GOD AND MAMMON

God and Mammon

Protestants, Money, and the Market,
1790–1860

Edited by
MARK A. NOLL

OXFORD
UNIVERSITY PRESS

2002

OXFORD
UNIVERSITY PRESS

Oxford   New York
Athens   Auckland   Bangkok   Bogotá   Buenos Aires   Cape Town
Chennai   Dar es Salaam   Delhi   Florence   Hong Kong   Istanbul   Karachi
Kolkata   Kuala Lumpur   Madrid   Melbourne   Mexico City   Mumbai   Nairobi
Paris   São Paulo   Shanghai   Singapore   Taipei   Tokyo   Toronto   Warsaw

and associated companies in
Berlin   Ibadan

Published by Oxford University Press, Inc.,
198 Madison Avenue, New York, New York 10016

Oxford is a registered trademark of Oxford University Press

Library of Congress Cataloging-in-Publication Data
God and Mammon : Protestants, money, and the market, 1790–1860 /
[edited by] Mark A. Noll.
p.   cm.
Includes bibliographical references and index.
ISBN 0-19-514800-2—ISBN 0-19-514801-0 (pbk.)
   1. Protestant churches—United States—History.   2. Christianity and economics—History.
3. United States—Economic conditions—To 1865.   I. Noll, Mark A., 1946–

BR525 .G63 2001
261.8'5'0973—dc21   2001023501

9 8 7 6 5 4 3 2 1

Printed in the United States of America
on acid-free paper

To

David E. Johnston

Acknowledgments

Many of the chapters in this volume originated at a conference in December 1998 on "Financing American Evangelicalism." The conference was held under the auspices of Wheaton College's Institute for the Study of American Evangelicals (ISAE), with generous funding provided by the Lilly Endowment. It is a privilege to thank Edith Blumhofer, Larry Eskridge, Jennifer Farmer, Bryan Bademan, and Katri Brewer of the ISAE for their help at the conference and in the preparation of this book. Conference participants engaged in unusually productive discussions, which have greatly assisted revision of the papers that were first presented there. Special gratitude is due to Richard Carwardine, David Hempton, Daniel Walker Howe, Richard Pointer, and John Walsh for writing new chapters expressly for this volume or for revising previously written material. P. J. Hill was an unusually helpful consultant on economic history. Rachel Maxson, Joel Moore, and Jeff Gustafson went far beyond the call of duty in helping the editor organize the chapters for publication. Some of the themes of this book overlap with the themes of a second volume that stemmed from the same Lilly project, *More Money, More Ministry: Money and Evangelicals in Recent North American History*, ed. Larry Eskridge and Mark A. Noll (Grand Rapids, Mich.: Eerdmans, 2000).

The book is dedicated to a longtime friend of the ISAE who for this project made a signal contribution as participant, as well as patron.

For permission to reprint revised versions of published material, we thank the University Press of Virginia (Daniel Walker Howe, "Charles Sellers, the Market Revolution, and the Shaping of Identity in Whig-Jacksonian America," and Richard Carwardine, "Charles Sellers's 'Antinomians' and 'Arminians': Methodists and the Market Revolution," in *The Market Revolution in America: Social, Political, and Religious Expressions, 1800–1880*, ed. Melvyn Stokes and Stephen Conway, 1996, pp. 259–307). We also thank *The Pennsylvania Magazine of History and Biography* for permission to reprint Richard W. Pointer, "Philadelphia Presbyterians, Capitalism, and the Morality of Economic success, 1825–1855" 112 (1988): 349–74.

Contents

Contributors

RICHARD CARWARDINE is Professor of American History at the University of Sheffield and the author of the *Evangelicals and Protestants in Antebellum America* (New Haven, Conn.: Yale University Press, 1993).

DAVID HEMPTON has written extensively on Methodists in the eighteenth and nineteenth centuries, including *The Religion of the People: Methodists and Popular Religion, ca. 1750–1900* (London: Routledge, 1996). He holds appointments in history and divinity at Boston University and in history at the Queen's University of Belfast.

DANIEL WALKER HOWE, Rhodes Professor of American History at Oxford University, has published books on moral philosophy at Harvard in the early national period, on the American Whig party, and on the understanding of self in America from the mid-eighteenth to the mid-nineteenth century.

ROBIN KLAY is a colleague of John Lunn in the Department of Economics and Business Administration at Hope College. She has published *Counting the Cost: The Economics of Christian Stewardship* (Grand Rapids, Mich.: Eerdmans) and writes regularly for *The Christian Century*.

KATHRYN T. LONG, who teaches in the History Department at Wheaton College, is the author of *The Revival of 1857–58: Interpreting an American Religious Awakening* (New York: Oxford University Press, 1998).

JOHN LUNN teaches economics and business administration with Robin Klay at Hope College. His articles have appeared in a number of journals, including *Industrial and Labor Review*, *European Economic Review*, and *Journal of Legal Studies*.

MARK A. NOLL is McManis Professor of Christian Thought at Wheaton College. He has published books and articles on religion in early America and on Protestant evangelicals in the twentieth century.

DAVID PAUL NORD is professor of journalism and adjunct professor of history at Indiana University. He has served as acting editor of the *Journal of American History*, and he has published widely on print culture and religious voluntary societies in the early Republic.

RICHARD W. POINTER is a professor of history at Westmont College. After writing on religious diversity in eighteenth-century New York, he is now completing a book on how the encounter with Native Americans shaped Euro-American religion.

KENNETH STARTUP is the academic dean of Williams Baptist College in Walnut Ridge, Arkansas, and the author of *The Root of All Evil: The Protestant Clergy and the Economic Mind of the Old South* (Athens: University of Georgia Press, 1997).

JOHN WALSH is retired as a fellow of Jesus College, Oxford, where he directed many doctoral dissertations and published widely on religion in England during the eighteenth century.

I

CONTEXTS

Introduction

Mark A. Noll

Money was close to the heart of all significant developments in the history of American Protestants between the War for Independence and the Civil War. It might, therefore, seem mysterious that there have been so few comprehensive efforts to explain connections between God and mammon in the United States' early history, or that monographs in which money and Protestant religion factor large for particular localities and specialized themes have only recently begun to appear. Much of the mystery as to why the history of Protestant-fiscal connections remains fragmentary, unsatisfactory, or unattended dissolves, however, when the different levels at which money entered into American Protestant history are considered. The main reason that scholarship falls so short of the reality is that the reality was extraordinarily complex.

The big picture can be framed by two sets of general observations about the United States during the early national and antebellum periods, the first by visitors from abroad, the second by respected historians of our own time. Prominent among the features of American life that most impressed outsiders then and historians now were its feverish commercial activity and the vitality of its churches.

In 1842 Charles Dickens concluded his account of a trip through the United States and Canada by noting that "all kinds of deficient and impolitic usages are referred to the national love of trade." Dickens was thinking mostly about trade in pirated literary property, but his comments reached further: "Healthful amusements, cheerful means of recreation, and wholesome fancies must fade before the stern utilitarian joys of trade."[1] By the 1840s, such judgments had become the foreigner's stock-in-trade. A Hungarian, Ágoston Mokcsai Haraszthy, whose stay in America overlapped Dickens's trip, concluded,

"There is one single point to which all are driving, and this is 'wealth.'"[2] Only a few years later, Frederica Bremer from Sweden attended a wedding where she was shocked to discover the bride dressed in an outfit for travel. To Bremer the incident spoke of a larger reality: "This marriage ceremony seemed to me characteristic of that haste and precipitation for which I have often heard the Americans reproached. Life is short, say they, and therefore they hurry along its path, dispensing with all needless forms and fashions which might impede the necessary business of life, and perform even this as rapidly as possible, making five minutes suffice to be married in."[3]

Such comment about American preoccupation with financial betterment was usually spoken censoriously. Other visitors, however, who were struck by the same kind of feverish entrepreneurship applied to the religious sphere, were more likely to end their voyages in admiration. In 1832 Achille Murat, an exiled Bonapartist whose religious ideal remained a unitary Christian state, nonetheless could not help but be impressed by "the thousand and one sects which divide the people of the United States. Merely to enumerate them would be impossible, for they change every day, appear, disappear, unite, separate, and evince nothing stable but their instability. . . . Yet, with all this liberty, there is no country in which the people are so religious as in the United States."[4] An English Methodist, James Dixon, who was in America at the same time as Frederica Bremer was more straightforwardly impressed: "There are no sects in America, no Dissenters, no seceders—or, whatever other term may be employed to designate the position and standing of a Christian society. They are all alike considered as Christians; and adopting, according to the judgment of charity, with equal honesty the common charter of salvation, the word of God, they are treated as equal, and as possessing similar and indefeasible rights. This is certainly a new aspect of living and visible Christianity."[5]

A surprising number of visitors brought together commentary on America's burgeoning economic life and its voluntary, democratic religion. For some, like a Russian visitor in 1815, the two forces were simply antagonistic: "Money is the American's deity; only his piety and the wealth of the country have until now sustained his morals."[6] Mrs. Frances Trollope, who spent two years (1828–29) in Cincinnati, never passed up a chance to compare America's free market in religion unfavorably with the Anglican establishment of her native Britain. In contrast to the healthy religion she felt an established church provided for all citizens, the United States' system was captive to low economic forces: "As there is no legal and fixed provision for the clergy, it is hardly surprising that their services are confined to those who can pay for them." Mrs. Trollope was equally dismayed at the skill of itinerant revivalists in separating hearers, especially gullible women, from their money. When she once inquired how a particular itinerant "got paid for his labours," she was told "that the trade was an excellent one, for that many a gude wife bestowed more than a tithe of what her gude man trusted to her keeping, in rewarding the zeal of these self-chosen apostles."[7] A few years after Mrs. Trollope published her opinions, a visitor from Sweden, rather than taking offense, was more simply impressed with a vitality that seemed to flow naturally between the worlds of commerce and religion: "Here [Rochester, New York], as in all of the new cities we passed through, we were surprised to notice the great number of *churches* and *banks*—evidence of the greater intensity of both spiritual and material activity here than in older communities."[8]

With prose that has captivated later historians, Alexis de Tocqueville also spoke eloquently of the United States' free trade in goods and faith. To him, the pursuit of gain, the exercise of liberty, and the flourishing of Christianity were all of a piece: "The love of wealth is . . . to be traced, as either a principal or an accessory motive, at the bottom of all that the Americans do." And, "there is no country in the world where the Christian religion retains a greater influence over the souls of men than in America." For this conjunction, Tocqueville thought he had an explanation: "The character of Anglo-American civilization . . . is the result . . . of two distinct elements, which in other places have been in frequent disagreement, but which the Americans have succeeded in incorporating to some extent one with the other and combining admirably. I allude to the *spirit of religion* and the *spirit of liberty*."[9]

Not everyone who read Tocqueville at the time, however, was entirely convinced. John Robert Godley, an Oxford Movement Anglican, who arrived in America shortly after the publication of Tocqueville's *Democracy*, agreed that attitudes toward wealth, religion, and politics were related, but he felt that Tocqueville had ascribed too much to "the operation of equality" as a grand explanation. Godley held, rather, that America's excessive commercialism and its ecclesiastical antinomianism could both be traced to a strategic religious-political defect: He was "inclined to think that the materialism thus admitted [by Tocqueville] to exist may chiefly be traced to the prevailing indifference with respect to religious creeds; and that this indifference, again, is intimately connected with the compulsory neutrality of the government in religious matters."[10]

Other foreign observers were more interested in describing the fiscal-religious connection than evaluating it. Such voices tended to combine bemusement with admiration as they traced the interpenetration of spheres that remained largely distinct in Europe. Anticipating the vocabulary of modern rational-choice theorists, the Austrian Francis Grund thought he knew why religion was faring better in the United States than in Britain during the 1830s: "In America, every clergyman may be said to do business on his own account, and under his own firm. He alone is responsible for any deficiency in the discharge of his office, as he is alone entitled to all the credit due to his exertions. He always acts as principal, and is therefore more anxious, and will make greater efforts to obtain popularity, than one who serves for wages. The actual stock in any one of those firms is, of course, less than the immense capital of the Church of England; but the aggregate amount of business transacted by them jointly may nevertheless be greater in the United States."[11]

The Swiss-born, German-trained Philip Schaff had been in America for ten years when he returned to Germany in 1853. That trip was the occasion for lectures in which he tried to explain his new home. As part of a warning to prospective immigrants, Schaff expatiated at length on the intimate links between American commerce and American religion:

> If you wish a calm and cheerful life, better stay at home. . . . The genuine American despises nothing more than idleness and stagnation; he regards not enjoyment, but labor, not comfortable repose, but busy unrest, as the proper earthly lot of man; and this has unspeakable importance for him, and upon the whole a most salutary influence on the moral life of the nation. The New York merchant is vexed, if stopped with a question on the street; because he loses a couple of minutes. The same zeal, the same parsimony of time, is employed by the minister, the missionary, the colporteur, the tract and bible societies, for higher ends. Even the business man, if in any degree religiously disposed, considers his pecuniary gain only a means "to do good"—as he expresses it; and though the Americans are not unjustly reproached with avarice and covetousness, yet they are entitled, on the other hand, to the praise of a noble liberality towards all sorts of benevolent objects,—a liberality unrivaled in modern history save by the extraordinary offerings of the Free Church of Scotland in the glow of her first love.[12]

Even though he did not make the kind of comprehensive assessment found in Tocqueville and Godley, Schaff joined these observers in asserting an intimate connection between American commercial practice and American religion.

The testimony of modern historians has a different character. Here the remarkable thing is the number of superlative judgments rendered about overlapping spheres of reality that only rarely have been considered together. First is a statement by Gordon Wood about the historical significance of the American Revolution:

> In 1760 America was only a collection of disparate colonies huddled along a narrow strip of the Atlantic coast. . . . Yet scarcely fifty years later these insignificant borderland provinces had become a giant, almost continent-wide republic of nearly ten million egalitarian-minded bustling citizens who not only had thrust themselves into the vanguard of history but had fundamentally altered their society and their social relationships. Far from remaining monarchical, hierarchy-ridden subjects on the margin of civilization, America had become, almost overnight, the most liberal, the most democratic, the most commercially minded, and the most modern people in the world.[13]

Second is a judgment by missiologist Andrew Walls about the spread of Christianity: "[Kenneth Scott] Latourette rightly calls the nineteenth century 'The Great Century of Missions.' But in no part of the world did that century see such a striking outcome as in North America. The main missionary achievement of the nineteenth century was the Christianizing of the United States."[14] Third, from chapter 3 of this book comes Daniel Walker Howe's conclusion about one of the most important factors that undergirded this Christianization: "The voluntary basis of American religion—economic, legal, and, in the dominant evangelical heritage, theological—was unique in the world. It probably explains why the modernization of American society was not accompanied by a corresponding secularization."[15] Fourth, the geographical historian D. W. Meinig adds a conclusion about the physical boundaries of the United States: "No federation before or since has enlarged itself by the almost routine sequential addition of so many new territorial units."[16] Finally, Nathan Hatch expresses a judgment about the Christian denomination that did the most to fill up the vast expanse of the new United States: "Quite simply, Methodism remains the most powerful religious movement in American history."[17]

It is not difficult to conclude that generalizations of such scope must somehow relate to one another. Nor is it a stretch to think that the religious dimensions of these relationships involved primarily Protestant considerations since in this era the Protestant churches by themselves came close to defining the formal religious life of the nation.[18] However, without, careful attention to specific situations, cases, and circumstances, it is difficult to say with any precision what those relationships actually were.

A Range of Questions (and a Thesis)

The purpose of this book is to offer more concentrated research and more sharply focused interpretations of the connections between Protestants and money. The previous difficulty in locating Protestants within American economic history may have had something to do with the assumptions of modern historians, some of whom, perhaps, find it difficult to take religion seriously as a primary motive for human action or to believe that religious speech can be anything but a screen for social or economic motives. On the other hand, historians predisposed to regard religion as a base, as well as a superstructure, may know of no alternative to such secular frameworks of reference except providential models of history that discourage actual research. Or they may be swayed by residual Victorian squeamishness that obscures or euphemizes the functions of money in human society.

A more likely explanation for this situation in which the subject dwarfs its historiography, however, is a combination of unexplored factual questions and profoundly complicated structural connections. Many separate issues bear on relationships between money and Protestants in this era. Each of them poses singular difficulties for research or narrative explanation, and because the issues are connected, the larger picture is even more obscure. As the following chapter by economists Robin Klay and John Lunn indicates, even basic questions about the economic dimensions of the Protestant churches and voluntary societies remain unanswered. Their effort to estimate the scale of religiously connected economic activity relative to the national economy addresses that problem, but it also indicates how much work needs to be done before an exact picture comes into view. The paragraphs that follow outline other dimensions of a complex nexus between the nation's dominant Protestant faiths and a full range of economic circumstances.

Despite the complexity they document, the studies that make up this book do point toward a defensible general thesis. Money in the first decades of the new American nation was everywhere on the minds of church leaders and many of their followers. Economic questions were almost everywhere important for religious self-definition, they figured regularly in preaching and pamphleteering, and they contributed greatly to perceptions of both public and private morality. Yet there was no one Protestant approach to money. Neither did there exist any grand system of economic practice that systematically governed the behavior, the speaking, or the life assumptions of the churches and church people. Single-cause explanations—whether from the Bible, Max Weber, Karl Marx, E. P. Thompson, or any other authority—simply do not work as a satisfying covering explanation for religious-economic connections. What does work is an integrated perspective that recognizes the fully connected relationship of religious faith and economic forces in this era, that highlights the multiplied negotiations between religious principle and economic practice (as well as religious practice and economic principle), and that stresses the contingent nature of historical development. The Methodists, as a prominent example explored in several of the chapters, were always engaged with economic questions, but the nature of that engagement differed between days of apostolic expansion and later periods of sorting out ownership of properties after the national schism of 1844; the religious conception of money took shape differently in the South than in the North; and the bearing of characteristic Methodist theological emphases on questions of money varied between good times and bad. In sum, the detailed historical studies that follow demonstrate the ubiquity but also the ambiguity of money in relation to Protestant religion. Money was always a religious question in these decades. Religion always had major economic implications. Along the way the chapters also demonstrate the folly of trying to write about broader cultural developments of the period—including political developments—without considering religion, economics, and religion and economics together.

Raising Money

Obscurity attends the means by which the religious bodies raised money. To be sure, the main procedures are known: For the first decades of national history, Congregational churches in New England relied on an establishment to collect tithes, oversee construction of meetinghouses, and fix ministers' salaries. Even before the official end of disestablishment (Connecticut, 1818; Massachusetts, 1833), many churches in the region, as well as churches founded in New York and further west in the great New England diaspora, operated under a system of dual control, where "congregations" of members who met spiritual criteria existed alongside "societies" of those (members and nonmembers) who supported the church financially.[19]

In addition, where colonial churches had been established by law (Congregationalist in most of New England, Anglican in much of the South), glebes continued as a funding source into the national period.[20] The glebe system, in which land was provided to ministers or church agents as a source of revenue, was a direct carryover of old-world patterns. At first, many ministers farmed or managed the land allotted to them, and churches often rented out their additional properties for income from pasturage, woodcutting, and farming. Later, many of these holdings were sold. One theoretical advantage of the glebe was that the minister's income could keep up with rising prices since the value of what land produced was supposed to rise along with inflation. Yet the same tie between produce and prices could lead to serious tensions when prices fell, as in the celebrated two-penny/pound tobacco conflicts in Virginia during the 1750s and 1760s.[21] The glebe's advantage of more stable real income was also offset by the time that farming took away from pastoral duties. One New England deacon, for example, is supposed to have complained, "Wa'll our minister gives so much attention to his farm and orchard that we get pretty poor sermons, but he is mighty movin' in prayer in caterpillar and cankerworm time."[22]

The collection of annual rents for pews seems to have become the first widely used substitute for state subsidies once establishment came to an end. Methodists and Baptists, however, usually attacked pew rentals in the early decades of the century, with Baptists using an entirely voluntary system of church support (along with much labor donated by ministers and elders) and Methodists trying to funnel voluntary giving into connectional channels in order to provide itinerants with their meager stipends. Later, however, urban congregations like First Baptist in Richmond and First Baptist in Savannah also came to use the pew rental system.[23] The refinement of Methodism that took place after the death of Francis Asbury in 1816 was also marked by an increasing use of pew rentals, as noted by Richard Carwardine in chapter 9. Yet Methodist itinerants long continued to augment their income by acting as colportuers for the Methodist book concern.

Free-will offerings also became more important early in the national period, often as a supplement to pew rentals. In many churches they were at first gathered only on special occasions, perhaps at Communion time for the support of poor families or at Thanksgiving Day services for similar causes. Typically, monies raised by free-will offerings were distributed for benevolent purposes and not used to pay the minister's salary. Once mission and benevolence programs of denominations and voluntary societies developed, a system of agents was introduced whereby organizations (like the American Bible Society) sent representatives for annual fund-raising visits to local congregations.[24] Eventually, this competitive drumming up of business for benevolent causes came to be a burden for some urban churches, which found too many agents knocking too often at the door.

In addition to pew rentals, glebes, and free-will offerings, some churches called for voluntary subscriptions, which represented an early form of the pledge system that has come into common use among many churches in the twentieth century.[25] At first, subscriptions were sought mostly to fund new construction. Methodists on the frontier, however, also put subscriptions to use in supporting their work generally, with in-kind contributions of food, building materials, and labor being quite common. Later, the Methodists began to use subscriptions for ministerial salaries. Baptists, as a rule, were reluctant to follow suit. At least early in the century, church subscriptions were regularly paid in commodities—wheat, beef, whiskey, linen, flour, and so on. The names of subscribers and amounts pledged were also sometimes circulated within the congregation, as a means of encouraging (or shaming) all into doing their part.

A final form of religious fund-raising that appeared briefly in the early republic, but memory of which may have been suppressed, was the lottery. Lotteries for public purposes had been common in Britain, and at least a few churches and religious-connected institutions employed them in the colonial and early national periods. St. Paul's Protestant Episcopal Church in Baltimore, for example, raised over $30,000 for the construction of a new building from a lottery in 1780.[26] But results were usually not so favorable. An example of the kind of distress that led most groups to give up lotteries was the experience of the (mostly Presbyterian) College of New Jersey (later Princeton). It sponsored a lottery in the late 1770s but then spent at least twelve years negotiating with an offended participant who felt that the college had withheld his rightful winnings.[27]

"Free churches" in the antebellum era were congregations whose buildings, ministers, and sometimes week-to-week maintenance were provided by philanthropists. In setting up such churches in New York City in the late 1820s, evangelical businessmen like Lewis Tappan and the Dutch Reformed minister James C. Bliss looked in part to the example of Thomas Chalmers of Scotland, who had established churches on this scheme for the Glasgow poor during the previous decade.[28] The difficulties in implementing such a "free" system were indicated by the range of issues an advocate for such churches felt it necessary to discuss in the Episcopal *Church Review* in 1857: whether the urban poor could make any contribution at all to churches, whether donated church buildings would make the poor uncomfortable, and whether ministers could ascertain the names and addresses of their congregants in the absence of a rental pew list.[29]

Despite what is known about the existence of these diverse ways of raising money for Protestant enterprises, serious studies with full details on the mechanics of fund-raising and on its meaning for the religious and cultural lives of the churches remain rare.[30] One of the exceptions to this generalization is the awareness, discussed below, that national fund-raising for religious causes was a major source of intra-Protestant strife.

Broader Cultural Considerations

In the absence of hard information on such basic questions as how religious organizations raised money, it is difficult to address the most far-reaching interpretive question, that is, the nature of connections between broad trends of Protestant development and broad trends of American political economy.

During the early history of the United States, Protestants were more successful than ever before or since at enrolling Americans as members of their churches. Even as the national population increased twelvefold from 1775 to 1860, the number of Americans affiliated with the churches rose by a multiple of perhaps twenty-five.[31] In proportionate terms, the period of most rapid population growth in U.S. history was also the period of most rapid growth in church adherence. In this religious surge, moreover, the fastest growing churches were the Methodists and the Baptists, the Protestants who most thoroughly rejected the once nearly universal system of church establishments and who most thoroughly embraced the revivalistic evangelicalism that began to emerge during the age of Whitefield, Edwards, and Wesley. The Methodists were especially noteworthy. Their development within two generations from a few scattered groups at the start of the War for Independence into the nation's largest denomination, with over one-third of the country's church adherents, was an achievement of staggering proportions.[32]

The era witnessed also the flourishing of Protestant voluntary societies, a form of religious organization that went outside denominational boundaries to mobilize money, personnel, and publicity for specific tasks. These societies originated with the continental pietists and early British evangelicals, mostly as special-purpose units within the state churches. They came into their own in America, however, during the second decade of the nineteenth century.[33] The first large voluntary agencies had narrowly religious purposes, like the American Board for Foreign Missions (1810) and the American Bible Society (1816). But soon somewhat broader goals were also being served by organizations like the Colonization Society for Liberated Slaves (1817), the American Society for the Promotion of Temperance (1826), and the American Antislavery Society (1833). The annual budgets for some of these associations equaled the yearly income of relatively large business firms. Whether providing books and Bibles through traveling vendors, organizing to improve the treatment of the insane, or reaching out to prostitutes and other social outcasts, the Protestant voluntary agencies aimed at the social conversion of the United States.

At the same time, all forms of Christianity in America—from the most radical evangelical sects to the largest Protestant churches, older European confessional bodies, and Roman Catholics—moved in the direction of what Nathan Hatch has called populist and sectarian "democracy."[34]

One cannot stress too greatly that these momentous religious developments were occurring alongside the nation's rapid economic growth and, moreover, in the very places that foreigners visited as they drew their conclusions about the American mania for commerce. The United States in which Methodism came to the fore was the America in which the value of cotton exports rose from $23 million per annum to $124 million, where steamship freight rose from virtually nothing in 1811 to 195,000 tons in 1860, and where miles of railroad track leapt from almost none in 1830 to over 30,000 in 1860. The age of the voluntary society was the age of the new American city, none of which contained even 50,000 residents in 1790 but two of which (New York and Philadelphia) had grown by 1860 to over half a million, with five others above 50,000.[35]

The difficulties in interpreting connections between these two spheres, which everywhere overlapped, are severe. On the one hand, serious study of the actual religious dynamics of the period are only just beginning.[36] On the other hand, as noted in more detail below, major efforts in charting religious and economic relationships have been fragmentary or misleading. In light of the magnitude of all factors in the problem, the relationship of money and Protestantism between the two defining wars of the United States' early existence must be regarded as a mountain where exploration has reached only to the foothills.

Intra-Protestant Divisions

Still, despite the absence of compelling answers to larger questions, several other aspects of the economic history of Protestants have been recognized as obviously important. One such reality is the prominence of money in creating the major Protestant divisions of the period. In the first instance, as a few historians are beginning to document, interdenominational strife, often exceedingly fierce, was a major feature of Protestant life throughout this period. Fueling that strife were regular charges that theological error led to the misuse of money and the arrogance of wealth.[37] As the contributions of Kenneth Startup and Richard Carwardine to this book document, and as the immense literature on abolition also indicates, issues concerning money and the control of property exacerbated the tensions between North and South that made the Civil War a religious conflict.

Sectional and interdenominational theological differences were instances of a larger fault-line within early American Protestantism that Curtis Johnson has helpfully described as dividing "formalists" from "antiformalists."[38] As Johnson describes them, formalists were mostly from the middle and upper classes; they featured orderly worship, an educated clergy, and Victorian decorum; they were most common among Congregationalists, Presbyterians, low-church Episcopalians, and some of the Reformed denominations; they were the mainstays of the voluntary societies; and they were often aligned with the Whig party. Johnson's account of religious formalism tallies nicely with Daniel Walker Howe's compelling description of religious contributions to Whig ideology.[39] By contrast, antiformalists arose from the middle and lower classes; they preferred emotional worship, ministry as a function of personal charisma, and plain living; they predominated among Baptists, Restorationists, and early Methodists; they were suspicious of voluntary societies; and they were often aligned with the Democratic party. Johnson's picture of antiformalists fits well with accounts by Nathan Hatch, Richard Hughes, and George Rawlyk of dynamic revivalist groups in the early national period.[40] For our purposes, the key matter is that money was always a central matter of dispute between formalists and antiformalists. Antiformalists distrusted its accumulation and those who sought to harness wealth for broad, national projects; formalists felt a responsibility to use money for such purposes but to guard against greed or the arrogance of power.

A significant dimension of this intra-Protestant division, however, is the fact that formalist and antiformalist stances were not fixed. Over the course of only a few decades, for example, the Methodists moved decisively from antiformalism to formalism, as indicated perhaps above all by altered financial interests—decreasing distrust of regular salaries for ministers in line with standard professional scales, greater willingness to solicit funds for building imposing church edifices, and growing participation in the evangelical united front of voluntary societies.[41]

Money was just as important for the black-white fissure in early American Protestantism. Slavery was never less than a statement about the sovereignty of capital and its rights in relation to humanity and its rights. The story of Richard Allen, founder of the African Methodist Episcopal Church, is far from untypical. Not only did Allen have to purchase his own freedom, but also his efforts at securing property for an independent African-American church and then at gaining clear titles for churches and black benevolent societies generated constant conflicts throughout his religious career.[42] In the South, economic restrictions on religious organization by black Christians was part and parcel of the racial system that undergirded slavery and the marginalization of free blacks. Important works by a number of historians on slavery and abolition do, however, touch significantly on religious-economic connections.[43]

Money also figures in the history of male and female trajectories. The voluntary societies, by opening up avenues for publishing and public speaking for women, like the sisters Sarah and Angelina Grimké, also opened up avenues for relatively independent economic activity. Religious convictions supplied the most powerful impulses for the women's rights movement, and that movement was much concerned about economic matters. The famous "Declaration of Sentiments" from the women's rights convention in Seneca Falls, New York, in July 1848, was shot through with economic, as well as political, social, and religious, concerns. It complained, for instance, that men had monopolized "all right in property" and "all the profitable employments" while closing to women "all the avenues to wealth and distinction which he considers most honorable to himself."[44] This Seneca Falls meeting, it is worth recalling, took place in a Wesleyan Methodist chapel, a new denomination founded only in 1843 as a protest against Methodist refusal to condemn the traffic in slaves.

Religious accommodation to the commercialization of American culture also had far-ranging implications for the religious roles of men and women. To the extent that a "women's sphere" developed as a refuge from, and alternative to, the expanding world of commerce, then Protestant faith would seem to be a major contributor, both by fueling the energy of the commercial sphere and by providing religious content for the domestic religion of the female sphere.[45]

In sum, all of the significant divisions within early American Protestantism had a prominent economic dimension. To study those divisions without considering the significance of money is to doom historical analysis to partiality in every sense of the term.

Money and Theology

The omnipresence of financial concerns in general Protestant development also raises questions about its possible role in other aspects of Protestant history. For example, did money or commercial activities play a role in the era's major theological expressions?

For the most part, historians have attempted only preliminary answers to such questions, yet even these are intriguing. The evolution of New England Calvinism, which Bruce Kuklick rightly describes as the nation's most sophisticated early intellectual tradition, was certainly an evolution that responded to altered social circumstances. The New Divinity of Nathaniel William Taylor and Lyman Beecher, for example, has been characterized as a form of religious thought in dialogue with "the judgments of Common Sense materialism and American republicanism."[46] Others, however, have perceived even more intimate connections with changing economic circumstances. Richard Rabinowitz, for example, depicts the theology of Taylor and Beecher as a response to "this commercial economic world" that was governed by "the language of moral experience—terms like investment, contract, and credit."[47]

Moving in the opposite direction theologically, some advocates of older confessions felt it was necessary to defend their traditions against the intellectual corruption worked by the marketplace. Among the new Dutch immigrants of midcentury, as James Bratt reports, the scornful rejection of American "Methodism" included a repudiation of "'the world,' whether of mass entertainment, of big business with its mergers and banks, or of charities with their assorted benevolences."[48] Questions of direct influence or antagonism do not exhaust the theological possibilities. For example, the holiness theology of Phoebe Palmer has been construed as theology for a domestic sphere self-consciously posing an alternative to aggressive commercial practices.[49] In short, although the bearing of economic issues on Protestant theology has only begun to be explored, it would seem to have played some kind of role.

Money and Protestant Reform

Financial factors were obviously also at work in the redirection of Protestant reform that started in the mid-1830s.[50] Charles G. Finney's 1835 move from New York City to Oberlin, Ohio, preceded by only a brief period the financial collapse of 1837–38, which seriously undercut the reformist philanthropy of active evangelicals like Lewis and Arthur Tappan.[51] In Finney's words, "I have been astonished to find that the pecuniary embarrassments of the few past years have so far crippled the movements of the great benevolent societies for want of funds."[52] Only shortly after that embarrassment, the antislave effort fragmented in disagreements over whether to expand reform to other inherited institutions and cultural conventions. The literal costs in attempting such reforms—whether against slavery, for women's rights, for reform of the workplace, or for urban development—were always a factor. By the time of the "businessmen's revival" of 1857–58, the price tag attached to the comprehensive reforms that once had been sought by the American Protestant mainstream became simply too much to bear. Kathryn T. Long's analysis of this revival, portions of which are contained in chapter 11, suggests that the revival's goal was a spiritual renewal directed more to inner human needs than to social relationships; that its characteristic method was the noon-hour prayer meeting, which could be attended without disrupting the working day; and that its final result was, in her phrase, "revivalism without social reform."[53] The tragic story that Donald Mathews has told so capably about the expansion of southern Protestantism alongside the evangelical acceptance of slavery presents a similar picture.[54] North and South, Protestant reform cooled as its market presence grew.

Money and New Religious Movements

Again, without lapsing into economic determinism, it is clear that answers to financial questions were part of what new American faiths provided. The Restorationist movement, spearhead by Alexander Campbell and Barton W. Stone, was resolutely antiformalist, even as it promoted the ideal of agrarian yeomanry beloved of Jeffersonian and Jacksonian rhetoric.[55]

The Millerites, who followed William Miller in fixing a definite date for the return of Christ, could be considered a typical evangelical movement of the 1830s and 1840s that became unusual only for the way it combined antiformalist and formalist characteristics.[56] Its antiformalism expressed itself in an eagerness to forsake all—including long-term leases and receivable accounts—for the coming Kingdom. Its formalism could be seen especially in Joshua Himes's skill at merchandising the ideas, sermons, and biblical charts of William Miller.[57]

The Mormons were even more complicated. Joseph Smith's early community combined an antiformalist rejection of America's competitive religiosity with a formalist effort to re-create a perfect Christian society.[58] By the end of Brigham Young's life, the Church of Jesus Christ of Latter-day Saints became even more complex. While maintaining Joseph Smith's new religious community, the church underwent the development that increasingly made it look like a Whig, formalist, market-embracing religious organization.[59]

The enduring new religious movements that arose in America after the War of 1812 were fundamentally what their founders said they were—that is, religious movements. At the same time, the engagement of these movements in the expanding American market was as thorough as it was complex. No single interpretation can adequately describe the range, depth, and variety of religious-economic relationships in early U.S. history. Yet interpretations that attempt to describe Protestant developments without full attention to questions related to money are just too simple.

Historiography

Despite increasingly sophisticated study of individual issues involving Protestants and money, satisfying general interpretations are still in short supply. The most important effort to link the main trajectories of American economic life with the main developments in the churches is Charles Sellers's volume from 1991, *The Market Revolution*.[60] Sellers's account, in turn, is a sharply formulated instance of the widely accepted historical metanarrative brilliantly limned for Britain in E. P. Thompson's *Making of the English Working Class* (1963) and then embraced, to one degree or another, by a number of influential historians.[61] In crude overview, this narrative pictures a colonial (sometimes colonial and early national) society largely composed of self-contained, organic communities of agrarian producers and urban artisans. These communities, if not idyllic, nonetheless enjoyed much of what was lost when "the market revolution" occurred. The earlier way of life was characterized by personal security reinforced by domestic stability, communal solidarity, anti-authoritarian ideology, valorization of play and leisure, and a face-to-face economy that sustained humane values. The new market world, by contrast, featured mobility, efficiency, individual self-exertion, specialization, productivity, expanding consumption, and a way of life that disrupted communities, uprooted relationships, and commodified family connections. Within this narrative, religion has been important, primarily for its role in either retarding or accelerating the turn to markets. To Thompson, for example, the late eighteenth century was critical in the history of British Nonconformity for creating "the extraordinary correspondence between the virtues which Methodism inculcated in the working class and the desiderata of middle-class Utilitarianism."[62] In more general terms, religion was useful if it resisted the spread of the market but retrograde if it facilitated the transition to modern capitalism.

Charles Sellers's particular construction of the role of religion is singular, but his functional understanding of religion is not untypical of the major interpretive pattern. His impressive effort to show how economic changes triggered a major shift in American cultural ethos includes a systematic statement about the behavior of Protestants. In his view, three antagonistic groups existed. "Antinomians" were committed to noncommercial, agrarian, local, patriarchal, and egalitarian values; as adherents to the Democratic party, they resisted the spread of national markets; as Baptists and Methodists, they embraced a radical, revivalistic, New Light folk evangelicalism. Opposed to "antinomians" were Sellers's "arminians" who promoted efficiency, production, the enterprising spirit, individualistic competition, and banks; as exploiters of the period's expanding markets, they were adherents to the Whig party and could be found among the progressive Protestants—especially Unitarians, Congregationalists, and New Light Presbyterians. A third group of Protestants, Sellers's "Moderate Lights," were ministers, led by Charles G. Finney, who won over "antinomians" to the "arminian" values of the market.

Since Sellers's interpretation is not only the most impressive but also the most comprehensive effort to link Protestants and the rising American economy, two full-scale responses to his picture are provided in chapters 3 and 4 of this book.[63] Both Daniel Walker Howe and Richard Carwardine appreciate the reach and creativity of Sellers's proposal, but both feel that it misconstrues critical aspects of what really took place.

As a parallel recognition of the immense influence of Thompson's *Making of the English Working Class* for creating a picture of Protestantism, especially evangelical Protestantism, in relationship to the market realities of modern economies, the book also includes an analysis by David Hempton and John Walsh (chapter 5) of Thompson's depiction of the Methodists, the central group in his construction of the Protestant-market relationship. By examining Thompson's more general stance toward Methodism, Hempton and Walsh show how much Thompson has stimulated historical inquiry but also how far short Thompson (and, by implication, his imitators) have fallen from recording a full, reliable history of Protestants and the economy in the early modern period. These historiographical chapters conclude that, for all their brilliance, the major statements by Sellers and Thompson fail to present a convincing general picture of the pertinent religious-economic relationships.

If the history of Protestants and money provided by the extant general narrative has been artificially limited, it is clear that help is on the way. The first instance of that help is the perceptive criticism of "*the* market revolution" thesis itself. Some of that criticism questions the whole idea of a cultural turning point such as Sellers and others have perceived in economic activity,[64] some has pointed out the fluidity and geographical variability that attend market changes,[65] and some has found much earlier evidence for the characteristics widely held to have been promoted by the market revolution.[66]

Much help on the Protestant-market relationship has also been provided by students of the internal character of colonial evangelicalism. As indicated by Kathryn T. Long in chapter 11, the depiction of George Whitefield as a pioneering marketer, as well as a pioneering revivalist, has raised important questions about the defining characteristics of early Protestant evangelization itself. Harry Stout and Frank Lambert, both drawing on the scholarship of T. H. Breen, picture Whitefield as one who adapted a historic Protestant message to the new dynamics of transatlantic merchandising.[67] The importance of their work is to show that at least some Protestants in some situations eagerly embraced some aspects of "the market" because they thought such a strategy could enhance the spread of their religion.

Of perhaps greater significance for historians focused on nineteenth-century developments are suggestions that the moral valences ascribed to pre-capitalist and capitalist culture—that is, the basic antithesis between the wholesome, humane virtues of pre-market communities and desiccated, mechanical practices of market individualism—whether in Britain or the United States, are flawed.[68] Responses to E. P. Thompson—from grand alternatives to his whole interpretive enterprise to careful research that modifies different aspects of his Methodist portrait—suggest the range of possible criticism.[69] For American religious developments, important demurrals to the dominant narrative include Daniel Walker Howe's convincing argument that many Protestants committed themselves to the values of the Whig party because they held that cautious embrace of opening markets would *strengthen* human flourishing and moral stability.[70] In addition, the tardy but welcome maturation of Methodist history holds out the hope that Methodist economic behavior can be analyzed in terms supplied by the Methodists, as well as by modern students of economic life. In particular, several historians have shown that many Methodists in early America seemed quite naturally to embrace, without a sense of contradiction, moral values that market revolution historiography posits as incompatible.[71]

Although Sellers's *Market Revolution* has been the most comprehensive effort at linking economic and religious spheres for American history, several other general interpretations have also contributed significant insights. George Thomas's *Revivalism and Cultural Change* (1989) offered an intriguing set of conclusions that link evangelical revivalism with large-scale political and economic realities. His key assertion was that where historians of the nineteenth century can identify a strong attachment to "entrepreneurship and the small capitalist enterprise," there they will also find "outbreaks of revival and the adherence to a revivalist world view." Moreover, where entrepreneurship dominates, there will also be found "support for Republicanism [i.e., the Republican party after 1856] and Prohibitionism." Similarly, where revivalism holds sway, there will again appear "strong support for Republicanism and Prohibitionism."[72]

R. Laurence Moore's *Selling God* (1994) is likewise broad in its concerns. The book is a thoughtful meditation on the many ways in which the story of religion in the United States can be written as a story of "commodification." Moore's thesis is that American Protestants have always sustained an active relationship with commerce, marketing, and the dynamics of American business. The book's great merit is to show that secularization in America has never meant a zero-sum struggle, in which the "world" and the "church" do battle over the same bit of turf, but rather that secularization has always been a much more nuanced question, with the gains and losses for religion attending the same set of circumstances. It is "religion's systematic and expansive complicity in mechanisms of market exchange" that gives Moore his argument and dictates the arrangement of his evidence.[73] That evidence begins with Protestant support for the great voluntary societies, the theatricality of urban revivalists, and Protestants' difficult acceptance of leisure, before moving on to how Protestant leaders self-consciously used religious controversy to "sell" their distinctive beliefs but also to distribute books and periodicals, market their meetings, finance their church buildings, and meet budgets.

Thomas and Moore both avoid the reductionism of Sellers's account, and both also are alert to religious matters as defined by religious actors in a way Sellers is not. Yet neither amounts to a magisterial interpretation of the 1790–

1860 period. Thomas's concern to frame a general theory for religious movements in the context of modern societies takes him away from the historical specifics of the period to the kind of abstract reasoning (and prose) that have never played well among historians. Moore's success lies in showing the constant complicity of Protestants in marketing strategies rather than in providing an explanation for how and why such complicity took place.

One other general interpretation that links Protestants and commerce must be mentioned. This is the effort by a few historians and a talented group of sociologists to explain the rise and decline of various religious groups in America by using marketplace categories. Sometimes described as rational-choice theory, this interpretation posits a market in religious goods that responds to the quality and quantity of competition. One of the first historians to use such reasoning was Terry Bilhartz in his study of religion in early national Baltimore. As Bilhartz describes the situation, the "pluralistic society" that existed after the Revolution meant that adherence to a particular church was a free decision. "But in any economy of exchange, individual choice is influenced by the supply of goods or services being offered and the marketing skills of the competing vendors." In the religious competition of the early national period, "large numbers of Baltimoreans bought into evangelical religion largely because the price was right and the streets were filled with vendors."[74]

Roger Finke, a leading sociologist in propounding rational-choice theory, carries the analysis even further:

> This surge in revivals and the growth of organized religion was due to a shift in supply, not demand. With the restrictions on new sects and itinerants reduced, a new wave of religious suppliers emerged—suppliers that aggressively marketed their product to the masses. Whitefield's crowds didn't materialize out of nowhere, they responded to the publicity, press releases, and sermons circulated in their communities up to two years in advance. . . . The so-called Second Great Awakening was nothing more (nor less) than the successful marketing campaigns of "upstart" evangelical Protestants. Thus early American religion flourished in response to religious *deregulation*.[75]

Although historians have received this interpretative scheme with indifference or even hostility, its ability to explain, at least on one level, many features of American religious ecology gives it considerable plausibility.[76] The great difficulty for historians in putting the theory to use, however, is that it divorces explanation from the lived worlds of the historical participants. Methodists may have flourished in early America because they successfully adapted to a deregulated religious market, but they thought they were winning souls to Christ, experiencing the witness of the Holy Spirit, correcting the doctrinal mistakes of the Baptists, and preparing converts to serve God. Moreover, for Protestant connections to the economy, rational-choice theory offers little direct interpretation of the theological or religious meaning of either money or the accommodation of religious commitments to commercial practices.

The absence of convincing overarching interpretations should not, how-ever, be taken to suggest that no works of enduring value have been published on aspects of the subject. To the contrary, a recent surge of studies has pro-vided fresh examination of several questions concerning money in relation-ship to Protestant religion. These include a small shelf of sophisticated studies on social transformation in antebellum communities, mostly in New England and New York;[77] several books on the surprisingly positive connections between revivalistic Protestantism and early labor movements;[78] Kenneth Startup's revealing examination of what a wide range of Southern clergymen actually wrote on economic questions in the decades before the Civil War;[79] some helpful articles on widely scattered subjects;[80] and illuminating reflec-tions touching on economic matters in works devoted to other issues.[81]

Yet even with much distinguished writing, no one picture has gained wide acceptance. Authors of the local studies agree on the interconnectedness of material, cultural, ideological, symbolic, gender, class, and religious reali-ties, yet they differ considerably on the relative weight to ascribe to such fac-tors. Such productive disagreements—as between Paul Johnson, who while disclaiming economic reductionism still considers the revivals "a crucial factor in the legitimation of free labor," and Curtis Johnson, who sees revival growing integrally out of people's beliefs that Christian "salvation" was "vitally im-portant" and that they could "have an impact on the salvation of themselves and others"[82]—may not touch the most difficult issues. Fresh study of differ-ent local circumstances or of different individuals in the same place might point to a variable landscape (sometimes economic motivation trumping re-ligious conviction, sometimes moral aspiration trumping material interest). It might call into question whether any interpretation that assumes simple motivation for religious or economic behaviors can be adequate.

The Shape of the Book

The chapters that follow seek clarity in describing relationships between Protestants and economic matters and sophistication in accounting for those relationships. They begin with two kinds of contextual studies. The first is a chapter by economists Robin Klay and John Lunn that sketches American economic history for the period, estimates the size of religion in the national economy, and provides a case study of the financial affairs of one (Albany, New York) church with an unusually full set of local records. The other con-textual chapters are historiographical. Daniel Walker Howe and Richard Carwardine subject Charles Sellers's account of religion and "the market revolution" to critical analysis, Carwardine by focusing on the Methodists and Howe on the more general Protestant situation in Sellers's period. These chapters are followed by the assessment of E. P. Thompson's influential views of English Methodism from David Hempton and John Walsh.

Part II features three case studies for the period when what might be called traditional Protestant economics gave way to the new market realities of the nineteenth century. David Hempton and David Paul Nord explore economic aspects of the great engines that drove the new century's evangelical surge—the Methodists, a movement which, as Hempton shows, interpreted money with slightly different emphases on either side of the Atlantic, and the voluntary societies, which, as Nord describes them, both welcomed and demonized the market they so thoroughly mastered. Richard W. Pointer's chapter on the economic attitudes of Philadelphia Presbyterians balances the treatments in these early chapters by focusing on a most influential middle state and a denomination, the Presbyterians, who with Congregationalists remained the era's recognized leaders in formal Protestant thought. Although Pointer carries his analysis into the 1850s, the heart of his concern is the adaptation of historic attitudes toward money by one of the country's most traditional denominations.

The three chapters of Part III address economic questions for the period of Protestant maturity when the earlier explosion of Protestant energy was being consolidated. Richard Carwardine's assessment of the fiscal strife that lingered long in the wake of the great Methodist schism of 1844 testifies to the pervasiveness of economic concerns in this largest, most representative antebellum denomination. Kenneth Startup's chapter emphasizes the importance of traditional Protestant attitudes toward money in Southern criticism of Northern mammonism, but also shows that Southern ministers could only be expert in criticizing the North because they had had so much practice in rounding on their own congregations for economic sins. In a chapter on fiscal aspects of the 1857–58 "businessmen's revival," Kathryn T. Long explores both the marketing of revival as news and the influence of business thinking on the understanding of revival itself. Only with this chapter and its focus on events in the late 1850s do we see for the first time market practices clearly outweighing indigenous Protestant values in the money-religion relationship.

The book's final section contains two efforts at grasping a larger picture. Mark Noll uses a survey of Protestant publications from throughout the period to argue that the kind of Protestantism that flourished in the United States (strongly evangelical and disestablishmentarian) explains much in religious-economic attitudes of the period. Daniel Walker Howe draws matters to a close by looking back over the salient findings of the book but also forward in specifying the many levels of research and sensibility that are required for addressing religious-economic questions satisfactorily.

Chapters in the book proceed more or less chronologically but are also linked by important themes. In keeping with recent first-order scholarship more generally, a number of them feature the Methodists as a key to the era's religious dynamism. They share the assumption that to understand the rise of Methodism is to understand not only the religious wellsprings of the period but also the economic place of religion. In addition, several authors, especially David Hempton and John Walsh, draw fruitfully on British sources to show how much history was shared across the Atlantic but also where American developments ran ahead of the Old World. By returning to such concerns, as also to practices of philanthropy, attitudes toward the spread of markets, responses to economic booms and busts, and other general matters, the chapters offer a substantial thematic treatment of the era.

By no means, of course, do they address all levels of the problem outlined at the start of this introduction. But they do push further into important cases and wider into important relationships. They remind us that Protestants were involved in the United States' new market realities to different degrees, with different attitudes, and in response to different visions in different areas of the country. Most Protestants were at ease with the burgeoning markets of early American history, but never unequivocally so. Although these chapters constitute the most thorough study of the subject yet attempted, they open many doors for further inquiry and draw attention to many subjects that demand equally serious attention. The book will succeed if, despite inevitable lacunae, oversights, and misperceptions, it convinces readers that connections among Protestants, money, and the market for the period 1790 to 1860 were multidimensional rather than unilinear, complex rather than simple, and driven at least as much by the religious as by the market forces at play.

Notes

1. Charles Dickens, *American Notes* (New York: St. Martin's, [1842] 1985), 225.

2. Ágoston Mokcsai Haraszthy, *A Journey to North America*, quoted in Marc Pachter and Frances Wein, eds., *Abroad in America: Visitors to the New Nation, 1776–1914* (Reading, Mass: Addison-Wesley, [1844] 1976), 43.

3. Frederica Bremer, *The Homes of the New World* (from a visit in 1849–1851), quoted in Pachter and Wein, *Abroad in America*, 123.

4. Achille Murat, *A Moral and Political Sketch of the United States of North America* (1832), quoted in Milton B. Powell, ed., *The Voluntary Church: American Religious Life, 1740–1860, Seen Through the Eyes of European Visitors* (New York: Macmillan, 1967), 50.

5. James Dixon, *Methodism in America* (1848–49), quoted in Powell, *Voluntary Church*, 174.

6. Pavel Svin'in, *A Picturesque Voyage through North America* (1815), quoted in Pachter and Wein, *Abroad in America*, 16.

7. Frances Trollope, *Domestic Manners of the Americans*, ed. Donald Smalley (New York: Knopf, [1832] 1949), 109, 125.

8. Gustaf Unonius, *A Pioneer in Northwest America, 1841–1858*, ed. Nils W. Olsson (1950), quoted in Marvin Fisher, *Workshops in the Wilderness: The European Response to American Industrialization, 1830–1860* (New York: Oxford University Press, 1967), 109.

9. Alexis de Tocqueville, *Democracy in America*, ed. Thomas Bender (New York: Modern Library, ([1835, 1840] 1981), 35–36, 182, 514.

10. John Robert Godley, *Letters from America* (1844), quoted in Powell, *Voluntary Church*, 163.

11. Francis Grund, *The Americans* (1837), quoted in Powell, *Voluntary Church*, 77.

12. Philip Schaff, *America: A Sketch of Its Political, Social, and Religious Character*, ed. Perry Miller (Cambridge, Mass.: Harvard University Press, [1855] 1961),

29–30. The balance in Schaff's judgment was echoed almost exactly in a defensive statement by Harvard's Francis Bowen in his widely used textbook, *The Principles of Political Economy*, 4th ed. (Boston: Little, Brown, 1865), 545: "We may be a speculating, but we are not a miserly people."

13. Gordon S. Wood, *The Radicalism of the American Revolution* (New York: Knopf, 1992), 6–7.

14. Andrew Walls, "The American Dimension of the Missionary Movement," in *The Missionary Movement in Christian History: Studies in the Transmission of Faith* (Maryknoll, N.Y.: Orbis, 1996), 227.

15. Daniel Walker Howe, chapter 3, p. 63, in this volume.

16. D. W. Meinig, *The Shaping of America, Vol. 2: Continental America, 1800–1867* (New Haven, Conn.: Yale University Press, 1993), 431.

17. Nathan O. Hatch, "The Puzzle of American Methodism," *Church History* 63 (June 1994): 177–78.

18. The 1850 national census found that Protestants accounted for 97% of the nation's churches, 95% of seating accommodations in those churches, and 89% of the value of church property. Roman Catholic churches (1,227 out of a total of 38,183) and Jewish houses of worship (37 in the entire country), which tended to be located in cities and to be larger than the average Protestant church structure, were effectively the only non-Protestant religious organizations in the country. *Statistical View of the United States. . . . Being a Compendium of the Seventh Census* (Washington, D.C.: Senate Printer, 1854), 132–39.

19. The tensions this dual system led to in Cortland County, New York, during the 1820s and following are well described by Curtis D. Johnson, *Islands of Holiness: Rural Religion in Upstate New York, 1790–1860* (Ithaca, N.Y.: Cornell University Press, 1989), 80–82.

20. For details, see Luther P. Powell, *Money and the Church* (New York: Association Press, 1962), 123–25.

21. For a succinct treatment, see Carl Bridenbaugh, *Mitre and Sceptre: Transatlantic Faiths, Ideas, Personalities, and Politics, 1689–1775* (New York: Oxford University Press, 1962), 208.

22. Powell, *Money and the Church*, 124.

23. Merrill D. Moore, "Church Finance," *Encyclopedia of Southern Baptists* (Nashville, Tenn.: Broadman, 1958), 1:284.

24. A full discussion of the expansion, management, and difficulties of the agency system can be found in Peter J. Wosh, *Spreading the Word: The Bible Business in Nineteenth-Century America* (Ithaca, N.Y.: Cornell University Press, 1994); see also chapter 7 by David Paul Nord in this volume.

25. Powell, *Money and the Church*, 139–42.

26. Ibid., 146.

27. The subject was discussed at almost every meeting of the college board from September 1780 to April 1791; Trustee Minutes, Princeton University Archives, Princeton, N.J.

28. On Chalmers's example, see *The Memoirs of Charles G. Finney: The Complete Restored Text*, ed. Garth M. Rosell and Richard A. G. Dupuis (Grand Rapids, Mich.: Zondervan), 296, n.70.

29. "What the Free-Church System Requires," *Church Review* 10 (April 1857): 88–105.

30. But see Conrad Edick Wright, *The Transformation of Charity in Postrevolutionary New England* (Boston: Northeastern University Press, 1992), with its

careful account of the multiplication of benevolent societies in the decade after the Revolution; and E. Brooks Holifield's detailed treatment of the relationship between urban ministerial salaries and the income of other professionals, "The Penurious Preachers: Nineteenth-Century Clerical Wealth, North and South," *Journal of the American Academy of Religion* 58 (Spring 1990): 17–36.

31. Church statistics are notoriously imprecise for all periods of American history, especially the revolutionary and postrevolutionary eras. The figures proposed here begin with a growth in national population from 2.6 million in 1776 to 31.5 in 1860. For church growth between 1775 and 1850, I use figures for church adherence (a larger number than formal membership) supplied by Roger Finke and Rodney Stark, "How the Upstart Sects Won America: 1776–1850," *Journal for the Scientific Study of Religion* 28 (March 1989): 27–44, esp. 31, and then multiply the 1850 figure by 30%, which is the average growth rate in *membership* calibrated for the nation's largest Protestant churches from 1853 to 1861 by Kathryn Teresa Long, *The Revival of 1857–58: Interpreting an American Religious Awakening* (New York: Oxford University Press, 1998), app. B, "Membership Statistics and Church Growth Rates, 1853–61," 144–50. If Finke and Stark's figure for adherence in 1775 is too low—as suggested by both Patricia U. Bonomi and Peter R. Eisenstadt, "Church Adherence in the Eighteenth-Century British American Colonies," *William and Mary Quarterly* 39 (1982): 245–86; and Richard W. Pointer, *Protestant Pluralism and the New York Experience: A Study of Eighteenth-Century Religious Diversity* (Bloomington: Indiana University Press, 1987), 29–31—then the rate of church growth from 1775 to 1860 is less impressive. If, on the other hand, their 1775 estimate is too high, as suggested by Jon Butler, *Awash in a Sea of Faith: Christianizing the American People* (Cambridge, Mass.: Harvard University Press, 1990), 191–92, then the rate of church growth from 1775 to 1860 is even more impressive. It is hard to believe that any fresh calculations would show church growth lagging behind the pace of population growth.

32. For excellent recent work on the Methodists, whose significance in the era was long matched by the insignificance of scholarship about them, see Russell E. Richey, *Early American Methodism* (Bloomington: Indiana University Press, 1991); A. Gregory Schneider, *The Way of the Cross Leads Home: The Domestication of American Methodism* (Bloomington: Indiana University Press, 1993); Russell E. Richey, Kenneth E. Rowe, and Jean Miller Schmidt, eds., *Perspectives on American Methodism* (Nashville, Tenn.: Kingswood, 1993); Christine Leigh Heyrman, *Southern Cross: The Beginnings of the Bible Belt* (New York: Knopf, 1997); John H. Wigger, *Taking Heaven by Storm: Methodism and the Rise of Popular Christianity in America* (New York: Oxford University Press, 1998); Sylvia R. Frey and Betty Wood, *Come Shouting to Zion: African American Protestantism in the American South and British Caribbean to 1830* (Chapel Hill: University of North Carolina Press, 1998), 105–08 and passim; Cynthia Lynn Lyerly, *Methodism and the Southern Mind, 1770–1810* (New York: Oxford University Press, 1998); and Dee E. Andrews, *The Methodists and Revolutionary America: The Shaping of an Evangelical Culture* (Princeton, N.J.: Princeton University Press, 2000).

33. See especially Walls, "American Dimension," 229–30, and "Missionary Societies and the Fortunate Subversion of the Church," in *Missionary Movement*, 241–54.

34. Nathan O. Hatch, *The Democratization of American Christianity* (New Haven, Conn.: Yale University Press, 1989). For an indication of these trends among Catholics, see Patrick W. Carey, *People, Priests, and Prelates: Ecclesiastical Democracy and the Tensions of Trusteeism* (Notre Dame, Ind.: University of Notre Dame Press, 1987). For how "democratization" could actually work to silence some groups in

American religious life, see Susan Juster, *Disorderly Women: Sexual Politics and Evangelicalism in Revolutionary New England* (Ithaca, N.Y.: Cornell University Press, 1994).

35. Jonathan Hughes and Louis P. Cain, *American Economic History*, 5th ed. (Reading, Mass.: Addison-Wesley, 1998), 147, 151, 156, 170.

36. On the historiographical aversion to studying those dynamics, see Hatch, "Puzzle of American Methodism," 175–89; Hatch, "Refining the Second Great Awakening: A Note on the Study of Christianity in the Early Republic," in *Democratization*, 220–26; and Michael Zuckerman, "Holy Wars, Civil Wars: Religion and Economics in Nineteenth-Century America," *Prospects: The Annual of American Studies*, 1992, 205–40.

37. See, for example, Johnson, *Islands of Holiness*, 45–47; Richard Carwardine, "Unity, Pluralism, and the Spiritual Market-Place: Interdenominational Competition in the Early American Republic," in *Unity and Diversity in the Church*, ed. Robert Swanson, *Studies in Church History*, No. 32 (Oxford: Blackwell, 1996), 297–335; Carwardine, *Evangelicals and Politics in Antebellum America* (New Haven, Conn.: Yale University Press, 1993), 228–29 and passim; and for Baptist-Methodist-Presbyterian strife in British North America, Daniel C. Goodwin, "'The Faith of the Fathers': Evangelical Piety of Maritime Regular Baptist Patriarchs and Preachers, 1790–1855" (Ph.D. diss., Queen's University, Ontario, 1997).

38. Curtis Johnson, *Redeeming America: Evangelicals and the Road to Civil War* (Chicago: Ivan R. Dee, 1993), with definitions, 7–8.

39. Daniel Walker Howe, *The Political Culture of the American Whigs* (Chicago: University of Chicago Press, 1979).

40. Hatch, *Democratization*; Richard T. Hughes, *Reviving the Ancient Faith: The Story of Churches of Christ in America* (Grand Rapids, Mich.: Eerdmans, 1996), 33, 112; and George A. Rawlyk, *The Canada Fire: Radical Evangelicalism in British North America, 1775–1812* (Montreal: McGill–Queen's University Press, 1994).

41. All of the more recent, sophisticated studies of Methodism (see note 32) stress the evolution of Methodism toward formalism; for particular sensitivity to the economic dimensions of that evolution, see Wigger, *Taking Heaven by Storm*, 175–77, 188–93. The best study of how rising income affected tastes in church construction is a book on Upper Canada, whose conclusions would also pertain in large measure to the United States: William Westfall, *Two Worlds: The Protestant Culture of Nineteenth-Century Ontario* (Montreal: McGill-Queen's University Press, 1989).

42. See Carol V. R. George, *Segregated Sabbaths: Richard Allen and the Emergence of Independent Black Churches* (New York: Oxford University Press, 1973).

43. See, for example, David Brion Davis, *The Problem of Slavery in the Age of Revolution, 1770–1823* (New York: Oxford University Press, [1975] 1999); Thomas Bender, John Ashworth, David Brion Davis, and Thomas L. Haskell, *The Antislavery Debate* (Berkeley: University of California Press, 1992); John Ashworth, *Slavery, Capitalism, and Politics in the Antebellum Republic, Vol. 1: Commerce and Compromise, 1820–1850* (New York: Cambridge University Press, 1995); Eugene D. Genovese, *The Slaveholders' Dilemma: Freedom and Progress in Southern Conservative Thought, 1820–1860* (Columbia: University of South Carolina Press, 1992); and James L. Huston, "Abolitionists, Political Economists, and Capitalism," *Journal of the Early Republic* 20 (Fall 2000): 487–521.

44. Elizabeth Cady Stanton, et al., eds., *History of Woman Suffrage*, 2nd ed. (Rochester, N.Y., n.p. 1889), 1:70–71.

45. For perceptive discussion of how the developing "women's sphere" reduced the scope for women's public ministries, see Catherine A. Brekus, *Strangers and Pilgrims: Female Preaching in America, 1740–1845* (Chapel Hill: University of North Carolina Press, 1998), 267–306.

46. William R. Sutton, "Benevolent Calvinism and the Moral Government of God: The Influence of Nathaniel W. Taylor on Revivalism in the Second Great Awakening," *Religion and American Culture* 2 (Winter 1992): 39. For Kuklick's caution about linking intellectual traditions to social conditions, see *Churchmen and Philosophers from Jonathan Edwards to John Dewey* (New Haven, Conn.: Yale University Press, 1985), 301–2. Fruitful reflection on the bearing of economic practice on theology in colonial New England is found in Mark Valeri, "Universal Signs: The Language of Theology and the Culture of the Market in Early New England," unpublished paper for the Institute of Early American History and Culture, Williamsburg, Va., June 1998.

47. Richard Rabinowitz, *The Spiritual Self in Everyday Life: The Transformation of Personal Religious Experience in Nineteenth-Century New England* (Boston: Northeastern University Press, 1989), 227–28.

48. James D. Bratt, *Dutch Calvinism in Modern America* (Grand Rapids, Mich.: Eerdmans, 1984), 59.

49. See Kathryn T. Long, "Consecrated Respectability: Phoebe Palmer and the Refinement of America Methodism," in *Methodism and the Shaping of American Culture*, ed. Nathan O. Hatch and John H. Wigger (Nashville, Tenn.: Kingswood, 2001), 281–307.

50. This assessment leans considerably on the arguments in James D. Bratt, "The Reorientation of American Protestantism, 1835–1845," *Church History* 67 (March 1998): 52–82.

51. See Keith J. Hardman, *Charles Grandison Finney, 1792–1875: Revivalist and Reformer* (Syracuse, N.Y.: Syracuse University Press, 1987), 337–38; and Bertram Wyatt-Brown, *Lewis Tappan and the Evangelical War Against Slavery* (Cleveland, Ohio: Press of Case Western Reserve University, 1969), 168–70, 174–75.

52. Charles G. Finney, "Lecture XIII: Being in Debt," *Oberlin Evangelist* 1, no. 17 (July 31, 1839): 130.

53. Long, *Revival of 1857–58*, 124. For an article that expands upon these features of Long's book, see Joel A. Carpenter, "Revivalism Without Social Reform," *Books & Culture: A Christian Review* (Nov./Dec. 1998), 26–27.

54. Donald G. Mathews, *Slavery and Methodism: A Chapter in American Morality, 1780–1845* (Princeton, N.J.: Princeton University Press, 1965); Mathews, *Religion in the Old South* (Chicago: University of Chicago Press, 1977).

55. See especially David Edwin Harrell, Jr., *Quest for a Christian America: The Disciples of Christ and American Society to 1866* (Nashvillem, Tenn.: Disciples of Christ Historical Society, 1966), chap. 3, "An American Economic Gospel."

56. On the movement's typicality, see especially Ruth Alden Doan, *The Miller Heresy, Millennialism, and American Culture* (Philadelphia: Temple University Press, 1987).

57. On that merchandising, see especially Hatch, *Democratization*, 145.

58. On the antiformalism, see Richard Bushman, *Joseph Smith and the Beginning of Mormonism* (Urbana: University of Illinois Press, 1984); on the formalism, see Jan Shipps, *Mormonism: The Story of a New Religious Tradition* (Urbana: University of Illinois Press, 1985).

59. See Daniel Walker Howe, chapter 3, p. 69, in this volume.

60. Charles Sellers, *The Market Revolution: Jacksonian America, 1815–1846* (New York: Oxford University Press, 1991).

61. For a summary statement of that metanarrative, including reference to Thompson, see Winifred Barr Rothenberg, *From Market-Places to a Market Economy: The Transformation of Rural Massachusetts* (Chicago: University of Chicago Press, 1992), 2–3. Important American instances of a picture that resembles Thompson's include Anthony F. C. Wallace, *Rockdale: The Growth of an American Village in the Early Industrial Revolution* (New York: Norton, 1972), "The Theory of Christian Capitalism," 394–97; and Sean Willentz, *Chants Democratic: New York City and the Rise of the American Working Class, 1788–1850* (New York: Oxford University Press, 1984), for example, p. 87: "Evangelicalism began to affect the shops directly only when the new workshop regime—and the social boundaries of class—that had begun to emerge in the Jeffersonian period matured."

62. E. P. Thompson, *The Making of the English Working Class* (New York: Vintage, 1966), 365.

63. These responses first appeared in a volume entirely devoted to assessing Sellers's proposals: Melvyn Stokes and Stephen Conway, eds., *The Market Revolution in America: Social, Political, and Religious Expressions, 1800–1880* (Charlottesville: University Press of Virginia, 1996), used by permission.

64. George M. Thomas, *Revivalism and Cultural Change: Christianity, Nation Building, and the Market in the Nineteenth-Century United States* (Chicago: University of Chicago Press, 1989), x.

65. Richard Lyman Bushman, "Markets and Composite Farms in Early America," *William and Mary Quarterly* 55 (July 1998): 351–74.

66. Alan Macfarlane, *The Origins of English Individualism: The Family, Property, and Social Transition* (Oxford: Blackwell, 1978).

67. See Harry S. Stout, *The Divine Dramatists: George Whitefield and the Rise of Modern Evangelicalism* (Grand Rapids, Mich.: Eerdmans, 1991); Frank Lambert, *"Pedlar in Divinity": George Whitefield and the Transatlantic Revivals, 1737–1770* (Princeton, N.J.: Princeton University Press, 1994); and as an example of Breen's work, "An Empire of Goods: The Anglicization of Colonial America, 1690–1776," *Journal of British Studies* 25 (October 1986): 467–99.

68. For example, Rothenberg, *From Market-Places to a Market Economy*; and Gordon S. Wood, "Inventing American Capitalism," *New York Review*, June 9, 1994, 44–49.

69. For the former, see J. C. D. Clark's effort to restore the religious credibility of the ancien regime in works like *English Society, 1688–1832* (Cambridge: Cambridge University Press, 1983); and *The Language of Liberty, 1660–1832: Political Discourse and Social Dynamics in the Anglo-American World* (Cambridge: Cambridge University Press, 1994). For the latter, see, for example, W. R. Ward, *Religion and Society in England, 1790–1850* (New York: Schocken, 1973); Ian R. Christie, *Stress and Stability in Late Eighteenth-Century Britain: Reflections on the British Avoidance of Revolution* (Oxford: Clarendon, 1984), 206–11; James E. Bradley, *Religion, Revolution, and English Radicalism: Nonconformity in Eighteenth-Century Politics and Society* (Cambridge: Cambridge University Press, 1990), 32–33; and David Hempton, "Motives, Methods and Margins in Methodism's Age of Expansion," *Proceedings of the Wesley Historical Society* 49 (1994): 189–207.

70. Howe, *Political Culture*.

71. See, for example, Richard Carwardine, chapter 4, pp. 85–86, in this volume.

72. Thomas, *Revivalism and Cultural Change*, chap. 3, with summary, 7.

73. R. Laurence Moore, *Selling God: American Religion in the Marketplace of Culture* (New York: Oxford University Press, 1994), 91.

74. Terry D. Bilhartz, *Urban Religion and the Second Great Awakening: Church and Society in Early National Baltimore* (Cranbury, N.J.: Associated University Presses, 1986), 139.

75. Roger Finke, "Supply Side Explanations for Religious Change," in *Rational Choice Theory and Religion: Summary and Assessment*, ed. Lawrence A. Young (New York: Routledge, 1997), 48. With essays by Finke, Rodney Stark, R. Stephen Warner, Nancy T. Ammerman, and others, this book is a valuable collection. Major statements about and illustrating this interpretive strategy are Finke, "Religious Deregulation: Origins and Consequences," *Journal of Church and State* 32 (1990): 609–26; Roger Finke and Rodney Stark, *The Churching of America: Winners and Losers in our Religious Economy* (New Brunswick, N.J.: Rutgers University Press, 1992); and R. Stephen Warner, "Work in Progress toward a New Paradigm for the Sociological Study of Religion in the United States," *American Journal of Sociology* 93 (1993): 1044–93.

76. See Martin E. Marty, Review of Finke and Stark *Churching of America*, in *Christian Century*, January 27, 1993, 88–89.

77. Pioneers in this effort were Wallace, *Rockdale*; Paul E. Johnson, *A Shop-keeper's Millennium: Society and Revivals in Rochester, New York, 1790–1865* (New York: Hill & Wang, 1978); and Mary P. Ryan, *Cradle of the Middle Class: The Family in Oneida County, New York, 1790—1865* (New York: Cambridge University Press, 1981). More recent examples include Randolph A. Roth, *The Democratic Dilemma: Religion, Reform, and the Social Order in the Connecticut River Valley of Vermont, 1791–1850* (New York: Cambridge University Press, 1987); John L. Brooke, *The Heart of the Commonwealth: Society and Political Culture in Worcester Country Massachusetts, 1713–1861* (New York: Cambridge University Press, 1989); Curtis Johnson, *Islands of Holiness*; Rabinowitz, *Spiritual Self in Everyday Life*; Alan Taylor, *Liberty Men and Great Proprietors: The Revolutionary Settlement on the Maine Frontier, 1760–1820* (Chapel Hill: University of North Carolina Press, 1990); David G. Hackett, *The Rude Hand of Innovation: Religion and Social Order in Albany, New York, 1652–1836* (New York: Oxford University Press, 1991); Rothenberg, *From Market-Places to a Market Economy*; and Alan Taylor, *William Cooper's Town: Power and Persuasion on the Frontier of the Early American Republic* (New York: Vintage, 1995).

78. Something of a pioneer was Bilhartz, *Urban Religion*. Confronting directly the standard picture of labor-evangelical antagonism are Teresa Anne Murphy, *Ten Hours' Labor: Religion, Reform, and Gender in Early New England* (Ithaca, N.Y.: Cornell University Press, 1992); Jama Lazerow, *Religion and the Working Class in Antebellum America* (Washington, D.C.: Smithsonian Institution Press, 1995); and William R. Sutton, *Journeymen for Jesus: Evangelical Artisans Confront Capitalism in Jacksonian Baltimore* (University Park: Pennsylvania State University Press, 1998).

79. Kenneth Moore Startup, *The Root of All Evil: The Protestant Clergy and the Economic Mind of the Old South* (Athens: University of Georgia Press, 1997).

80. For example, Mark Valeri, "The Economic Thought of Jonathan Edwards," *Church History* 60 (March 1991): 37–54; David Paul Nord, "Free Grace, Free Books, Free Riders: The Economics of Religious Book Publishing in Early Nineteenth-Century America," *Proceedings of the American Antiquarian Society* 106 (1996): 241–72; Nathan O. Hatch, "The Second Great Awakening and the Market Revolution," in

Devising Liberty: Preserving and Creating Freedom in the New American Republic, ed. David Thomas Konig (Stanford, Calif.: Stanford University Press, 1995), 243–64; Linda Pritchard, "The Spirit in the Flesh: Religion and Regional Economic Development," in *Belief and Behavior: Essays in the New Religious History*, ed. Philip R. Vandermeer and Robert P. Swierenga (New Brunswick, N.J.: Rutgers University Press, 1991), 88–116; Mark S. Schantz, "Religious Tracts, Evangelical Reform, and the Market Revolution in Antebellum America," *Journal of the Early Republic* 17 (Fall 1977): 425–66; and Robert Wuthnow and Tracy L. Scott, "Protestants and Economic Behavior," in *New Directions in American Religious History*, ed. Harry S. Stout and D. G. Hart (New York: Oxford University Press, 1997), 260–95.

81. For example, Daniel Calhoun, *The Intelligence of a People* (Princeton, N.J.: Princeton University Press, 1973); Stout, *Divine Dramatist*; Lambert, *"Pedlar of Divinity"*; Carwardine, *Evangelicals and Politics in Antebellum America*; Mark Y. Hanley, *Beyond a Christian Commonwealth: The Protestant Quarrel with the American Republic, 1830–1860* (Chapel Hill: University of North Carolina Press, 1994); Wosh, *Spreading the Word*; Richard R. John, *Spreading the News: American Postal System from Franklin to Morse* (Cambridge, Mass.: Harvard University Press, 1995); and Long, *Revival of 1857–58*.

82. Paul Johnson, *Shopkeeper's Millennium*, 141; Curtis Johnson, *Islands of Holiness*, 86.

2

Protestants and the American Economy in the Postcolonial Period: An Overview

Robin Klay and John Lunn

Attempts to write the economic history of the United States for the period between the ratification of the Constitution and the outbreak of the Civil War are handicapped by a scarcity of precise aggregate data. The carefully verified and expertly tabulated summaries of economic information that have been prepared by the United States Department of Commerce, for example, are much fuller for decades beginning in the late nineteenth century than for the early 1800s.[1] Efforts aimed at the economic history of American religious groups are even more scantily served by collections of aggregate financial data. To be sure, the fact that the Census Bureau during parts of the nineteenth century collected systematic information on churches means that certain data exist for the earlier period that have not been collected throughout the twentieth century. For example, a clearer picture of the total value of church property is present for 1850 and 1860 than for a century later. Yet for religious bodies, as well as for the nation as a whole, the general trend is for the quality and quantity of well-verified data to increase the closer we come to the present.

Despite measurement problems, an attempt to outline a general picture is necessary as context for the more specific studies that follow in this book. For that purpose, an elementary sketch of national economic history will review the main conclusions of economic historians for the period. That sketch is followed by an effort to grasp the scale of religious participation in the national economy. The chief finding of this preliminary survey from the sketchy evidence available is that the Protestant churches and religious societies, although never an overwhelming force, did exert more weight in the early American economy than has been the case in later American history. Finally, to impart some specificity to an otherwise general account, a relatively full set of financial records from one church in Albany, New York, will be explored to illustrate the way in which individual institutions functioned as significant economic, as well as religious, entities in their particular localities.

The Postcolonial Economy: From Regional Markets to a National Economy

During the period 1790–1860, the United States grew from a small country on the eastern edge of the continent to one that eventually encompassed the forty-eight contiguous states.[2] Table 2.1 details the changes in the size of the country from the days just after the ratification of the new Constitution to just before the Civil War. Population increased eightfold, and territory more than tripled in size. For the economy, agriculture increased absolutely but decreased in relative importance because of the burgeoning manufacturing and trade. From 1790 to 1860, the proportion of American workers in agriculture declined from approximately five out of six to about one out of two. Important economic forces at work in this early period included land tenure decisions, westward expansion, and improvements in transportation. Regional differences were also important for the interrelationship between economic activity and church finances.

The United States in 1790 did not have a national economy. High transportation costs and a thinly dispersed population meant that most economic activity was local. Only from the 1810s, as canal building expanded and national monetary institutions began to develop, did a national economy (and improved economic record keeping) emerge. Even in an increasingly national economy, however, the country's three main regions differed from one another

TABLE 2.1. Population and Territory 1790-1860

Year	Total Resident Population (thousands)	Land Area (square miles)	Total Territory (square miles)
1790	3,929	864,746	888,811
1800	5,297	864,746	888,811
1810	7,224	1,681,828	1,716,003
1820	9,618	1,749,462	1,788,006
1830	12,901	1,749,462	1,788,006
1840	17,120	1,749,462	1,788,006
1850	23,261	2,940,042	2,992,747
1860	31,513	2,969,640	3,022,387

Source: *Historical Statistics of the United States, Colonial Times to 1957* (Washington, D.C.: U.S. Bureau of Census, 1960), Series A7 and J1–2.

significantly and continued to do so through the era of the Civil War. One main area was the North, consisting of New England and the former "middle colonies"—New York, New Jersey, Pennsylvania, and Delaware. Although there were differences between New England and the middle states, key similarities were more important. In particular, farming tended to be on a smaller scale and similar patterns of industry developed in both subregions. A second main area was made up of the southern states, where plantation agriculture played an unusually large role. A final region was the West, consisting of the lands beyond the Appalachians, which originally were sparsely populated by Americans of European descent. Of the 3.9 million people enumerated by the first census in 1790, only about 200,000 lived in the West, with the remaining divided between the North and the South. Approximately 700,000 were slaves. During the seven decades prior to the Civil War, all regions moved westward, with the "economic North" and the "economic South" adding new states and territories and the "economic West" moving further outward, toward the Mississippi River and beyond.[3]

An important difference between the West and the other two regions was that more of its economic activity did not involve markets. The difficulty of transporting heavy goods into or out of the frontier necessitated a considerable degree of self-sufficiency. There was also a greater tendency to barter when exchanging goods and services rather than to use money. Not coincidentally, organized church activity in the West lagged behind levels in the settled North and South; during these years there were, for example, far fewer full-time clergy per unit of population in the West than in the other two regions.

Land Tenure and Distribution

A critical early question in the economic history of the United States was the disposition of western lands.[4] Several of the original colonies claimed lands west of their borders, and these claims often conflicted. Furthermore, some whites had purchased land from Native Americans directly, and these claims often clashed with other settlers' claims. The new country needed a way to allocate these lands in a legally recognized manner. From early times families had moved into the wilderness, cleared land, and farmed without bothering to secure clear titles. The first step toward resolving this issue was that the new states turned over their western land claims to the national government. These lands became part of the public domain of the United States.

The second step involved transferring the public lands to private citizens. The Land Ordinances of 1785 and 1787 established the principles that led ultimately to the distribution of land in 31 of our present 50 states. Thomas Jefferson was the major force behind these principles. First, tenure acquired by private persons from the federal government would not be subject to further government claim. Second, the distribution should be orderly and sales would take place only after the land was surveyed. Third, the new lands would become eligible to form new states on an equal footing with the original states.[5]

The Ordinance of 1785 called for selling the land at auction. The land was surveyed and arranged in 6-mile-square townships with thirty-six sections of 640 acres each. One section was set aside in each township for the support of public schools, and four sections were set aside for government use. A proposal to set aside land for church use—which would have maintained something like the historic glebe system—was defeated in congressional debate. Thus, even before the passage of the First Amendment in 1789, the federal government indicated that it would not be providing financial support for churches. Although several of the original thirteen states continued to provide such maintenance or support of some kind for certain churches, this historic pattern did not develop in the new states.[6]

Another factor that affected land distribution was the federal government's aggressive pursuit of new land. The most famous example is the Louisiana Purchase of 1803; by 1860, the United States had acquired all the territory of the contiguous forty-eight states. Table 2.2 details these acquisitions. Because of the presence of resistant Native Americans, not all this new land could immediately be settled by white Americans.

For an age preoccupied by questions of land tenure, it is pertinent to ask why the United States did not recognize the property rights of the original occupants of the land. The answer is that the new national government followed European practices concerning the establishment of property rights. An important dimension to this philosophy was that the purpose of property rights was to develop the land. Since it was widely held that the Indians used the land but did not develop it, they were seen as not having proper claims to the property.[7] The significance of land development continued as the western lands were settled. In general, when disputes arose concerning the use of specific parcels of land, the courts tended to side with the person who was using the land to enhance economic development. It fits with the conclusions drawn in other chapters of this book to note that although some Protestants did protest white treatment of Native Americans, their protests almost never concentrated on these widespread conventions about what it meant to "own" land.

TABLE 2.2. Area and Acquisition of the Public Domain, United States: 1790–1860

Year	Acquisition	Acquired Area (square miles)	Total Area of U.S. (square miles)
1790			888,811
1803	Louisiana Purchase	827,192	1,716,003
1819	By treaty with Spain	72,003	1,788,006
1845	Texas	390,143	2,178,149
1846	Oregon	285,580	2,463,729
1848	Mexican Cession	529,017	2,992,747
1853	Gadsden Purchase	29,640	3,022,387

Source: *Historical Statistics of the United States, Colonial Times to 1957.* (Washington, D.C.: U.S. Bureau of the Census, 1960), Series J1–2.

Land sales did not follow a steady path, as might be expected, but sales occurred in three waves that coincided with upswings in the business cycle. Figure 2.1 shows the annual sales figures from 1800 to 1860. There are peaks around the booms of 1818, 1836, and 1854–55, with an incredible 20 million acres sold in 1836 alone (greater than the combined land area of Connecticut, Massachusetts, New Hampshire, and Vermont). The steep declines in land sales after the booms coincide with business depressions and monetary crises. Since immigration was not a significant factor before 1840, the first two peak periods reflect demand for land from people already settled in the United States. Stanley Lebergott is also able to explain a considerable amount of the annual variation in land sales in the South on the basis of cotton prices, improvements in river transportation, and cessions of Indian lands.[8] For his part, Douglass North has shown that land sales in seven western states followed price movements of corn and wheat fairly closely.[9] He argues that rising prices led to sales of new lands in western regions, then (after the land was cleared) to large increases in the acreage planted. The resulting large increase in supply then led to declining prices. The pattern was evident both in the South with cotton and in the West with corn and wheat.[10]

Improvements in Transportation

The major ports along the Atlantic coast were important trading centers from the earliest days of the nation. Yet other parts of the country were more isolated from world trade and even from other parts of the country. Transporta-

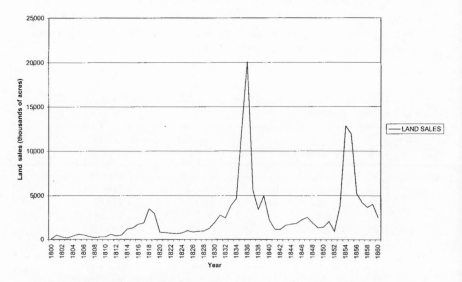

FIGURE 2.1 Land Sales, 1800–1860

tion costs were high, which made it difficult for specialized manufacturers in one place to supply the needs of other places. Douglass North provides an example of the need for local specialization from a census of manufacturing interests in and around Mt. Pleasant Ohio in 1815, which numbered about 500 residents at the time:

> 3 saddlers, 3 hatters, 4 blacksmiths, 4 weavers, 6 boot and shoemakers, 8 carpenters, 3 tailors, 3 cabinet makers, 1 baker, 1 apothecary, and 2 wagon makers shops—2 tanneries, 1 shop for making wool carding machines, 1 with a machine for spinning wool, 1 manufactory for spinning thread from flax, 1 nail factory, 2 wool carding machines. . . . Within the distance of six miles from the town were—9 merchants mills, 2 grist mills, 12 saw mills, 1 paper mill with 2 vats, 1 woolen factory with 4 looms and 2 fulling mills.[11]

This system of production can be called household manufacture because of its small scale and because it often took place literally in the household.

Rapid growth of manufacturing began in the 1830s. Manufacturing (with textiles in the lead) was located mostly in the Northeast, although the West did have some industry tied to its particular resources. An important source for the expansion of manufacturing was the reduced costs of transport. In the early years of the republic, farmers in the West did not sell much of their produce because it was very costly to transport to the eastern population. Often westerners used agricultural produce to feed cattle and then walked the cattle to market or converted the crops into whisky, which had a higher value relative to the cost of shipping. Similarly, manufacturers could not exploit scale economies because of the high cost of shipping. It was often easier to ship between the United States and Europe than from the East Coast of the nation to the interior.

Several factors led to reduced costs for transportation, especially the building of canals, roads, and railroads, although the effect of railroad building became important only toward the end of this period. State governments were responsible for most of these improvements in transportation.[12] The Erie Canal was the most famous project, but numerous other canals were important during the period. By the 1850s, railroads began to supersede the canals. From 1815 to 1860, 4,254 miles of canals were constructed. By 1860, there were over 30,000 railroad miles. Canals were especially important for linking the Great Lakes to the North, and thus strengthening economic links between the West and the North, whereas the Mississippi River aided trade between the South and West. For example, between 1836 and 1860, trade on the Erie Canal from the western states increased from 54,219 tons to 1,896,975 tons.[13] This increase in trade was accompanied by a dramatic drop in unit cost per mile. Whereas transporting freight from Buffalo to New York in 1817, before the Erie Canal was opened, cost 19.12 cents per ton-mile, the cost on the Erie Canal for the period 1857–60 was 0.81 cents per ton-mile.[14] Initially, such dramatic economic gains put special strain on the churches, especially along the Erie Canal. The most pressing issue concerned what position to take on the movement of goods and the use of the postal system (which was greatly stimulated by economic development as well) on Sundays.[15]

By the end of this period, the railroad was performing the same function of increasing tonnage and cutting cost. By 1860, railroad mileage was seven times greater than canal mileage. As a natural consequence of these developments, the Northeast became more dependent on western food while at the same time adding to its development of manufacturing. The reduction of freight costs allowed western farmers to become a part of the national economy and to engage in greater specialization of labor themselves.

Periods of Economic Activity

During its earliest days (1790–1814), the new American economy was dominated by international concerns. Thereafter, domestic factors predominated, although a constant factor in the growth of the American economy was always trade with Europe.

During the period 1815–23, after the hostilities of the War of 1812 were ended, the South flourished, whereas the North went through a painful period of readjustment. Cotton became the most important export crop, accounting for about 39 percent of all exports.[16] Cotton prices rose from 1815 to 1818, promoting prosperity in the South, and then fell sharply from 1818 to 1820, resulting in a sharp depression over those years. The high prices had stimulated cultivation of new lands, financed partly by debt and speculation. When the new sources of supply became productive, cotton prices fell and many borrowers could not repay their debts.

The West also expanded during these years, and population grew rapidly as many settlers moved into the region from the eastern seaboard. From 1815 to 1819, 7.5 million acres of land were sold in the Northwest Territories. Demand was high for farm products from the region. Improvements in transportation benefited the region greatly, benefits that were, however, considerably slowed by a financial crisis in 1821–22.

In the next general period, 1823–43, critical changes occurred in American economic history, which Douglass North has helpfully summarized as follows:

> If one were to date the beginning of acceleration in the economy's growth and the years when industrialization began, it would be during this period. It was also a period of tremendous westward expansion. Underlying both these developments was the cotton trade. During this period, cotton played its greatest role in inducing growth in the size of the market for manufacturers and in influencing the pace of the westward movement. When this era of expansion came to a close with the severe depression of the early 1840s, the cotton trade had irrevocably done its part as the proximate prime mover in quickening the pace of the country's growth.[17]

Prices for almost all goods and services declined during most of the 1820s before rising the following decade, except for the years 1834 and 1837–38. The depression of 1839–43 was one of the most severe in the nation's history prior to the Great Depression of the 1930s. Each of the regions experienced fundamental transformation, including greater interregional dependence

among them all. The North came more and more to be the center for manufacturing, especially textile manufacturing, whereas the South continued to
concentrate on cotton.[18] A shift in population was also underway from the
Old South (along the Atlantic Coast) to the New South (states further to the
west such as Alabama, Mississippi, and Louisiana). During the decade of
the 1820s, while the still much more populous states of North Carolina, South
Carolina, and Georgia grew by not quite one-quarter, the populations of Alabama, Mississippi, and Louisiana increased by 85 percent. The West prospered as it traded more with the South, especially those western areas where
there was ready access to waterways.

Manufacturing was also increasing with unusual speed. From 1821 to 1825,
the annual production of cotton spindles in the Northeast grew from 230,000
to 800,000.[19] Shipping declined in importance as an industry, and the domestic market became more important than the international market. Service industries to facilitate manufacturing also appeared with increasing frequency.

The 1830s saw increasing prosperity for all three regions, with each continuing previous economic patterns. The South relied on cotton, the West grew
food products for the North and the South, and the North produced manufactured goods for the South and the West. The rising prosperity of each region
stimulated growth in the other regions. Rising incomes meant fresh demand
for products like clothes, shoes, and leather goods. Machinery for canal building, steamboats, and farm implements was also produced in dramatically increased quantities. The surge of growth in the market for manufactured products spurred specialization. As one very important instance, the rapid growth
of textile manufacturing stimulated increased production of steam engines,
not only for the textile sector but also for ships and railroads. Capital markets
that had earlier developed in connection with cotton and foreign trade were
now opened to the financing of domestic manufacturing.[20]

The depression that began in the fall of 1839 brought dramatic changes.
Hardest hit were the South and the West. Several states defaulted on interest
payments on their debts. Cotton farmers reduced acreage for their main crop
in order to grow their own food supplies and so seek self-sufficiency. Capital
flow from the North to the West fell sharply, causing severe depression in
many western states. The Northeast was harmed least because of flexible
wages, which allowed employment to remain high and firms to continue operating. Once the nation had pulled out of the depression, growth returned
but cotton was no longer as central to the economy.[21]

The next period in the nation's economic history, 1844–1860, witnessed
vast acquisitions of land—Texas, the Oregon territory, California, and the
Southwest. The California gold rush of 1849 and the following year led eventually to California's entry into the Union as a state. During the years 1844 to
1855, manufacturing grew more rapidly than at almost any other time in the
nation's history. The Northeast became a full-fledged manufacturing region.
From 1844 to 1859, while the national population was growing by 57 percent,
value added by manufacturing increased 127 percent (whereas the value added
by agriculture increased 58 percent over the same period). Ties between the

North and the West strengthened, although those between the West and the South weakened. In the northern region the market had grown large enough to support a substantial manufacturing base on its own. Railroad construction was becoming important in the economy, both as a stimulus to manufacturing and as a factor in reducing transportation costs.[22] Foreign capital became a critical source of investment for financing railroad construction.

By this time, the business cycle was firmly in place. A severe recession in 1854 affected the West badly, but recovery was rapid. The financial crisis in 1857 affected all regions of the country. But again the crisis was not prolonged, though the expansion from 1857 to 1861 was relatively modest. The long period of relative prosperity and peace came to an end in 1861 with the beginning of the Civil War. By the time the war began, the American economy had expanded tremendously; it had been transformed from its local, agriculturally based beginning into a truly inter-regional economy fueled by growing industrial capacity. Per capita income had approximately quadrupled (in real terms) from 1790 to 1860, making the United States one of the wealthiest nations in the world on a per capita basis.[23] After the diversion of economic resources occasioned by the war, the process of industrialization continued.

For the entire period between the Revolution and the Civil War, the economics of slavery has always been an obviously important, but also highly contentious, issue.[24] Modern economic historians have studied with great sophistication such questions as whether the economic gain to the Old (eastern) South from the sale of slaves to the New (Western) South was offset by declining land values in the former, or whether the systemic malnourishment of slaves was a result of a deliberate decision or mere inadvertence. Most economic historians now believe that it was still functioning productively on the eve of the Civil War, but there is more disagreement on whether the slave system was more efficient than free agriculture or whether the material (as opposed to moral and psychological) condition of slaves was better than that of industrial laborers in the North.[25] Whatever the consensus on such questions, there is complete agreement on the great economic importance of the slave system. When the economic aspects of slavery are considered along with its moral aspects, and when the churches' own deep involvement in the national economy is grasped, there is even more reason for understanding why slavery became the most controversial issue for the churches in the decades before the war.

Protestants and the National Economy

Because only a few American denominations systematically collected and reported comprehensive financial information—and those few not until the years shortly before 1860—it is difficult to calculate with precision the exact role of Protestants in the early American economy. Studies of individual denominations or voluntary societies, such as that by David Paul Nord, chapter 7 in this volume, do suggest that Protestant organizations raised significant sums, but what those sums meant in comparative social, cultural, or economic terms has not yet been fully clarified. Money given to the churches and voluntary societies, as well as property owned by religious groups, does not seem to have been unusually large relative to the wealth being generated by the incredibly rapid expansion of private enterprise. Yet by comparison to the money controlled and spent by the national government, which of course was much smaller in the antebellum period than it has become since the Great Depression, Protestant religious groups occupied a surprisingly large position.

The eighth census of the United States, for 1860, provided figures for the annual value of various products manufactured in the country.[26] In 1860, contributions to Protestant churches and voluntary societies may be estimated in the range of $35–$40 million.[27] By comparison, the value of flour and meal produced was $224 million, of cotton goods $115 million, and of lumber $96 million.[28] Contributions to Protestant causes were comparable to the production value of machinery and steam engines ($47 million); printing, including books and newspapers ($42 million); and sugar refining ($38.5 million).

What such aggregates mean individually can be suggested by figures from the Reformed Church in America (RCA), the denomination descended from early Dutch immigration, which in 1857 became one of the first American denominations to provide information about total church giving alongside its annual reports of membership. In 1860, with 51,740 members from 33,130 families, the total contributions to the denomination's individual churches was $527,210. This figure represented $10.19 per member, or $15.49 per family.[29] Also in 1860, skilled artisans on the Erie Canal made from $1.75 to $3.00 per day.[30] If the income of RCA members met those levels, annual giving to their churches represented one to two weeks of labor, or perhaps 2% to 4% of annual income. As a point of comparison, this percentage would rank toward the high side of giving to American churches in the early 1990s.[31]

Comparative assessment suggests that such levels of religious giving in the antebellum period meant a great deal in the national economy. The comparison that points most strongly to such a conclusion concerns religious contributions versus the monies of the federal government. The more conservative of the two national Presbyterian denominations, the "Old School," began to record total denominational giving somewhat earlier than even the RCA. Its figures show that this one denomination, whose membership never exceeded 1% of the national population and whose adherents would never have numbered more than 3% to 4% of the population, collected annual sums equal to between one-thirty-sixth and one-eighteenth of the total revenues of the federal government (see Table 2.3). By the end of the 1850s, a decade marked by steadily shrinking federal revenues, the total annual contribution to Protestant denominations and agencies (estimated at $35 to $40 million) was coming close to equaling the total income of the federal government.

A modern comparison puts those ratios into perspective. The *Yearbook of American and Canadian Churches* reported that in 1995, 55 American denominations, representing over 48 million adherents (or 18% of the national

TABLE 2.3. Old School Presbyterian Giving in Relation to Federal Income

Year	Members	% of National Population	Total Giving	Federal Income	Government: Presbyterian
1851	210,306	0.9	1.46m	52.56m	36:1
1854	225,404	0.8	2.05m	73.80m	36:1
1857	244,825	0.8	2.76m	68.96m	25:1
1860	292,926	0.9	3.18m	58.06m	18:1

Source: Herman C. Weber, *Presbyterian Statistics: Through One Hundred Years, 1826-1926* (Philadelphia: General Council of the Presbyterian Church in the U.S.A., 1927).

population), contributed $21.4 billion to their churches. In that year the federal government took in $63 for every dollar donated to all of those churches.

By the 1850s, Methodists had constructed about as many churches as the U.S. government had built post offices, and the two main Methodist denominations employed about as many ministers as the post office did employees, and this at a time when postal workers constituted the vast majority of federal employees. In addition, each of the large Protestant denominations was raising about as much money each year as the postal service took in for its business. (See Table 2.4 for these and other comparisons.)

The proper conclusion from these comparisons is not that the big Protestant denominations were necessarily a giant force in the antebellum economy, for the U.S. government was a much smaller enterprise than it later became. Similarly, the fact that Old School Presbyterian annual contributions in the early 1850s (about $1.5 million) came close to matching the total annual income of all colleges and universities in the United States (about $2 million)[32] indicates mostly how very tiny higher education still was at that time. Still, these kinds of comparisons do show that antebellum churches were not marginal economic forces. In the absence of a dominant economic presence like the modern federal government, churches exerted considerable economic influence.

The social, as well as the economic, reach of the Protestant denominations was also extended many times over by the extraordinary activities of the

TABLE 2.4. Churches and the National Economy, 1850 and 1997

	1850	1997	Multiple
Value of church buildings	87m	????	
National savings deposits	43m	227b	5279
Number of churches	38,183	350,000	9.2
Number of post offices	18,417	38,019	2.1
Number of banks	872	83,911	96.2

Source: Figures from 1997 are from *Statistical Abstract*; for 1850 from *Historical Statistics to 1970,* and *Statistical View of the United States, Being a Compendium of the Seventh Census* (Washington, D.C.: Senate Printer, 1854), 133–35.

voluntary societies. Although it is important to remember that not all voluntary agencies of the period were Protestant or even religious, nonetheless the "epidemic of organization" arising from the rapid expansion of evangelical Protestant churches in the national period meant a great deal in economic, as well as in other, terms.[33]

Comparisons are again illuminating. From its beginnings to 1828, the United States spent nearly $3.6 million on internal improvements (roads, canals, and communications). In that same span of years, the thirteen leading benevolent societies, overwhelmingly Protestant in constituency and purpose, spent over $2.8 million to further their goals.[34] No broad-based movement, not even the political parties, brought together so many people committed to so much social construction as did the national meetings of the benevolent societies. From the mid-teens, the societies were sponsoring major local and regional gatherings. By the late 1820s, many of the societies were staging their annual conventions as impressive public spectacles. The high point of this visible demonstration of evangelical social construction may have occurred in the first week of May 1834, when at least sixteen different societies gathered in New York City for reports and exhortation relating to their work.[35] The widely ranging interests of these societies spoke for wide-ranging economic influence as well:

- Six agencies for social benevolence at a time when neither the states nor the federal government sponsored programs of social welfare
- Four educational societies offering scholarships and other coordinated support for higher learning at a time when only the most progressive states had just begun to consider publicly sponsored education
- Three societies for distributing the Bible and religious literature throughout the nation at a time when American publishing was otherwise mostly carried out by local firms for local audiences
- Two home missionary societies at a time when there existed almost no other altruistic attention to frontier settlers
- One large foreign missionary society at a time when, after immigration and foreign trade, missionaries represented the United States' primary window to the world at large and, for the non-Western world, the most important window

The "evangelical united front" did not long survive the great convention week of May 1834. The societies were badly affected by the national depression at the end of the 1830s. Their very success had also engendered significant opposition from locally minded Protestants, especially in the West and South, who resented what they regarded as Yankee intrusions on their own liberty. Protests from such groups were illustrated, for example, by Baptist and Restorationist publication of Bible translations in order to compete with the American Bible Society. The mounting sectional crisis, which witnessed North-South schisms over slavery among both the Methodists and Baptists in 1844, further reduced the national impact of unitary Protestant movements. Yet all such circumstances notwithstanding, the economic presence of the voluntary societies remained very large.

Exactly how large is still an important question. Hints on questions of magnitude can be supplied by a brief consideration of the American publishing industry, which by 1860 had become a major factor in the American economy.[36] The census for that year found publishing (including newspapers and job printing) to be the ninth most productive industry in the United States (after flour and meal, cotton goods, lumber, boots and shoes, leather goods, clothing, woolen goods, and machinery and steam engines).

American publishing was an industry built on the Protestant religion. The industry leaders—until about 1830, Carey & Lea of Philadelphia; from the mid-1830s, Harper & Brothers of New York—became preeminent through their skill in marketing religious books. Numerous indications suggest the close ties between the expansion of Protestantism and the growth of publishing. Although Matthew Carey was a Roman Catholic, his firm came to prominence in no small part because of its supersalesman, the Rev. Mason Weems. Weems is now remembered as the hagiographer of George Washington, but he was also renowned in his own day as an extraordinarily effective salesman of the Bible.[37] The four Harper brothers, who managed the nation's largest publishing company from the mid-1830s, maintained a dedicated commitment to the Methodism in which they had been raised and to the Methodist luminaries after whom two of them (Wesley and Fletcher, for Wesley's successor John Fletcher) were named.[38] Moses W. Dodd, of Dodd, Mead, and Co., and Charles Scribner both began their publishing firms in large part as a way of expressing their own serious commitment to Presbyterian causes.[39]

If American commercial publishing arose as a result of skill in marketing religious books, that was still far from the whole story. Many of the denominations, with the Methodists in the lead, established large publishing ventures. The $612,636 worth of business generated by the Methodist Book Concern in 1848 illustrated its size as one of the nation's largest publishing ventures.[40] The size of this sum also suggests why the struggle among Methodists to control the publishing concern after the schism of 1844 was so rancorous, as explored by Richard Carwardine in chapter 9 in this volume.

For its part, the American Bible Society (ABS), founded in 1816, was always a major player in the publishing world.[41] By the mid-1840s, the society was represented by between thirty and forty agents in the field, and it employed a large staff at its printing operations in New York. By 1860, there were also between 4,000 and 5,000 local auxiliary Bible societies that recruited volunteers to work on behalf of the ABS. From its founding, the ABS maintained a large presence among American publishers. Its annual receipts regularly amounted from 2% to 3% of the total value of books manufactured and sold in the United States (for example, in 1830, the industry value was $3.5 million and ABS receipts were $117,000; in 1850 the industry took in $12.5 million, the ABS $277,000).[42] The quantity of Bibles produced was perhaps an even truer indicator of its presence. By the late 1840s, Harper & Brothers had become one of the largest publishers in the world; in that period the firm was printing 2 million books per year. From 1842 to 1848, the ABS distributed an average of 426,293 Bibles and testaments annually, with over 600,000 each in 1847 and 1848.[43]

Much more painstaking research would be required to gauge the exact place of the Protestant economy in the national economy during this early period. What even a preliminary survey suggests, however, is that the churches and societies were nearly as important for the era's economic history as they were for the nation's cultural, intellectual, and social development.

A Case Study of First Church, Albany

Because of the relative lack of comprehensive economic information for American religion as a whole in the years between the Revolution and the Civil War, it is all the more interesting to find local situations where complete records do exist. One local body for which such records are available is the First Church of Albany, New York, a Dutch Reformed congregation founded in 1642 and still existing today. The financing of First Church was not necessarily representative of broader trends, but the details of its financial history do make an informative case study of religion and finance.[44] To enhance the meaning of this one run of good records, historians are also fortunate to have a fine recent monograph on religion and society in early Albany by David Hackett, a work that provides a local context for First Church's financial history.[45]

First Church's extant records show a well-rounded picture of its finances—especially how money was raised and spent and what impact the local and national economies had on the church. Several unusual features set the financial history of First Church apart, but at least by the middle decades of the nineteenth century its position was probably quite typical for urban Protestant churches of the era. Hackett's monograph shows that by this period First Church, although still strongly identified by its ethnic Dutch heritage, no longer functioned as an established church, as it had during the early period of Hudson River settlement by feudal Dutch families. In 1835, First Church's approximately 320 members made it Albany's seventh largest of over twenty churches, following three Presbyterian congregations (the largest with about 640 members), two Baptist, and one Methodist. Where political allegiance can be determined, the members of First Church and its daughter RCA congregation were divided between Martin Van Buren's Regency party (supportive of the Democratic president, Andrew Jackson) and the local Workingmen's party, made up of rising artisans and smaller merchants (whose program resembled the national Whig party platform). First Church's congregation leaned toward a traditional, proprietary, or formal kind of Protestantism but also nurtured surprisingly broad sympathy for the newer, more aggressive evangelicalism of the era. In other words, the membership of First Church seems to have been more thoroughly representative of the major Protestant constituencies of the day and of their theological and practical interests than the church's patrician past might have suggested.[46]

The special circumstance that made First Church's financial situation unusual was its ownership of key properties in the city of Albany. Real estate sales and rental incomes were thus relied on to cover a substantial portion of the church's operating budget. Apart from rent and land sales, regular church income came from two sources—pew rentals and Sunday collections for the Deacon's Fund. (In addition, members also made special donations—for example, to pay for new windows or construct an organ—for which no adequate financial records were kept.) In churches without an endowment, presumably church leaders would have had to make adjustments over time in pew rental rates to cover rising operating costs, but for a long time First Church enjoyed the luxury of not depending so heavily on their members' present contributions. In the early years of record keeping, operating costs were mostly a matter of the pastor's salary, but later expenses included gas for lamps, other fuel for heating, and the services of a sexton. It was not until 1857 that First Albany's budget was balanced entirely from its members' yearly contributions.

In the eighteenth century, family pew rentals were set (depending on the location or size) at $2 to $4 per year, which equaled 1.5 to 3 days of the typical artisan's daily wage.[47] In modern terms, if we suppose an annual income in 1998 of $52,000, 1.5 to 3 day's earnings would be equivalent to an annual contribution of roughly $313 to $624.[48] First Church's pew rentals seem to have been standard for the period. Figures available for the First Reformed Church of Bethlehem, New York, are comparable.[49] In 1797 it asked members for a flat $2.25 per year. By 1869, the Bethlehem church was asking members for $6 to $28 per year. (As late as 1922, it was still financing its operations by pew rental with rates of $10 to $100 per pew per year.) For wage comparison, in 1830 the average daily wage of an artisan in Philadelphia was $1.73;[50] in 1875, white laborers in Massachusetts were earning from $605 to $790 per year.[51] The same survey of Massachusetts workers in 1875 showed that one-third of families contributed an average of $14.60 per year to religious causes. It is difficult to bring together Massachusetts income and New York pew rental fees, but if something like the Massachusetts figures applied in Albany and Bethlehem, New York, it would indicate that many families not only paid pew rents but also contributed offerings above and beyond this flat fee.

At First Church, Albany, whatever increases may have come from contributions after the Revolutionary War were offset by the steep inflation of the postwar period. By 1812, the pew rates were so nominal that many families paid for pews not only in the original church but also in the daughter congregation (North Church), which had been founded in Albany by the Dutch Reformed. At this time, pew rental income covered only 20% of the church's operating expenses. When the agreed-upon payment for the church's pastors fell behind rising prices, the church sometimes assessed its members *Domine gelt* ("minister's money") as a way of supplementing salaries. Incomplete records from the time often do not distinguish between the *Domine gelt* raised by such assessments and pew rent collected from individuals.

The main source of income during this early period, however, was the money from rentals and lot sales. But these revenues were naturally subject to the vicissitudes of the local and national economies. During the eighteenth century, more than half of the operating budget was covered by income from pasturage fees. When income did not match expenses, the church took out bank loans. By 1791, its total debt was twice the size of the operating budget. In 1761, the church bought 150 acres of house lots. When these lots were later sold at auction, typically more than three-fourths would be purchased by church members who considered them good investments. If the rise in prices after the Revolution put a strain on the contributions given to First Church, it had the opposite effect on its substantial real estate holdings, which became more valuable during this inflationary cycle. On the basis of the increase in property values, the church founded a charity school. Over the years 1797–99, it also built a new daughter church building (North Church), for which the cost was $32,500 ($7,500 for the land, $25,000 for the building). Using figures from about the same time for wages in Philadelphia, the cost of the new church building in Albany represented the annual income of about 63 artisans.[52] In terms of wages for the late 1990s, these figures would equal a building project of something over $3 million.

In 1799, the Consistory of elders and deacons voted to erect a new church building (Middle Church) as a replacement for the old stone structure in the center of town. Before the new building could be finished, however, a severe depression in the local economy—caused by disruption in trade, resulting from warfare between France and England—halted construction from 1806 to 1809. The completed building was finally opened for worship in 1811. By 1813, it was possible to hire a second pastor to help in serving the two structures. In 1814, the church Consistory authorized the construction of a separate house for their meetings; its second story was designated as quarters for a sexton whose task was to look after the two church buildings.

During the early part of the nineteenth century, the Albany church started to lease its land rather than sell lots. This change in financial practice resulted in a steadier cash flow and reduced the need for taking out temporary loans (from banks, ministers, and well-to-do members) between lot sales. (The church was also able to gain significant new money by commuting annual quit rents to one-time payments.) This approach proved successful until the depression of 1815, which hit Albany hard. A majority of tenants defaulted on their payments. The Consistory offered to renegotiate leases for those who could pay their arrears. Because few had the money to do so, most of the property was simply abandoned and repossessed by the church. Not only did property income plummet, but also many members defaulted on their nominal pew rents. This financial crisis became the occasion for abandoning the collegiate organization of the churches; in 1815, the North and Middle Churches divided into two separate congregations.

Good times returned to Albany with the construction of the Erie Canal, which was begun in 1817 and completed in 1823. The prosperous local economy led North Church in 1820 to make plans for refurbishing its sanctuary. This time, however, financial reforms were put into place: (1) There would be an effort to keep down church expenditures; (2) a drive would be conducted to collect pew rental arrears; (3) pew rental rates would be raised; (4) the first canvas would be held for pledges to meet maintenance needs (e.g., repainting in 1825); and (5) a new policy would allow the Consistory to draw on any surpluses in the Deacons' Fund to meet the operating budget. To this time, the Deacons' Fund, which was designated for the assistance of poor members of the congregation and community, had been managed separately from the operating budget. (In its early days the fund had been used to hire poor laborers to clean and service the church facilities.) The Deacons' Fund had built up large surpluses over time, probably because of the social stigma attached to the receiving charity. This cash fund had become a sort of bank from which money was loaned to Dutch entrepreneurs, and occasionally even to the church for construction or operating expenses. (Records do not always show whether these internal loans were repaid.) The financial reforms of 1820 allowed the North Church to elevate its nonproperty income to more than 50% of the annual operating budget.

During the economic boom of the late 1820s, the North Church embarked on new construction, which was to be financed through lot sales and pledges. But these plans had to be set aside when fraud by a church treasurer was discovered. The resulting shakedown led to temporary insolvency, but also a major restructuring of North Church's financial practices. In 1833 its Consistory, staring at ruin, adopted several important new measures: (1) Pew rents were doubled; (2) subscriptions were requested from key members to rescue the church; (3) it was decided that the entire Deacons' Fund could be used to cover the operating budget, except what might be necessary to help poor families in need of emergency assistance; and (4) a decision was made to sell off more house lots. This last decision, however, encountered immediate disappointment since a national economic crisis meant that the value of the lots had sunk so low as to be hardly worth selling. Partially in response, pew rents were tripled again in 1834, which led to a doubling of income from this source between 1834 and 1837. (The fact that income from pew rents only doubled, when the rate increased threefold, leads to the conclusion that the national economic crisis made it difficult to collect the full increase in rental.) In 1837, North and Middle Churches were reunited, and together they became known as First Church.

In the late 1830s, Albany experienced both good times and bad. On the one hand, traffic on the Erie Canal was at its peak. On the other hand, the financial panic of 1837 and the depression that followed in 1839 put a strain on local church finances. Several special collections were taken to deal with increased economic need, and the Consistory of First Church again raised pew rents by 33%. Financial problems even led First Church to first consider (but not implement) a trustee system that would have removed financial affairs from the purview of the Consistory.

The decade of the 1840s brought strong economic growth to the Albany area. During these years the efforts of the previous decade's experience bore fruit since pew rents now covered two-thirds of the operating budget. The rest was supplied from sales of land whose value increased as population growth again picked up. Sales of lots continued into the 1850s. In 1850, a new plan for refurbishing the church was initiated, paid for at first by lot sales but after 1853 by member subscription. In 1852, income from pew rents had diminished to only 35% of the operating budget.

In 1854, the finances of Albany's Dutch Reformed churches entered a new era, for in that year its last house lot was sold. Faced abruptly with a sharp change from the situation dating from its earliest years, the Consistory dithered. Evidently the board expected the tide of prosperity to roll on, for without bothering to institute new forms of financial support, the consistory nonetheless voted to raise the minister's salary by 25%, from $2,000 to $2,500. (By comparison, the national average earnings of nonfarm employees in 1860 was $363.[53]) When word got out to the congregation about these actions, its members voted to replace the Consistory with a Board of Trustees as the church's financial managers. Soon thereafter, in 1857, the trustees doubled the pew rents and initiated efforts to collect them in a timely way. Since even this effort did not raise enough new money to rescue the church, the trustees chipped in funds from their own pockets and solicited donations from others to keep the church solvent.

As this new pattern of church finances took shape, First Church was also developing a new approach to benevolences. Money received from actual collections at the Sunday worship service had formerly been designated entirely to the Deacons' Fund. But as the Deacons' Fund came less and less to be used for support of the poor (and finally was absorbed into the general operating fund), new appeals for benevolence took its place. Sometime during the late 1830s or 1840s, it became the practice to assign receipts from the Sunday collection to different programs—one Sunday a month to the Poor Fund, one to foreign missions, and other Sundays to a miscellaneous range of benevolences. The following figures indicate the benevolent purposes that were supported by First Church in 1857:

Deacon's Poor Fund	$443.81
Foreign missions	535.61
Domestic missions	388.60
Albany Tract and Missionary Society	376.80
Board of Education (of RCA)	207.71
Board of Publication (of RCA)	175.14
Church Sunday school	137.81
American and Foreign Christian Union	50.43
American Colonization Society	47.37
American Sunday School Union	41.09
Total	$2,404.40

Since the operating budget for that year was $3,623, these benevolences represented about 40% of First Church's annual receipts. (The pastor's salary was probably still $2,000; it was taken from the operating budget.) Since by this time the last lot had been sold, payments of pew rents by church members covered an estimated 88% of the operating budget. It was in this same year, 1857, that the budget was balanced for the first time from the year's actual receipts. Because of the various new strategies developed in the 1850s, the church's financial standing remained on a firm footing for many years to come.[54]

Several observations about the financial history of this one church are pertinent. Most obviously, First Church's land endowments set it apart from the financial situation of most other American churches. Whereas the ability to collect rents and sell house lots did sustain the church's finances for a long time, it also meant that the congregation probably grew overly complacent in periods of rising real estate prices. By contrast, the collapse of real estate markets threw the church's finances into turmoil.

First Church's unusual situation did, however, make for a most interesting history. In 1799, for instance, the church hired Philip Hooker, who became the leading architect of Albany, to design North Church (as it was then called), and it spared little expense on other occasions when this structure and Middle Church were remodeled. More generally, it is clear that financial decisions were made primarily with a view to good real estate business. One of the church treasurer's ongoing tasks was to look into tax sales of prime properties that would make good investments for the church. The church also arranged several land swaps with the city to consolidate its holdings in the most advantageous way. These swaps spoke to the close relationship between this church and the city of Albany. In 1809, the city agreed to pay the church for ringing its bell to mark the hours. As an indication of First Church's centrality in the community, it often rented out its funeral hearse to other churches in the city. At one time (1816), the church even contemplated going into the wharfage business but then decided that the best arrangement was to swap its riverfront property for 120 house lots and let the city construct and administer the new wharves.

In many particulars, however, the financial history of this church was probably quite typical of at least those Protestant churches that had become well established in urban areas. Church financial records reveal, for example, attitudes toward liquor that were not unusual for Protestants until about the mid-1840s, when temperance became a great cause. At First Church, the treasurer paid bonuses to all construction workers in the form of rum. In addition, wine, ale, and spirits were regularly billed to the treasurer, along with food and tobacco, as normal expenses for church suppers. These wares were also commonly part of ministers' salaries in those days. Alcoholic beverages were, at the time, regarded as a safe alternative to drinking contaminated city water. Drunkenness was a problem that warranted discipline of members, and on at least one occasion the dismissal of a pastor from Albany, but not moderate drinking.

More generally, the financial history of the Albany church shows the increased importance of national religious mobilization. As the 1857 North Church benevolence receipts indicate, the local RCA congregation was contributing to denominational purposes, as it no doubt had been doing since its earliest days. The seven different national voluntary societies on the church's list of benefactions would, however, have appeared only shortly before. They suggest that national religious movements had come to be as closely related to the Albany church as national financial developments had long been to its economic life.

What First Church also shared with almost all the other Protestants of the era was its interconnections with local and economic developments. It is clear that both the church's operating budget and its facilities were strongly affected by changes in the local and national economy, such as population growth, expanding business, depressions, and inflation. Also, Albany's First Church and similar churches made an impact on the local economy. They supported welfare funds for the good of the community. Their financial commitment to missionary movements, charity schools, and hospitals contributed significantly to the social infrastructure of the new nation.

In its many economic activities, however, it is difficult to find First Church's leaders acting on the basis of any explicitly religious conviction. Rather, participation in the economy seems to have been largely unreflective. In keeping with conclusions suggested by Mark A. Noll in chapter 12 in this volume, it is hard to see that the management of the operational budget of the First Church (or of most other Protestant churches of the time) represented self-conscious economic choice. Even when Albany's RCA congregations created boards of trustees to run the churches' financial affairs, this seems to have been an instinctive move to use an organizational form developed in the wider society to meet the immediate needs of the congregations. "Ecclesiastical economics," at least as represented by this one example, seem, therefore, to have meant church participation in broader economic domains rather than sharp-edged, intentional *action* on the basis of religious perspectives.

Whatever particular insights can be gained from financial records like those available from First Church, Albany, the general picture they reveal corresponds well with conclusions from preliminary sketches of economic and religious history in the period 1790–1860. During an era when the Protestant churches were expanding rapidly, the churches occupied considerable space in the American economy. The description and interpretation of salient circumstances that resulted from the existence of Protestants in both religious and economic realms are the concerns of the rest of this book.

Notes

1. For example, *Historical Statistics of the United States to 1970* (Washington, D.C.: Bureau of the Census, 1975); *Statistical Abstract of the United States 1998* (Washington, D.C.: Bureau of the Census, 1998).

2. We have found especially useful for reliable summaries of this period Jonathan Hughes and Louis P. Cain, *American Economic History*, 5th ed. (Reading, Mass.: Addison-Wesley, 1998); Douglass C. North, *The Economic Growth of the United States 1790–1860* (New York: Norton, 1966); Sidney Ratner, James H. Soltow, and

Richard Sylla, *The Evolution of the American Economy: Growth, Welfare, and Decision Making* (New York: Basic Books, 1979); and the data available from sources mentioned in note 1.

3. An interesting recent survey of economists and historians suggests that for the antebellum period there is widespread scholarly agreement on the stimulus provided to the economy by export trade (especially cotton) and on the fact that the tariff system favored the North while harming the South. Opinion is more divided on whether national land policy stimulated net productivity or whether the cotton textile industry was dependent on protective tariffs. Robert Whaples, "Where Is There Consensus Among American Economic Historians? The Results of a Survey on Forty Propositions," *Journal of Economic History* 55 (1995): 139–54, with 145–46 on the antebellum period.

4. For a general discussion of land issues and land laws, see Paul W. Gates, *History of Public Land Law Development* (Washington, D.C.: Public Land Law Review Commission, 1968); and Vernon Carstensen, ed., *The Public Lands: Studies in the History of the Public Domain* (Madison: University of Wisconsin Press, 1962).

5. Jonathan Hughes, "The Great Land Ordinances: Colonial America's Thumbprint on History," in *Essays on the Emerging of the Old Northwest*, ed. David C. Klingman and Richard K. Vedder (Athens: Ohio University Press, 1987).

6. For the extent of that support, see Thomas J. Curry, *The First Freedoms: Church and State in America to the Passage of the First Amendment* (New York: Oxford University Press, 1986), 193–222. By the 1830s all the states had ended direct payments to churches.

7. Chief Justice Marshall outlined some of this theory in *Johnson v. M'Intosh*, 21 U.S. (8 Wheaton), 543. Further elaboration on the philosophical background can be found in Jesse Dukenminier and James E. Krier, *Property* (Boston: Little, Brown, 1993); and Stanley Lebergott, "The Demand for Land: The United States, 1820–1860," *Journal of Economic History* 45 (June 1985): 181–212.

8. Lebergott, "Demand for Land."

9. The states were Illinois, Indiana, Iowa, Michigan, Missouri, Ohio, and Wisconsin.

10. North, *Economic Growth*, 129, 140.

11. Ibid., 157.

12. The classic work here is Carter Goodrich, *Government Promotion of American Canals and Railroads, 1800–1890* (New York: Columbia University Press, 1960).

13. North, *Economic Growth*, 251.

14. Ibid., 146–47.

15. See Richard R. John, *Spreading the News: The American Postal System from Franklin to Morse* (Cambridge, Mass.: Harvard University Press, 1995).

16. North, *Economic Growth*, 179.

17. Ibid., 189.

18. For a discussion of the rise in manufacturing, see Peter Temin, "Manufacturing," in *American Economic Growth: An Economist's History of the United States*, ed. Lance E. Davis et al., (New York: Harper & Row, 1972); Jeremy Atack, Fred Bateman, and Thomas Weiss, "The Regional Diffusion and Adoption of the Steam Engine in American Manufacturing," *Journal of Economic History* 40 (1980): 281–308; and Thomas Cochran, *Frontiers of Change: Early Industrialization in America* (New York: Oxford University Press, 1981).

19. North, *Economic Growth*, 193.

20. Lance E. Davis, "The New England Textile Mills and the Capital Markets: A Study of Industrial Borrowing, 1840–1860," *Journal of Economic History* 20 (1960): 1–30.

21. For a general discussion of the business cycle during this period, see Peter Temin, "The Anglo-American Business Cycle, 1820–1860," *Economic History Review*, 2nd ser., 27 (1974): 207–21; Hugh Rockoff, "Varieties of Banking and Regional Economic Development in the United States, 1840–1860," *Journal of Economic History* 35 (1975): 160–81; and Richard Sylla, "American Banking and Growth in the Nineteenth Century: A Partial View of the Terrain," *Explorations in Economic History* 9 (Winter 1971–72): 197–227.

22. The overall importance of the railroad has been greatly debated by economic historians. Much of the focus of the debate has been on the postbellum period, but is also relevant for this period. The standard works are Robert W. Fogel, *Railroads and American Economic Growth* (Baltimore: Johns Hopkins University Press, 1964); and Albert Fishlow, *American Railroads and the Transformation of the Antebellum Economy* (Cambridge, Mass.: Harvard University Press, 1965).

23. Douglass C. North, Terry L. Anderson, and Peter J. Hill, *Growth and Welfare in the American Past: A New Economic History* (Upper Saddle River, N.J.: Prentice Hall, 1983). Measures of stature generally confirm the growth in income for this period but raise doubts about whether incomes continued to grow after the Civil War. See Dora Costa and Richard Steckel, "Long-term Trends in Health, Welfare, and Economic Growth in the United Sates," in *Health and Welfare During Industrialization*, ed. Richard Steckel and Roderick Floud (Chicago: University of Chicago Press, 1995).

24. Important works in a vast bibliography include Alfred Conrad and John Meyer, "The Economics of Slavery in the Antebellum South," *Journal of Political Economy* 66 (1958); Robert William Fogel and Stanley Engerman, *Time on the Cross: The Economics of American Negro Slavery*, 2 vols. (Boston: Little, Brown, 1974); Gavin Wright, *The Political Economy of the Cotton South* (New York: Norton, 1978); and Richard H. Steckel, "A Peculiar Population: The Nutrition, Health, and Mortality of American Slaves from Childhood to Maturity," *Journal of Economic History* 46 (1986): 721–41. For a recent summary, see Hughes and Cain, *American Economic History*, chap. 10, "The Debate Over Slavery."

25. See Whaples, "Consensus Among American Economic Historians," 141, 146–47.

26. *Preliminary Report on the Eighth Census, 1860* (Washington, D.C.: Government Printing Office, 1862), 191.

27. This figure is, admittedly, an estimate. It is based on extrapolating from the full record of contributions to Old School Presbyterian churches and the fairly complete giving records for other Presbyterian churches. The Old School was the more conservative denomination that emerged from a schism in the main American body of Presbyterianism in 1837–38. By the early 1860s, contributions to Old School Presbyterian churches (including money used locally and money transmitted outside the local congregation for benevolent purposes) had passed the $3 million mark (1860: $3,175,304; 1861: $3,167,425). Other Presbyterian bodies (New School, Cumberland, United, and a number of smaller Reformed denominations) together employed more ministers and sponsored about as many churches as the Old School. The giving of these churches has been estimated at something over $2 million (ascertained from some extant records with some estimations). For these Presbyterian numbers, we rely especially on Herman C. Weber, *Presbyterian Statistics: Through One Hundred Years,*

1826–1926 (Philadelphia: General Council of the Presbyterian Church in the U.S.A., 1927); and Lewis G. Vander Velder, *The Presbyterian Churches and the Federal Union, 1861–1869* (Cambridge, Mass.: Harvard University Press, 1932), 7, which draws on Joseph M. Wilson, ed., *Presbyterian Historical Almanac and Annual Remembrancer of the Church* (Philadelphia: Wilson, 1862). To extrapolate from Presbyterian giving of about $5.5 million in 1860 to a national Protestant total, we first multiplied by six. According to the 1860 census Presbyterians accounted for 12.5% of all Protestant churches, 14.5% of all Protestant accommodations, and 18.7% of the value of all Protestant church properties; Jewish and Catholic places of worship accounted for 4.9% of the churches, 7.5% of all accommodations, and 16.3% of the value of all church property—*Statistics of the United States . . . 1860; Compiled from the . . . Eighth Census* (Washington, D.C.: Government Printing Office, 1866), 497–501. Our multiplication yields a sum of about $33 million as the total contributions to Protestant churches. To this figure we added a conservative estimate of $2 to $7 million for income to Protestant voluntary societies—the American Bible Society, for example, had receipts in its fiscal year ending in 1860 of almost $436,000; see note 41 for the society.

28. Revenues generated by the railroad industry were $130 million in 1861, with no figure available for 1860; *Historical Statistics to 1970*, 734.

29. *Minutes of the General Synod, RCA.*

30. *Historical Statistics to 1970*, 164.

31. Dean R. Hoge, Charles Zech, Patrick McNamara, and Michael J. Donahue, *Money Matters: Personal Giving in American Churches* (Louisville, Ky.: Westminster John Knox, 1996).

32. *Census of Population Compendium* (Washington, D.C.: Bureau of the Census, 1854), 141.

33. The phrase is from Donald G. Mathews, "The Second Great Awakening as Organizing Process," in *Religion in American History*, ed. John M. Mulder and John F. Wilson (Upper Saddle River, N.J.: Prentice Hall, 1978), 208.

34. Charles I. Foster, *An Errand of Mercy: The Evangelical United Front, 1790–1837* (Chapel Hill: University of North Carolina Press, 1960), 121. Foster's account remains the most helpful general study of the American voluntary societies.

35. Ibid., 148–53.

36. General information in the following paragraphs comes from John Tebbel, *A History of Book Publishing in the United States: Vol. 1, 1630–1865* (New York: Bowker, 1972), passim, with 184–90 on the religious press itself; and Charles A. Madison, *Book Publishing in America* (New York: McGraw-Hill, 1966).

37. See David Kaser, *Messrs. Carey & Lea of Philadelphia: A Study in the History of the Booktrade* (Philadelphia: University of Pennsylvania Press, 1957), with 31–34 on Weems.

38. Especially helpful on the deep economic, as well as religious, commitment to the Bible publishing of Harper & Brothers is Paul C. Gutjahr, *An American Bible: A History of the Good Book in the United States, 1777–1880* (Stanford, Calif.: Stanford University Press, 1999), 70–78 and passim.

39. See Madison, *Book Publishing*, 32–33.

40. Ibid., 132.

41. For a well-rounded general history, see Peter J. Wosh, *Spreading the Word: The Bible Business in Nineteenth-Century America* (Ithaca, N.Y.: Cornell University Press, 1994). For details on the income and production of Bibles, helpfully tabu-

lated by year, see Henry Otis Dwight, *The Centennial History of the American Bible Society* (New York: Macmillan, 1916), 576–77.

42. Tebbel, *History of Book Publishing*, 221; Dwight, *American Bible Society*, 576.

43. Tebbel, *History of Book Publishing*, 279; Dwight, *American Bible Society*, 577.

44. We are much indebted to Robert S. Alexander, *Albany's First Church and Its Role in the Growth of the City: 1642–1942* (Albany, N.Y.: First Church in Albany, 1988); and to personal correspondence from Mr. Alexander of January 24, 1998.

45. David G. Hackett, *The Rude Hand of Innovation: Religion and Social Order in Albany, New York, 1652–1836* (New York: Oxford University Press, 1991).

46. Ibid., 106, 125, 146–47, 153. First Church's daughter congregation, Second, or North, Church, had about 280 members in 1836.

47. *Historical Statistics to 1970*, 165.

48. See *Yearbook of American Churches 1998* (Nashville, Tenn.: Abingdon, 1998), 324.

49. H. S. VanWoerst, *History of the First Reformed Church, Bethlehem, New York 1763–1963* (Bethlehem, N.Y.: Dutch Church of Bethlehem, 1913), 29, 42, 64.

50. *Historical Statistics to 1970*, 163.

51. Daniel Horowitz, *The Morality of Spending: Attitudes Toward the Consumer Society in America, 1875–1940* (Baltimore: Johns Hopkins University Press, 1985), 14f.

52. *Historical Statistics to 1970*, 163.

53. Ibid., 165.

54. In the year after the Civil War, a controversial decision was made to automatically raise pew rentals to keep ahead of inflation. With this decision implemented, pew rental incomes increased fourfold between 1854 and 1874, rising annually from $2,000 to $8,000. Considering the prior history of precarious financing, Albany's North Church set itself on a sustainable path for the rest of the nineteenth century. (Pew rents continued to be collected until 1938.)

3

Charles Sellers, the Market Revolution, and the Shaping of Identity in Whig-Jacksonian America

Daniel Walker Howe

Like a number of prominent historians before him, Charles Sellers waited until he had retired before publishing his magnum opus, a summation of the results of his lifetime of research and teaching. Such a work is *The Market Revolution: Jacksonian America, 1815–1846.*[1] When I read this book, two things struck me: First, the interpretive framework was much the same as I had heard it thirty years before as a graduate student listening to Professor Sellers's lectures at the University of California; second, for some reason I *liked* the old lectures better. Of course one couldn't help wondering why: What was different? Fortunately, there was a published version of the lectures, so one did not have to rely simply on memory and old notes to make the comparison.[2]

Thirty years ago, Sellers was arguing that the antebellum era had been characterized by the formation of a national market economy based primarily on commercial agriculture and that this national market economy constituted a transition between the colonial staple-exporting economy and the industrial economy of modern America. In the colonial economy, he pointed out, only those people who lived within easy access to water transportation had been able to market their crops, and the rest of the population had had to be economically self-sufficient, if not within their families, then within their local communities. The national market economy that came into being during the antebellum period was made possible by improved transportation and was characterized (as Sellers put it) by "a vast extension of the division of labor or, in other words, specialization of economic activities. Areas and individuals that formerly had been self-sufficing," or nearly so, "began to concentrate on the one product or service they could produce most efficiently."[3] The expansion of cotton cultivation was particularly momentous. It contributed to such varied and important developments as the rise of textile manufacturing and the American merchant marine, conflict between the "five civilized tribes" and white settlers, and political disputes over economic issues: whether to have a protective tariff, how to provide credit to commercial farmers, and how to fund the internal improvements necessary to transport crops to market. Ultimately, this extension of commercial agriculture produced sectional competition over whether slave or free farming would be expanded. The explanatory model that Sellers described in those lectures, the causal dynamic of commercial agriculture, carried conviction.[4]

Building on the strengths of these lectures, *The Market Revolution* adds an interpretation of the cultural consequences wrought by economic change and relates these, too, to politics. To examine "the expansion of the market, the intensification of market discipline, and the penetration of that discipline into spheres of life previously untouched by it" is a necessary and demanding undertaking that has been addressed by a number of recent historians.[5] Recognizing the importance of religion to antebellum American life, Sellers casts his discussion of the cultural impact of the market largely in terms of religion. Much of the time, his analysis of the interactions among economic, religious, and political history dovetails nicely with the more specialized findings of Richard Carwardine and others.[6]

In *The Market Revolution*, Sellers is at his best in his evocative, sympathetic descriptions of the culture of the subsistence farmers who lived outside the market.[7] Although somewhat romanticized, his presentation also takes account of the dark side of that culture: the oppression of women and children within the household, the bitter hatred of African Americans and Native Americans, and the vulnerability to certain kinds of demagogy. (Joseph Smith, the founding prophet of Mormonism, is portrayed as one of the worst of demagogues.[8]) But, despite its virtues, the presentation of cultural history in *The Market Revolution* is so tendentious, so heavily freighted with the determination to show that the market was undemocratic and inhumane, that it fails to convey a comparable understanding of the culture of the new middle class that was created by the market and embraced by many people. Because we are not made aware of this culture's attractions, we are left even at the end of this massive book wondering why the market revolution succeeded and the opposition to it failed. A more balanced assessment of the costs and benefits entailed by the market revolution would help explain this outcome. Besides critically analyzing the presentation of cultural history in Sellers's book, I will suggest in this chapter how we can understand the appeal of the market revolution to Americans living between 1815 and 1846.

Invoking the language of theology, Sellers describes a commercial worldview that he calls "arminianism" and a noncommercial, agrarian worldview that he calls "antinomianism." Here is how he explains their opposition:

"While arminian moralism sanctioned competitive individualism and the market's rewards of wealth and status, antinomian new birth recharged rural America's communal egalitarianism in resistance. A heresy of capitalist accommodation confronted a heresy of precapitalist cultural revitalization in a *Kulturkampf* that would decide American destiny on the private battle-grounds of every human relationship."[9]

Historically, Arminianism was not defined in opposition to antinomianism; it was defined in opposition to Calvinism. In Christian theology, Arminianism is the doctrine that, even after the Fall, human beings retain sufficient free will to choose whether to accept God's saving grace. The doctrine was advanced among Reformed Protestants in the seventeenth century as an alternative to the High Calvinist position that humanity was utterly depraved and divine grace was irresistible.

Among Dutch, Swiss, and New England Protestants, Arminianism flourished most within prosperous mercantile communities, which indeed raises the question of whether their merchant capitalism fostered a sense of human capacity and moral responsibility. Sellers puts it this way: "As God seemed kindlier, the environment more manageable, and their fate more dependent on their own abilities, [Arminians] could no longer see themselves as sinners helplessly dependent on the arbitrary salvation of an all-powerful God."[10] This is a plausible sociological explanation for the rise of what was manifestly a religion that affirmed a greater measure of human dignity. Another possible explanation is that a higher standard of living diminished the appeal of the austere Calvinist lifestyle. Of course, the two hypotheses are not mutually exclusive. Whatever the causal mechanism, the Calvinist doctrines of total depravity, unconditional election, and irresistible grace were losing their appeal during the course of the nineteenth century, and not only in the United States. It seems likely that even if early Calvinism fostered what Max Weber termed "the spirit of capitalism," in later generations a market economy fostered higher notions of human dignity and liberty than were compatible with orthodox Calvinism.[11] This much may be conceded by way of overall plausibility to Seller's association of Arminianism with the market. Whether the connection between theology and economic development was as straightforward and politically manifest as he makes out is something else.

By the nineteenth century, Arminian-like doctrines were growing in popularity among several denominations of American Protestants, having been embraced not only by rationalists like the Unitarians and Universalists and liturgicals like the Episcopalians but also by such important evangelical groups as the Methodists, "New School" Presbyterians, and Freewill Baptists. However, these assorted kinds of Arminians did not constitute a united front in support of common economic, political, or even religious issues. One convenient test of support for the market revolution was support for the Whig party, but only certain versions of Arminianism were associated with Whig voting. Arminian Unitarians certainly were predominantly Whigs, but Arminian Universalists and Baptists were probably more often Democrats. The politics of the Arminian Methodists are only now being sorted out by

THE SHAPING OF IDENTITY IN WHIG-JACKSONIAN AMERICA 57

Richard Carwardine.[12] Episcopalians—despite their Arminianism and high socioeconomic status—were often Democrats, especially in New England, where they, along with Baptists and Quakers, shared the status of religious dissenters from the Congregational establishments and consequently embraced Jeffersonian principles on the relationship of church and state.[13]

Although many Whigs may have been Arminians, there is no doubt that many other Whigs were Calvinists. The typical Whig reformers, such as the famous Tappan brothers, were evangelical Calvinists. The New England Calvinism of Andover Theological Seminary, founded for the express purpose of resisting Arminianism, was quintessentially Whig. The most uncompromisingly anti-Arminian of all American Calvinists, the Old School Presbyterian faculty of Princeton Theological Seminary, were "Cotton Whig" in their politics. In short, evidence that antebellum theological Arminianism was a stalking-horse for Whig politics or promarket economic policies is difficult to find.

The socioeconomic origins of Arminianism are probably multiple, and the Arminianism of the New England commercial elite, which resulted in Unitarianism and best illustrates Sellers's thesis, is only one aspect of the story. The growth of Arminianism among the common people has been studied less, although Steven Marini and Nathan Hatch have made an exciting start.[14] It needs to be added that Arminianism, like other theological and philosophical positions, has an intellectual appeal on rational grounds that is independent of socioeconomic predilections, and a balanced historical account will recognize this. The debates carried on in this period over the nature of free will among theological exponents of various schools of thought, Calvinist and Arminian, were remarkable for their intellectual sophistication. Sellers shamelessly reduces these arguments to fronts for economic interest, often distorting them in the process.[15]

Antinomianism, the other of Sellers's categories, also dates from the time of the Reformation. Antinomianism is the doctrine that the moral law does not bind God's elect since they enjoy the indwelling presence of the Holy Spirit. It has not been a popular teaching because it seems to license anarchic behavior by fanatics who are claiming the direct guidance of the Spirit. Sellers does not so much as mention this definition, however; instead he describes antinomianism as the "heresy that God visits ordinary people with the 'New Light' of transfiguring grace and revelation."[16] This is confused: first, because antinomianism is not the same as New Light Calvinism; second, because the doctrine Sellers has stated is not heretical. What would make it heretical would be if the revelations of God's will that the believer received were considered to apply to anyone else or to substitute for Scripture or the doctrine taught by the church.

The only Christian body in antebellum America that subscribed more or less to the heretical form of antinomian doctrine was the Quakers. Yet the Quakers were conspicuously *not* a backwoods denomination or one whose members practiced subsistence agriculture; what is more, they were generally part of the Whig voting bloc. The founders of certain other sects, notably

Ann Lee of the Shakers and Joseph Smith of the Mormons, claimed to have received revelations that supplemented the Judeo-Christian Scriptures. Instead of accepting traditional agricultural society, however, they both founded utopian communities, highly regimented, with novel economic and sexual practices. One group of antebellum American Christians did indeed approximate the kind of antinomian peasant *mentalité* Sellers describes, scrupulously rejecting the market and all modernity. These were the Mennonites, including the Amish. But Sellers never mentions them in his book. All in all, antinomianism seems to work even less well as a cultural category in Sellers's book than Arminianism does.

In a footnote, Sellers explains that he does not intend the terms *arminian* and *antinomian* to be taken in their accustomed theological senses but rather as designating "clashing cosmologies," of which the "narrower theological distinctions" are symbols.[17] By not capitalizing *arminian* and *antinomian*, it is possible that the author may intend to remind us that the terms are not being used in their usual senses. But what is the relationship between "arminianism" and Arminianism, and why use terms in special, esoteric senses? The footnote goes on to say, "The evangelical movement galvanized by the antinomian New Light also contained doctrines susceptible to arminian drift under market pressures, especially the moralism of the Methodist *Discipline*, and the free grace for all believers of Methodists, Free-Will Baptists, and Universalists." This seems to mean that whatever sincere conviction motivated the evangelicals is to be attributed to their antinomianism, whereas their arminianism (or Arminianism) was an unfortunate lapse "under market pressures." But why should belief in human dignity rather than depravity be presumed hypocritical? And how could one test such a claim?

Sellers evidently adopted his bipolar interpretation of American Christianity from Alan Heimert's book on the Great Awakening of the eighteenth century. Like Heimert, he opposes the Arminianism of the merchant capitalists to the New Light evangelical faith of the rural democracy.[18] In his retrospective look at the Great Awakening, Sellers follows Heimert's interpretation. He and Heimert are alike devoted to Perry Miller's most unreliable work, his 1949 intellectual biography of Jonathan Edwards, long since superseded. Sellers goes out of his way to endorse its argument in a summary of Edwards's career.[19] However, Heimert tried to be more sensitive to theological distinctions, denying emphatically that the New Light Calvinists were antinomians. What is more, Heimert argued that the evangelical Calvinists were not hostile to the market but were simply concerned with democratizing it.[20]

Like Heimert, Sellers identifies New Light theology with communitarian values and Jeffersonian politics. But he exaggerates the extent to which there was an American political ideology that was celebrating and seeking to protect the noncommercial, agrarian way of life. He attributes such an ideology to Thomas Jefferson, distinguishing between the purity of Jefferson's own devotion to noncommercial agriculture and the compromises with commercialism made by his follower James Madison. "Jefferson was anxious about the corrupting effect of the market on American farm families," he writes.[21]

It is true that Jefferson and his followers thought that the ownership of property gave a citizen-householder a measure of independence that was politically valuable. The wage earner, like the apprentice, the slave, the minor, or the woman, was by their definition not fully independent and therefore not qualified for political participation. Thus a society consisting of family farms enjoyed, in their eyes, a strong economic basis for free political institutions. However, what was important to this model was that the farm be held in fee simple and be clear of mortgage indebtedness, not that it practice subsistence agriculture. Provided that Americans lived on family farms, they would derive the social and political advantages of Jeffersonian theory regardless of whether those farms produced crops for local consumption or for export. The Jeffersonian model of property-owning households was readily applicable to commercial farmers and urban artisans who were producing articles for sale. Indeed, the laissez-faire of Jeffersonian political thought seems to have been intended more as a means to commercial expansion than as a means to the preservation of premarket conditions.[22]

To enlist Thomas Jefferson on behalf of the kind of folk culture Sellers attributes to the noncommercial antebellum farmers is to misunderstand him. Jefferson, who was committed to the right to the pursuit of happiness and defined it as the fulfilling development of one's faculties, would hardly have wished the meager, subsistence way of life on his countrymen. What Sellers describes as the antinomian cosmology—a blend of superstitious supernaturalism, patriarchal repressiveness, and limited intellectual horizons—had zero appeal for Jefferson. Instead he wished to encourage education, literacy, and religious toleration—in short, a broadening of horizons. Jefferson was a spokesman of the Enlightenment, not of a provincial peasant *mentalité*. His vision of human beings as self-directed, autonomous free agents had much more in common with what Sellers calls arminianism than with what he calls antinomianism.

The merits of Sellers's *Kulturkampf* interpretation cannot definitively be settled by an analysis of the true content of Arminianism, antinomianism, and Jefferson's political philosophy. Ultimately, the issue is the way in which Sellers opposes democracy to commercial life, "the people" to what he reifies as "the market." In his account, American democrats—those who work with their hands, whether in the countryside or in cities—courageously resist the temptations of internal improvements, bank credit, and competition. As Sellers sees it, the conflict between the people and the market came to a climax in Andrew Jackson's war on the Bank of the United States: "While collective repression pushed bourgeois hegemony across the North, antinomian democracy—and racist slavery—empowered the patriarchal Hero of New Orleans to bring *Kulturkampf* to a showdown in national politics."[23]

In the remainder of this chapter, I accept Sellers's challenge to locate political conflict within a broader cultural context. In place of adopting the notion of an apocalyptic *Kulturkampf*, however, I wish to show how the new capitalist market was welcomed by a wide variety of people, how it broadened their horizons and offered them more control over their own lives than they had ever enjoyed before. The market greatly facilitated what Jefferson had called "the pursuit of happiness," something which, as he so appropriately recognized, most Americans valued next to life and liberty.

It is an oversimplification to suppose that "subsistence" and "the market" posed themselves as stark alternatives to most Americans of the antebellum period. In rural areas, farm families supplemented their incomes with various market activities that might include, depending on the time and place, taking raw cotton from a storekeeper to spin into yarn, distilling grain into whiskey, churning milk into butter, or sending a daughter off to work in a mill town. As Christopher Clark has shown so well for western Massachusetts, market activity presented itself as an additional resource within a complicated family survival strategy.[24] In small towns, goods and services were for sale, yet the prosperity of the community depended on that of the neighboring farms. The ways of the market were not perceived in sharp opposition to customary rural ways. There were means of combining and synthesizing them: For example, keeping account-book records of credits and debts was an intermediate step between a cash economy and a barter economy. The struggle of which rural families were aware was the struggle for a livelihood, not necessarily a struggle between two alternative kinds of livelihood.

Sellers's presentation of a political conflict between market and nonmarket regions probably works best in explaining conditions in the South.[25] In the North, commercial versus noncommercial conflict was often overshadowed by ethnic or class conflict. In the cities, the transition between two alternative commercial livelihoods—artisan labor and wage labor—could be more dramatic and traumatic for those who underwent it than the transition between subsistence and commercial farming.[26] Industrialization and the technological changes that went with it, rather than the expansion of the market per se, were responsible for the replacement of the old form of artisan manufacturing with factory production and the proletarianization of the work force. Class conflict emerged in nineteenth-century industrial cities that was unlike anything in eighteenth-century cities, where merchants and artisans, masters and apprentices, had often shared common political interests. Yet the most violent conflicts in Jacksonian America involved racial and religious hostilities: white against black, abolitionist against antiabolitionist, Protestant against Catholic, Mormon against Gentile. Often they pitted workers against workers. In such cases, cultural conflict cut across economic class lines rather than coinciding with them.[27]

Just as the market provided new economic resources for people, it also provided new cultural resources as well. Along with new ways for farmers to supplement their incomes, the market also multiplied schools and colleges and provided cheaper newspapers, magazines, and books; a better postal system; and, beginning with the invention of the telegraph, communication at the speed of light. The market revolution provided the opportunity for people to make choices on a scale previously unparalleled: choices of goods to consume, choices of occupations to follow, educational choices, and choices of lifestyles and identities. The ultimate test of autonomy is to be the kind of person one chooses—that is, to be self-constituted. The market brought people new power to shape their own identities.[28]

Sellers conceives of the cultural history he is relating as the subordination of the people to the market: "Where nobilities and priesthoods left folk cultures little disturbed, capital feeding on human effort claimed hegemony over all classes."[29] But the way in which Sellers is here using the neo-Marxist concept of hegemony implies that a false consciousness is being imposed. What if people really were benefiting in certain ways from the expansion of the market and its culture? What if they espoused middle-class tastes or evangelical religion or (even) Whig politics for rational and defensible reasons? What if the market was not an actor (as Sellers makes it) but a resource, an instrumentality, something created by human beings as a means to their ends?

Sellers is right to see that conflicts over culture were of crucial importance to the antebellum period. Whether all of them can be reduced to one titanic *Kulturkampf* is less clear, although modernization and its discontents were arguably central to the whole Victorian epoch.[30] But Sellers makes no conscientious effort to take account of the advantages, as well as the costs, of the market revolution and its cultural consequences. The economic costs, the dislocations, the loss of skills and the proletarianization of workers, the concentration of wealth—all these were real enough. So were the cultural hypocrisies and contradictions of middle-class ideology. But the market revolution also brought real advantages to large numbers of people.

Some of these advantages were material, such as a more varied diet, cheaper clothing, better housing, and new ways of earning a living. Of course, most of the comforts, conveniences, and improvements in medicine that technological progress would bring still lay in the future. In the period Sellers treats, up to 1846, it was not at all clear to contemporaries that the material benefits of the industrial revolution outweighed the costs. But the market revolution was broader than the industrial revolution and probably affected rural and small-town people more. The material benefits of the expansion of the market were more readily apparent than those of industrialization because they could give farmers extra income without drastic social dislocations.

Moreover, in assessing the advantages entailed by the market revolution before 1846, the nonmaterial ones seem particularly significant. The market revolution increased individual autonomy, the opportunity to exercise choice. Nineteenth-century Atlantic civilization celebrated the ideal of "careers open to talents," but this ideal required an appropriate variety of careers from which to choose. The market revolution and urbanization multiplied occupations, especially those making use of formal education, far beyond what subsistence husbandry could provide. The market also broadened people's horizons and therefore enhanced their quality of life. Besides more schools and colleges, better communication, and more printed matter to read, the market revolution provided more (and more beautiful) places of worship, more humane child-rearing practices, more opportunities to encounter the arts.

Precisely because American frontier life was so anarchic and violent, the indigenous social controls to which it gave birth were equally harsh. An uncompromising fundamentalism struggled to contain frontier hedonism like a stern patriarch disciplining unruly sons and daughters or like a lynch mob confronting cattle rustlers. Andrew Jackson typified the authority figure of the West: violent, passionate, and vindictive, unconstrained by the rules of law (or spelling). Sellers accurately characterizes the culture of the American frontier as patriarchal. The market revolution brought a new middle-class culture that, for all its psychological tensions, held out the promise of deliverance from this narrow world of patriarchy and violence. The historian Louise Stevenson, calling this new culture by its political name, Whiggery, describes it well: "Whiggery stood for the triumph of the cosmopolitan and national over the provincial and local, of rational order over irrational spontaneity, of school-based learning over traditional folkways and customs, and of self-control over self-expression. Whigs believed that every person had the potential to become moral or good if family, school, and community nurtured the seed of goodness in his moral nature."[31] Those who opted for the market and its culture of modernization were not necessarily victims of a false consciousness.

As Sellers well recognizes, party loyalty was strong in the period he treats, and political parties were vehicles for voter self-consciousness. To be a Democrat or Whig could be an important facet of a man's identity. To Abraham Lincoln, for example, being a Whig was part of his self-image and chosen way of life, which was based on education, temperance, self-control, and self-improvement.[32] Lawrence Frederic Kohl has proposed that we look at the Whigs as "inner-directed" personality types and the Democrats as "tradition-directed." This terminology, borrowed from social psychology, is not without its problems (Kohl ignores the way in which these terms were originally defined by David Riesman), but it seems to make the point Sellers wants to make in a way that is less misleading than his own theological terminology of "arminian" and "antinomian."[33]

What the terminologies of both Kohl and Sellers ignore is the importance of the Enlightenment to the Jeffersonian-Jacksonian political orientation. The Enlightenment, with its celebration of natural rights, its insistence on the separation of church and state, and its enthusiasm for laissez-faire and low taxes, must surely be recognized as the ideological fountainhead of the program of the Democratic party. It is not easy to reconcile—or figure out how contemporaries reconciled—these Enlightenment principles with the Democratic party's penchant for racism, patriarchy, and violence, but the problem is not solved by ignoring the Enlightenment.[34] A somewhat different strand of Enlightenment thought, emphasizing accommodation to classical learning, Christianity, and polite culture, was important to the Whigs.[35]

One important kind of opportunity for voluntary activity in America was represented by religion. In colonial and antebellum America, the growth of organized religion went hand in hand with the extension of the market. The founding of new churches kept pace with the founding of new towns, and the percentage of the population that was joining churches increased in parallel with the growth of towns and cities.[36] Sellers recognizes the connection between the market and the expansion of religious revivalism, but not the element of individual autonomy involved. Evangelical Protestantism celebrated the moment of "new birth," or conscious commitment to Christ, as a time when the believer was liberated from sinful dependence by God's grace. Among evangelical Arminians, it was seen as the moment when a person took charge of his or her own life by making a transforming decision. Because society was pluralistic and religion voluntary, even the religious bodies that were not Arminian in principle provided in practice a form of voluntary self-definition.

The wide range of religious options available, both indigenous and imported, had never been equaled in any previous human society. Religion could be a way of preserving an ancestral identity for immigrant ethnic minorities; it could also be a vehicle for upward social mobility. To make a decision for Christ for many young people meant to embrace a new self-definition as sober and responsible members of adult society. On the other hand, a religious commitment might represent the deliberate choice of a counterculture. People could create new religions or synthesize them through syncretistic processes. Religions proliferated for much the same reasons that unconventional schools of thought sprang up in science, medicine, and philosophy: There were Grahamites and Fourierites, phrenologists, mesmerists, and spiritualists. Many people, we sometimes forget, deliberately chose no religion at all—an option that was unprecedented in premodern society but one antebellum Americans were free to exercise without fear of reprisal.

The voluntary basis of American religion—economic, legal, and in the dominant evangelical heritage, theological—was unique in the world. It probably explains why the modernization of American society was not accompanied by a corresponding secularization.[37] Instead, Americans volunteered for service in the army of Christ, just as they volunteered for service in fire brigades, the militia, the Freemasons, the Antimasons, labor unions, sewing circles, political parties, or Washingtonians. This example of a religious yet individualistic and modernized population has persisted in the United States from antebellum times to the present. It is a distinctively American phenomenon, one that has survived the discrediting of so many other alleged cases of "American exceptionalism."[38] In antebellum America, religion was on its way toward what it would become more obviously in our own day, a consumer good that members of the public are free to purchase or not in any of a wide variety of brands. Revivalism in such a society can be interpreted as an effective form of the mass marketing of religion to the public.[39]

But the identity that went with a religious commitment was not merely a label to be pinned on. With the identity, there went a lifestyle. The parallel between religion and the consumer society of the present day should not mislead one into dismissing religion in antebellum America as merely a form of recreation. Most of the kinds of religion embraced by antebellum Americans required some form of self-discipline. Sellers calls such religious self-

discipline "collective repression," and he sees it as an example of the bourgeoisie imposing its values in the interests of "class needs for work discipline,
social order, and cultural hegemony."[40]

Paradoxically, however, an increase in social organization provided an
arena within which enhancement of personal autonomy could take place. A
market society and economic development presuppose a sense of corporate
trust and social responsibility in the population just as much as a respect for
individual rights.[41] In antebellum America, it was evangelical Protestantism
that provided most of the impulse toward social organization. The benevolent empire of voluntary associations provided a model of national, local, and
international organization in a society where little else was organized, in which
there was no nationwide business corporation except the Bank of the United
States and no nationwide government bureaucracy except the post office. The
Second Great Awakening promoted both social organization and optimism
about progress, sometimes expressed theologically in postmillennialism.[42]

Although one should recognize the economic function of a disciplined work
force within a market society, the history of religious self-discipline must take
account of much more. Since the Reformation, the members of Protestant sects
had recognized the right of their communities to impose church discipline.
The "watch and ward" that lay members exerted over each other represented,
within their voluntarily constituted communities, a substitute for the hierarchical subordination that maintained order in traditional society. It was a
part of the Protestant program of empowering the laity. Even in Jacksonian
America, evangelical discipline, both individual and communal, had its liberating and empowering side, as well as its restrictive one. The reforms undertaken by the evangelicals of the time were typically concerned with redeeming persons who were not functioning as free moral agents: slaves, criminals,
the insane, drunkards, children, and even—in the case of the most logically
rigorous of the reformers—women. Redemption and discipline went hand in
hand as aspects of personal autonomy.[43]

Nineteenth-century self-discipline came to a focus in the temperance movement. Temperance represented the confluence of religious discipline with the
needs of the marketplace. The economic incentive to temperance involved
more than the interests of industrial employers; commercial farmers were just
as prominent as manufacturers among the businessmen who joined with New
England clergy in originating the movement. Besides religious discipline and
workplace efficiency, there was also a third argument for temperance, which
was based on secular science. Americans in 1815 drank a great deal more
alcohol than was good for them, and the medical profession gave temperance
important backing. The temperance movement undoubtedly improved national health standards, just as its proponents claimed it did.[44] Nor was recognition of the evils of intemperance confined to the middle class. Even Sellers
notes that the working class had its own temperance movement, exemplified
by "indigenous working-class Methodism" and the Washingtonians—an
admission that would seem to embarrass, at least a little, his interpretation of
temperance as bourgeois "cultural imperialism."[45] Temperance may be taken

as an example of how the quest for personal improvement became a prominent feature of American life during this period. Eventually, the movement
for personal temperance became a movement for the improvement of society
as a whole, employing the coercive power of the state.

Originally, the temperance movement was part of a much larger trend
toward the reformation of manners that was by no means simply religious.
There was also a new secular code of conduct promoted by the marketplace
and practiced by people in their everyday lives. This was middle-class politeness. Given his commendable commitment to merging the public and private
sides of history, Sellers curiously neglects the history of politeness. As Paul
Langford has shown in his work on England, polite culture was one of the
great consequences of the commercial revolution of the eighteenth century
and the concomitant rise of the middle class.[46] The rise of politeness was a
giant undertaking of voluntary self-reconstruction made possible by the market
revolution. As obedience to the discipline of a church defined its membership, good manners defined a person as a member of the middle-class "polite
society," as it was termed. Besides self-discipline, the new ideology of politeness also encouraged the cultivation of good taste. Concern with taste was a
response to the wider consumer choices and rising material standard of living
made possible by the market revolution.[47] Why should we not regard the increasing ability to make such consumer choices as a form of personal empowerment? The principle of voluntarism, which the American separation of
church and state had secured for religion, was extended by the market revolution to patterns of consumption.

Unlike the hierarchical code of traditional European society, the new polite
culture was inclusive rather than exclusive, open to all who would adopt it. It
was part of a program of self-improvement whose patron saint was Benjamin
Franklin, a program that included education and economic efficiency. Etiquette
books, promoting politeness, were part of a massive and remarkably successful
didactic effort to remake the face of America. They taught a prudential basis
for social ethics in which the reason for treating others politely was to demonstrate that one deserved to be treated with reciprocal politeness.[48] The new
code of politeness provided a framework for interaction with strangers at a
time when the market revolution was bringing people increasingly into
contact with strangers. Accordingly, politeness became the cosmopolitan
(or metropolitan) code of behavior, as opposed to the folk practices of the
provinces.

Middle-class polite culture brought new dignity for women. It did not
simply teach men to pay courtesies to women ("ladies first"); it enjoined the
subordination of sexuality, along with other forms of emotional indulgence,
to principled self-discipline. Sellers is at his most intemperate in his denun-
ciations of nineteenth-century middle-class prudery. (He does not seem to
realize that condemnation of masturbation and homosexuality did not origi-
nate with nineteenth-century evangelicals.[49]) But what we think of as Victorian prudery can also be seen as a clumsy effort to make men regard women
as something other than sexual objects. Although polite culture put women

on a pedestal to avoid challenging the prerogatives of men, it represented in important respects an advance over the subjugation of women common in premodern society. In contrast to the patriarchy of folk tradition, middle-class polite society gave women a new role. It made them guardians of polite sensibility, religion, morality, and the home—roles that women themselves gradually amplified and politicized. Much of the early undertaking to provide education for women was conceived in terms of fitting them for this domestic role.[50]

Out of the sphere accorded them by Victorian middle-class culture and the wider scope of economic opportunities outside the home brought by the market revolution, women forged a new autonomy for themselves. Historians have universally acknowledged that women's liberation began in the urban middle class, within Protestant denominations that had encouraged a measure of gender equality (especially Unitarians and Quakers), and among women who had had experience in such other religiously motivated reform movements as those for temperance and antislavery.[51] Economically, the market revolution offered women alternatives to their traditional subjugation within the household. Politically, the early feminist movement occupied, together with abolitionism, what might be called the extreme left wing of the Whig-Republican political orientation.[52] If we are to view antebellum history in terms of a *Kulturkampf* between the market and the subsistence economy, there can be no doubt which side offered the more attractive option to women. The contribution of the market revolution to increasing autonomy for women is one of the ways in which it was a force for modernity and liberation, in antebellum America as in other times and places.[53]

Polite culture and evangelical religion had much in common: discipline, perfectionism, and the impulse to improve the individual and the world. Like the evangelical reform movements of the time, polite culture undertook to make the world a better place by reshaping individuals into better people. Like evangelical religion, politeness taught people to control their appetites and passions and practice consideration for others. In the largest sense, the rise of politeness was linked with the rise of humanitarianism, both of which involve the restraint of antisocial impulses.[54] Together, politeness and evangelical moral reform helped reshape the world into a place where violent behavior was discouraged and commercial relations between strangers would be facilitated. Not surprisingly, both movements often appealed to the same people: the middle class or those who aspired to middle-class identity.

Accordingly, many evangelical Christians embraced politeness and allied with it. The synthesis of evangelical religion and politeness produced the respectable evangelicalism that accepted market society but tried to reform and sensitize it—the type of evangelicalism Sellers calls "the Moderate Light." *Moderate Light* is the term he invented to distinguish bourgeois evangelicals from New Light folk evangelicals. So far, so good: There were different kinds of evangelicals in antebellum times. Some were ecumenical, polite, reformist, and Whig. Others were particularistic and stern, seeking not so much to reform the larger society (which had been written off as corrupt) as to preserve the distinctiveness of a saving remnant. What Sellers calls Moderate Light revivalism produced the Second Great Awakening, which I have argued was an important component of the political culture of the Whig party.[55]

Sellers's bifurcation of evangelicals into New Lights and Moderate Lights makes for confusion, however. The Quakers disapproved of middle-class politeness, yet they strongly supported perfectionist social reforms. Does this make them New Light or Moderate Light? Sellers himself finds many evangelicals who have characteristics of both; in such cases he attributes their good qualities to the New Light and their bad ones to the Moderate Light in ways that make for little real illumination.[56] And what is the precise relationship between his brand of antinomianism and the New Light? Upon analysis, New Light and Moderate Light seem to be sociopolitical rather than religious categories. Sellers wants the New Light to serve as a Jacksonian counterpart to the Moderate Light of the Whigs, but the category does not really work. It does not take account of the enormous variety of religious positions taken by the Jacksonian Democrats, many of whom were not any kind of evangelical. They could be Catholics, Lutherans, Episcopalians, Old School Presbyterians, freethinkers, or Freemasons. The best way to characterize the Democrats in religious terms is as outsiders. Sean Wilentz, writing of the workingmen of New York City, concludes: "It would be foolish to try to impose unity upon such diverse currents of artisan piety, irreligion, and apathy."[57] Democrats who demonstrated great religious diversity could unite in defense of the separation of church and state, in common opposition to the cultural onslaught of polite, perfectionist, Moderate Light Whiggery—but not in any common "antinomian" New Light theology.

One of the important movements of the period Sellers treats was Antimasonry. As a popular cause with both religious and political dimensions, Antimasonry provides an appropriate test case for the interpretation of politics in terms of religious categories. It has been the subject of considerable recent historical interest.[58] One might expect Antimasonry, therefore, to figure prominently in Sellers's account. But Sellers's treatment of Antimasonry is brief and conventional. He identifies it, in passing, with antinomianism at one point and with the middle-class Moderate Light at another.[59]

One of the principal features distinguishing the Moderate Light revival—accepting for the moment Sellers's terminology—was its broad sense of social responsibility.[60] Moderate Lights were committed to an ecumenical, national, and even international perspective. This commitment reflected a variety of influences, including long-standing Puritan philanthropy and the new, market-fostered, humanitarian concern for suffering outside one's personal encounter that Thomas Haskell has identified.[61] One of the ways in which the market revolution broadened people's horizons was in relation to their racial attitudes. Racism was, of course, pervasive in antebellum America, but there are degrees of racism, and they were reflected in the variety of political positions taken on the slavery question. The position taken by the bourgeois Whig party on slavery varied from one geographical region to another, but in each area the Whig party generally took a weaker proslavery position than

did the Democratic party. (Henry Clay, the party's architect and leader, supported gradual compensated emancipation and colonization as late as the
Kentucky constitutional convention of 1849.) Blacks had no difficulty in per-
ceiving which major party was the lesser evil; wherever they could vote, they
voted as a bloc against the Democrats.[62] Indian policy was an important political issue when the second party system was taking shape in the 1830s,
and policy toward Mexico was important throughout its duration. On both
these issues, too, the Whigs espoused a distinctly less racist (and less violent)
position than did the Democrats.

Sellers's treatment of abolitionism is idiosyncratic. In general, he regards
the reforms of the period as examples of sinister bourgeois social control;
unlike most of the social control school of historians, however, he makes an
exception for abolitionism. In the case of abolitionism, he sees the reformers
as genuinely enlightened and their cause as liberating. To distinguish abolition from other reform movements, he does all he can to link it with
"antinomian" and New Light influences.[63] Actually, abolitionism is better
understood as an extreme version of the perfectionist, improving impulse that
was so prominent a feature of Whig-market-postmillennial culture. Support
for abolitionism has been convincingly correlated with both the practice of
commercial farming for the market and the adoption of marketplace values.
As John Quist has put it, "The moral crusade to abolish slavery was dominated by people who were also preoccupied with self-improvement and better-
ing the circumstances of their families."[64] That abolition had connections to
New Light divinity (e.g., through Samuel Hopkins) is really an indication that
the New Light and Moderate Light versions of evangelicalism were not so
different from each other as Sellers makes out.

What Sellers calls New Light revivalism would appear to be more or less
what Nathan Hatch has described in his book *The Democratization of American Christianity*—that is, the Christianity of American popular culture. There
is a major difference between their descriptions of popular revivalism, however. Hatch does not see his New Lights as defenders of tradition against
modernity; he sees them as innovators, particularly in their assault on traditional social deference: "Rather than looking backward and clinging to a moral
economy, insurgent religious leaders espoused convictions that were essentially modern and individualistic. . . . Religious movements eager to preserve
the supernatural in everyday life had the ironic effect of accelerating the
breakup of traditional society and the advent of a social order of competition, self-expression, and free enterprise."[65]

The American exponents of the Moderate Light developed a creative synthesis of evangelical Christianity with polite culture, the intellectual basis of
which was their adaptation of Scottish moral philosophy to the needs of
American denominational colleges.[66] Perhaps Sellers's term *New Light* fits
those who rejected politeness as a form of worldly corruption and imposed
their own discipline on members as a godly alternative code of behavior. The
austere lifestyle of such evangelical outsiders did make them less vulnerable
to the seductive consumption patterns of the market, though it seldom de-

terred production for the market. Generally, the discipline of the groups Sellers calls antinomian was sterner than bourgeois polite discipline. Sellers's insistence on blaming all repression of impulse and passion on the bourgeois elite does not take account of this fact.[67] For example, temperance began as a polite and elite movement that was concerned with promoting moderation in the consumption of alcohol and endorsing wine as an alternative to whiskey. As it became less elite-dominated, the movement became more extreme in its demands, which culminated in total abstinence and legal prohibition. On the other hand, the "primitive" Baptists were not only antimission in an evangelistic sense but also antitemperance, and so they seem to illustrate a version of the antireform, antipolite kind of religion more in keeping with what Sellers calls antinomianism.[68]

The competition among different forms of evangelical, polite, and traditional cultures provides some of the most interesting historical examples of *Kulturkampf*, not only in the United States but also throughout the Western world.[69] Sometimes denominations started out resisting the market revolution and its culture only to reach accommodation with it—the most dramatic reversal being that of the Mormons, who switched at the turn of the last century from defending an autonomous enclave of polygamy and socialism to embracing monogamy, capitalism, and the Republican party. The struggles over culture, however, are not simply more complicated but also more morally ambiguous than is conveyed in Sellers's book.

In conclusion, we need to articulate why some Americans were suspicious of the market revolution and even hostile to aspects of it, like banks and bank currency. This Sellers does for us, eloquently. It is also important, however, to know why such suspicions and hostilities were overcome. In the United States during the Whig-Jacksonian period, white people were achieving greater autonomy and opportunity for self-definition than ever before. Sellers is right to characterize belief in human autonomy—what he calls arminianism—as basic to the culture of the market. What he does not do is make clear how the market enhanced autonomy and why so many contemporaries found this attractive. Ultimately, most Americans welcomed the economic benefits of the market and rose to the opportunities presented by it, often on a piecemeal basis. But it would take a civil war and tumultuous reconstruction for African Americans to be incorporated into market participation. As and when it happened, it formed part of the national recognition of their full humanity. For them, even more than for other Americans, market participation was indeed a revolution and a means of liberation.

There is a tendency today among some intellectuals in the Western world to romanticize traditional society and undervalue modernity. Sellers's book is an example of this tendency. His conception of antinomianism seems to contain both the anthropological concept of a traditional society awaiting modernization and a vision of fraternal, participatory democracy that can serve as a model for our own time.[70] But American frontier society and religion do not really serve his didactic purposes well. Like most premarket societies, the American hinterland was even more ethnocentric and patriarchal than bourgeois society, and Sellers is too good a historian not to notice this. Nor did the antinomian religion that Sellers celebrates constitute such a marked traditional contrast to that of the modernizing Moderate Lights. The New Light denominations, like those of the Moderate Lights, were descended from the Reformation. A sect like the Baptists, when compared with, say, Finneyite revivalism, simply represented an earlier historical version of the growth of individual autonomy associated with modernization.

Sellers's political sympathies lie with the Jacksonian Democratic party, and the history he writes properly reflects this. But, even given this commitment, the intellectual—and, yes, the moral—utility of Sellers's concept of antinomianism is open to question. The most constructive, admirable aspect of the Jefferson-Jackson political impulse in the antebellum United States was not its defense of traditional society in the face of an evil market revolution; it was its support for individual freedom and equality, which it expressed in Enlightenment terms of natural rights. All in all, then, Sellers's analysis of the impact of the market revolution on the United States was sounder before he added to it the anachronistic, ambiguous, and misleading categories "antinomian" and "arminian."

Notes

1. Charles Sellers, *The Market Revolution: Jacksonian America, 1815–1846* (New York: Oxford University Press, 1991).

2. The lectures are available in the form of a superb brief textbook that Sellers wrote with a colleague: Charles Sellers and Henry F. May, *A Synopsis of American History* (Chicago: I. R. Dee, 1963). Sellers wrote the antebellum and May the postbellum chapters.

3. Ibid., 110.

4. One may question, however, whether the "market revolution" contributed as much to economic growth, as distinguished from social, political, and economic change, as Sellers implied. See Thomas Weiss, "The American Economic Miracle of the Nineteenth Century," paper presented to the American Historical Association, San Francisco January 1994; and Paul David, "The Growth of Real Product in the United States Before 1840," *Journal of Economic History* 27 (1967): 151–97.

5. Quotation from Thomas Haskell, "Capitalism and the Origins of the Humanitarian Sensibility," *American Historical Review* 90 (1985): 339–61, 547–66. See also the illuminating "Forum" discussion among Haskell, David B. Davis, and John Ashworth, *American Historical Review* 92 (1987): 797– 878.

6. Richard J. Carwardine, *Evangelicals and Politics in Antebellum America* (New Haven, Conn.: Yale University Press, 1993).

7. For example, Sellers, *Market Revolution*, 8–9.

8. Ibid., 217–25.

9. Ibid., 31.

10. Ibid., 30.

11. See Daniel Walker Howe, "The Decline of Calvinism: An Approach to Its Study," *Comparative Studies in Society and History* 14 (1972): 306–27. If I were to

rewrite this article, I would complicate its line of argument, taking account, among other things, of a resurgence of Calvinism in the eighteenth century.

12. Carwardine, *Evangelicals and Politics*, esp. 113–126, 222–29.

13. After the disestablishment of religion by the New England states (a process not completed until 1833), Episcopalians still found the Democratic party a congenial refuge from evangelical moral reform crusades. See Robert Bruce Mullin, *Episcopal Vision/American Reality: High Church Theology and Social Thought in Evangelical America* (New Haven, Conn.: Yale University Press, 1986), 200.

14. Steven Marini, *Radical Sects of Revolutionary New England* (Cambridge, Mass.: Harvard University Press, 1982); Nathan Hatch, *The Democratization of American Christianity*, (New Haven, Conn.: Yale University Press, 1989).

15. For example, Sellers, *Market Revolution*, 210.

16. Ibid., 30.

17. Ibid., 452–53, n44.

18. Alan Heimert, *Religion and the American Mind: From the Great Awakening to the Revolution* (Cambridge, Mass.: Harvard University Press, 1966).

19. Sellers, *Market Revolution*, pp. 203–4, is a précis of Perry Miller, *Jonathan Edwards* (New York: Sloane, 1949). For a more reliable understanding of Edwards's ideas, see Norman Fiering, *Jonathan Edwards' Moral Thought and Its British Context* (Chapel Hill: University of North Carolina Press, 1981).

20. Heimert, *Religion and the American Mind*, 55, 56.

21. Ibid., 39.

22. See Joyce Appleby, *Capitalism and a New Social Order: The Republican Vision of the 1790s* (New York: New York University Press, 1984); John R. Nelson, *Liberty and Property: Political Economy and Policymaking in the New Nation, 1789–1812* (Baltimore: Johns Hopkins University Press, 1987); and Richard Vernier, "Political Economy and Political Ideology: The Public Debt in Eighteenth-Century Britain and America" (D.Phil. diss., Oxford University, Oxford, 1993).

23. Sellers, *Market Revolution*, p. 268.

24. Christopher Clark, *The Roots of Rural Capitalism: Western Massachusetts, 1780–1860* (Ithaca, N.Y.: Cornell University Press, 1990). See also James Henretta, "Families and Farms: *Mentalité* in Pre-Industrial America," *William and Mary Quarterly* 35 (1978): 3–32; John Mack Faragher, *Sugar Creek: Life on the Illinois Prairie* (New Haven, Conn.: Yale University Press, 1986); and Lacy K. Ford, *Origins of Southern Radicalism: The South Carolina Upcountry, 1800–1860* (New York: Oxford University Press, 1988).

25. See Charles Sellers, "Who Were the Southern Whigs?" *American Historical Review* 59 (1954): 335–46; J. Mills Thornton III, *Politics and Power in a Slave Society* (Baton Rouge: Louisiana State University Press, 1978); Harry Watson, *Community Conflict and Jacksonian Politics* (Baton Rouge: Louisiana State University Press, 1981).

26. The best account is Sean Wilentz, *Chants Democratic: New York City and Rise of the American Working Class, 1788–1850* (New York: Oxford University Press, 1984).

27. See, for example, Michael Feldberg, *The Turbulent Era: Riot and Disorder in Jacksonian America* (Oxford: Oxford University Press, 1980). Sellers tries to blame this violence on bourgeois instigation; it would be more accurate to say that employers took advantage of ethnic hostilities among workers than that they created them. See *Market Revolution*, 386–89.

28. See Richard D. Brown, *Knowledge Is Power: The Diffusion of Information in Early America, 1700–1865* (New York: Oxford University Press, 1989), chap. 9. On the importance of creating the power of choice in the modern world, see David Apter, *The Politics of Modernization* (Chicago: University of Chicago Press, 1965), 9–11.

29. Sellers, *Market Revolution*, 237.

30. As I have argued in *Victorian America* (Philadelphia: University of Pennsylvania Press, 1976).

31. Louise Stevenson, *Scholarly Means to Evangelical Ends: The New Haven Scholars and the Transformation of Higher Learning in America, 1830–1890* (Baltimore: Johns Hopkins University Press, 1986), 5–6.

32. Sellers is good on this. See his *Market Revolution*, 395.

33. Lawrence Frederick Kohl, *The Politics of Individualism: Parties and the American Character in the Jacksonian Era* (New York: Oxford University Press, 1989).

34. On why people might legitimately welcome both market economics and the political program of the Enlightenment as enhancing their autonomy, see Joyce Appleby, *Liberalism and Republicanism in the Historical Imagination* (Cambridge, Mass.: Harvard University Press, 1992).

35. On the contrasting ideological positions of the Democratic and Whig parties, see John Ashworth, *"Agrarians" and "Aristocrats": Party Political Ideology in the United States, 1837–1846* (London: Royal Historical Society, 1983); Major L. Wilson, *Space, Time, and Freedom: The Quest for Nationality and the Irresistible Conflict* (Westport, Conn.: Greenwood, 1974); and Daniel Walker Howe, *The Political Culture of the American Whigs* (Chicago: University of Chicago Press, 1979).

36. On urbanization and Christianity, see Steve Bruce, ed., *Religion and Modernization: Sociologists and Historians Debate the Secularization Thesis* (Oxford: Oxford University Press, 1992); Patricia Bonomi, *Under the Cope of Heaven: Religion, Society, and Politics in Colonial America* (New York: Oxford University Press, 1986); and Timothy Smith, *Revivalism and Social Reform in Mid-Nineteenth Century America* (New York: Abingdon, 1957).

37. See Roger Finke, "An Unsecular America," in Bruce, *Religion and Modernization*, 145–69.

38. See Byron Shafer, ed., *Is America Different: A New Look at American Exceptionalism* (Oxford: Oxford University Press, 1991); Ian Tyrell, "American Exceptionalism in an Age of International History," *American Historical Review* 96 (1991): 1031–55; and Michael Kammen, "The Problem of American Exceptionalism," *American Quarterly* 45 (1993): 1–43.

39. See Terry D. Bilhartz, *Urban Religion and the Second Great Awakening: Church and Society in Early National Baltimore* (Rutherford, N.J.: Fairleigh Dickinson University Press, 1986); and R. Laurence Moore, *Selling God: American Religion in the Marketplace of American Culture* (New York: Oxford University Press, 1994).

40. Sellers, *Market Revolution*, 259, 266–68.

41. See Nathan Rosenberg and L. E. Birdzell, Jr., *How the West Grew Rich: The Economic Transformation of the Industrial World* (New York: Basic Books, 1986), 123–26.

42. See Daniel Walker Howe, "The Evangelical Movement and Political Culture in the North During the Second Party System," *Journal of American History* 77 (1991): 1216–39.

43. See ibid.

44. See Ian R. Tyrell, *Sobering Up: From Temperance to Prohibition in Ante-bellum America, 1800–1860* (Westport, Conn.: Greenwood, 1979); and W. J. Rorabaugh, *The Alcoholic Republic: An American Tradition* (Oxford: Oxford University Press, 1979).

45. Sellers, *Market Revolution*, 267.

46. Paul Langford, *A Polite and Commercial People: England, 1727–1783* (Oxford: Oxford University Press, 1992), chap. 3.

47. See Richard Bushman, *The Refinement of America: Persons, Houses, Cities* (New York: Knopf, 1992).

48. See John F. Kasson, *Rudeness and Civility: Manners in Nineteenth-Century Urban America* (New York: Hill & Wang, 1990).

49. Sellers, *Market Revolution*, 246–52.

50. On the education of women for their role as "republican mothers," see Linda Kerber, *Women of the Republic: Intellect and Ideology in Revolutionary America* (New York: Norton, 1980).

51. See Ross Paulson, *Women's Suffrage and Prohibition* (Glenview, Ill: Scott, Foresman, 1973); Nancy Cott, *The Bonds of Womanhood: "Woman's Sphere" in New England, 1780–1835* (New Haven, Conn.: Yale University Press, 1977), 126–59; Mary Patricia Ryan, *Cradle of the Middle Class: The Family in Oneida County, New York, 1790–1865* (Cambridge: Cambridge University Press, 1981); Anne C. Rose, *Transcendentalism as a Social Movement, 1830–1850* (New Haven, Conn.: Yale University Press, 1981); Barbara Epstein, *The Politics of Domesticity: Women, Evangelicalism, and Temperance in Nineteenth-Century America* (Middletown, Conn.: Wesleyan University Press, 1981); and Nancy A. Hewitt, *Women's Activism and Social Change: Rochester, New York, 1822–1872* (Ithaca, N.Y.: Cornell University Press, 1984).

52. Cf. Ellen DuBois, *Feminism and Suffrage: The Emergence of an Independent Women's Movement in America* (Ithaca, N.Y.: Cornell University Press, 1978), 75.

53. An excellent treatment of the connections among women's rights, party politics, and the new culture of respectability is Norma Basch, "Marriage, Morals, and Politics in the Election of 1828," *Journal of American History* 80 (1993): 890–918.

54. See the wonderfully wide-ranging work of Norbert Elias, *The Civilizing Process*, trans. Edmund Jephcott (1939, first published in German, Oxford: Blackwell, 1982).

55. Howe, *Political Culture of the Whigs*, esp. 150–80, and "Evangelical Movement."

56. Sellers, *Market Revolution*, 210–12, 225–36.

57. Wilentz, *Chants Democratic*, 83.

58. Recent analyses of Antimasonry include Paul Goodman, *Towards a Christian Republic: Antimasonry and the Great Transition in New England, 1826–1836* (New York: Oxford University Press, 1988); Robert O. Rupp, "Parties and the Public Good: Political Antimasonry in New York Reconsidered," *Journal of the Early Republic* 8 (1988): 253–79; Kathleen Smith Kutolowski, "Antimasonry Reexamined: Social Bases of the Grass-Roots Party," *Journal of American History* 71 (1984): 269–93; William Preston Vaughn, *The Antimasonic Party in the United States* (Lexington: University of Kentucky Press, 1983); and Donald Ratcliffe, "Antimasonry and Partisanship in Greater New England, 1826–1836," *Journal of the Early Republic* 15 (1995): 197–237.

59. Sellers, *Market Revolution*, 282, 405.

60. See George M. Thomas, *Revivalism and Cultural Change: Christianity, Nation-Building, and the Market in the Nineteenth-Century United States* (Chicago: University of Chicago Press, 1990); George M. Marsden, *Religion and American Culture* (San Diego, Calif.: Harcourt, 1990); and Randolph A. Roth, *The Democratic Dilemma: Religion, Reform, and the Social Order in the Connecticut River Valley of Vermont, 1791–1850* (Cambridge: Cambridge University Press, 1987).

61. Haskell, "Capitalism."

62. Howe, *Political Culture of the Whigs*, 17–18, 133–37.

63. For example, Sellers, *Market Revolution*, 23, 402.

64. John Quist, "The Michigan Abolitionist Constituency of the 1840s," *Journal of the Early Republic* 14 (Fall 1994): 325–58; see also Eric Foner, *Politics and Ideology in the Age of the Civil War* (New York: Oxford University Press, 1980), 57–76.

65. Hatch, *Democratization of American Christianity*, 16.

66. See Mark A. Noll, *Princeton and the Republic, 1768–1822* (Princeton, N.J.: Princeton University Press, 1989); Donald Harvey Meyer, *The Instructed Conscience: The Shaping of the American National Ethic* (Philadelphia: University of Pennsylvania Press, 1972); Robert M. Calhoon, *Evangelicals and Conservatives in the Early South* (Columbia: University of South Carolina, 1988); and Brooks Holifield, *The Gentlemen Theologians: American Theology in Southern Culture, 1795–1860* (Durham, N.C.: Duke University Press, 1978).

67. Sellers, *Market Revolution*, 245–59.

68. Rorabaugh, *Alcoholic Republic*, 209.

69. See, for example, Dickson D. Bruce, *Violence and Culture in the Antebellum South* (Austin: University of Texas Press, 1979); and Leigh Eric Schmidt, *Holy Fairs: Scottish Communions and American Revivals in the Early Modern Period* (Princeton, N.J.: Princeton University Press, 1989).

70. Implicit throughout, this becomes explicit in the last sentence of the book. See Sellers, *Market Revolution*, 427.

4

Charles Sellers's "Antinomians" and "Arminians": Methodists and the Market Revolution

Richard Carwardine

No one can doubt the profound significance of the flowering of evangelical religion in the United States during the early national period. Intellectual and church historians have long regarded the waves of revival that made up the Second Great Awakening as expressing and accelerating a major ideological and theological reorientation, as well as demonstrating the American churches' extraordinary practical energy and enterprise in the years following dis-establishment.[1] More recently, exponents of the "new social history" have contributed a variety of painstaking and impressive local studies that have done much to anatomize the religious life of particular communities in the early Republic.[2] Even those who favor a primarily ideological explanation of the awakening see the value of pursuing the linkages between evangelical-revivalist culture and an expanding market economy whose salient features included better communications, urban growth, incipient industrialization, increasing class tensions, changing gender roles, and a mobile population.[3] In both the United States and Great Britain, the early decades of the nine-teenth century saw social and economic changes that transformed an older rural, preindustrial order, touching the lives of hundreds of thousands of men, women, and children; both societies simultaneously witnessed so massive a burgeoning of evangelical Protestantism that the relative strength of their churches was greater than at anytime before or since.[4]

In his grand interpretation of the era, Charles Sellers builds on these specialized studies to offer the most ambitious attempt yet at integrating what we know of Jacksonian Americans' religious experiences and the realities of their economic and political lives. Sellers contends that American society in the early republic was fundamentally divided by the experience of capitalist market revolution. On one side in this *Kulturkampf* stood the traditional rural, democratic order of subsistence farmers, marked by communal cooperation, family obligation, patriarchal authority, honor, and independence. Rapidly making inroads into this culture by 1815 was the world of market production, characterized by specialization, division of labor, competitive individualism, enterprise, and boom and bust.[5]

Sellers makes religion functionally pivotal in this analysis: Only religious fervor, he argues, could "nerve" Americans for their "stressful passage from resistance [to the capitalist revolution] through evasion to accommodation." He explains the ecstatic "New Light" revivals over the period of a century— from the mid-eighteenth-century Great Awakening through the turn-of-the-century Great Revival to the culminating Second Great Awakening of the 1820s and 1830s—as an "antinomian" reaction against the advance of capitalist enterprise by a subsistence world that valued egalitarianism, democracy, plainness, and communal love; the market, on the other hand, with its competitive ethic and its rewards of wealth and status, was nourished by an "arminian" heresy that related salvation to human capability and effort. Some disaffected evangelicals (the Shakers, for example), driven by "millennial fantasy," chose to "come out" of a corrupted world altogether; but far more chose to mobilize against the new economic order through the primary agency of the New Light churches, especially those of the Baptists and Methodists, plebeian movements of subsistence farmers and working people. New Lights found political expression in Jacksonianism, discovering in the hero of New Orleans a "moral preeminence" that could be asserted against cosmopolitan elites.[6]

The merchants and fashionable congregations who made up those elites were often Unitarians, believing in a God who had given them power to fashion their earthly success. But, Sellers continues, far more influential among Yankee merchants, millers, and other entrepreneurs was "a Moderate Light tenuously blending the self-discipline of arminian effort with antinomian love." Here the agent of transformation was the inspirational Charles Grandison Finney. The upstate New York attorney-turned-revivalist took the theological inheritance of Jonathan Edwards, Joseph Bellamy, and Samuel Hopkins—that is, their New Divinity of moderate antinomianism and "disinterested benevolence"—and among the transplanted Yankee communities of the "burned-over district" "met head-on the climactic antinomian challenge to the culture of capitalist accommodation." Finney jettisoned conventional Calvinism, "refocus[ing] the Moderate Light by appropriating the . . . universalism" of the antinomian sects, especially the Methodists. In due course, in a celebrated progress, he was to reach New York's Broadway Tabernacle and achieve unimagined fame, preaching capitalist discipline to the working class and asceticism to their employers. Ultimately he would retreat, disillusioned by the social conservatism of capitalist culture, but not before he had "nerved Americans for the personal transformation required by a competitive market."[7]

Sellers's integrative approach is deeply imaginative, provoking us to look anew at familiar materials. But his interpretation of popular religion raises as many questions as it tries to solve. A minor issue is terminology since he deploys familiar labels in unfamiliar ways. A more serious difficulty arises over the way he implicitly attaches significance to religion only insofar as it operates as a function of economic change and as a validation of economic self-interest: To this issue, and its restrictive implications for analyzing electoral behavior, we shall return. More problematic, however, is that the evidence on which Sellers builds his bipolar cultural conflict between "antinomianism" and "arminianism" is less emphatic than he seems to suppose. At the heart of his antimarket coalition sit the Methodists, institutionally nonexistent in America at the time of Jackson's birth but the fastest-growing and largest denomination in the country by the time he reached the presidency. Sellers describes the Methodists as "the main bearers of antinomian universalism" outside New England in the aftermath of the panic of 1819; they appear as core members of the democratic insurgency that took Jackson to the White House in 1829; only later (Sellers offers no date) did the Methodists' "ethical athleticism" open up their denomination to capitalist imperatives and "obliterate from the[ir] memory . . . their origins in a massive cultural mobilization against the market and its ways."[8]

But how far, in practice, were the Methodists the defenders of a "parochial and fatalist" subsistence culture against the developing momentum of the capitalist market? Were the Methodists' values really at odds with the competitive individualism of "arminian" entrepreneurs? How substantial was the philosophical distance between Methodist revivalists and the refocused Moderate Light religion of Finney? These questions bear investigation. Methodist "antinomianism" is essential to the purity of Sellers's model of cultural bipolarity: Without it, his interpretation of Jacksonian religion would lose much of its force.

The Economic Transformation of Methodism

The origins of Methodism as a church of the poor and common people are so well established they need little rehearsal here.[9] Early Methodist preachers held meetings anywhere they could secure a hearing: in barns, yards, private houses, schoolrooms, and public streets. To plebeian audiences of rural laborers, subsistence farmers, mechanics, journeymen, and artisans, they proffered a gospel of salvation and the means of finding "liberty." Usually farmers and artisans themselves and lacking formal theological training, these preachers spoke the language of the common folk they addressed, openly and implicitly criticizing worldly ostentation and the love of money and display. John Sale, a Virginia-born presiding elder in early Ohio, rejoiced at having moved to a part of the country "where there is so much of an Equallity & a Man is not thought to be great . . . because he possesses a little more of this Worlds rubbish than his Neighbour."[10] Self-denial in matters of dress was the primitive Wesleyan rule for both preacher and people. One early southern preacher, Nicholas Snethen, spoke contemptuously of slaveholding urbanites as "soft, effeminate and sensualised, the willing captives of every delusion which is covered with a thin veil of external decency."[11]

Convinced that secular trading corrupted devotional exercise, presiding elders worked strenuously to prevent the selling of goods at camp-meeting grounds. As the commercial revolution advanced, reflective and prayerful Methodists grieved over Sabbath trading. Freeborn Garrettson, from the vantage point of Rhinebeck, New York, lamented "the many boats on the Lordsday scraping the shores of the great Canal, and the north river gathering up the sabbath-breakers, either for business or pleasure."[12] Alarmed at the profane influences at work on those laboring in the multiplying workshops, northeastern Methodist ministers scheduled factory preaching into their itineraries.[13] When cotton looms were stopped for revivals, as in Otsego and in Troy in 1828, or when protracted meetings were held thrice daily over a period of several weeks, the tension between the demands of enthusiastic Methodist religion and the needs of productive industry was plain enough.[14] According to historians Teresa Murphy, William Sutton, and Jama Lazerow, Methodism provided evangelical artisans and other workers with secure congregational communities; an ethic that condemned excessive profits, the abuse of wealth, and the exploitation of the powerless; and the inspiration to oppose the injustices associated with the social relations of emerging capitalism.[15]

Looking back on the early days of the movement through which some of them had lived, many mid-nineteenth-century Methodists reflected warmly on the church's original simplicity of spirit, manners, and forms of worship. Critics of change like Beverly Waugh extolled the communalism and "brotherly love" that the early Methodists had fostered in the intimacy of class meetings. Even the unsentimental George Peck, writing in 1866, agreed that "there was more equality of social position, and, of course, more familiarity, in former times than at present." As Methodism won converts among the wealthy, the educated, and the socially influential, and as its own members rose in society and began to wear "gold and costly apparel," so it seemed that its primitive character evaporated. By the 1850s, the earlier indifference to formal education had yielded to institution building on an elaborate scale: In their Address of 1856 the bishops of the Methodist Episcopal Church proudly listed nineteen colleges and universities, sixty-eight academies offering "literary advantages" to both sexes, and two theological seminaries.[16]

The status that Benjamin Tefft was pleased to claim for his denomination in his triumphalist *Methodism Successful* was reflected in the building of more ornate churches whose steeples and bells proclaimed Methodist splendor; at the same time, the members of existing churches were spending thousands of dollars "remodeling and beautifying" their meeting places. The introduction of organs and other musical instruments alarmed those fearful that the expense of sustaining music would mean that "rich men become necessary to us." Symbolizing the denominations' increasing status was the advance of the pew rent system, which was introduced in the 1840s in Baltimore, Richmond, and other major urban centers of Methodist strength and which by the late 1850s was evident in towns of all sizes from the East Coast to the Mississippi Valley. Conscious that even the Free Methodist churches in New York and other cities might seem too luxurious for the very poor, Abel Stevens and other church leaders grew alarmed that the general population might have ceased to see Methodism as "the common religion of the common people." Midcentury Methodists could speak without embarrassment about riches, adopting a tone apparently alien to Wesleyan or Asburyan primitivism. Phoebe Palmer explained that she had "no sympathy with that querulous spirit which is ever denouncing the rich, merely because they are so; or perhaps oftener than otherwise, because the denouncer is not possessed of the same means, and who, were he possessed of them, would be less faithful than those whom he denounces. . . . There are gradations in society which always have been, and doubtless always will be, till the end of time."[17]

The Complexities of Methodists and Markets

The stark contrast between turn-of-the-century Methodism and its midcentury gentility seems to suggest that the movement had experienced a huge cultural shift and that Sellers is correct in pointing to a betrayal of its antimarket origins as it succumbed to "capitalist imperatives." But this interpretation raises the question, why did the process of embourgeoisement provoke so little resistance in the church? The great Methodist schisms of the first half of the century concerned episcopal authority, the rights of laymen, the relationship of whites and free blacks, and the place of slavery within the denomination. Only one minor schismatic group, the Stillwellites, can be said to have originated primarily in resistance to the growing respectability of Methodism. In New York City, the issue of lay independence was compounded in the 1820s by a conflict over the rebuilding of the mother church on John Street. Its lavish refurbishing, which included the addition of a carpeted altar, appeared to one of the trustees, Samuel Stillwell, and his nephew William to represent a betrayal of the primitive simplicity of Methodism. The resultant new sect grew rapidly for a while, and William Stillwell, along with a number of other radical Methodists, developed a "producerist" philosophy and a scriptural critique of industrial capitalism that he propagated through the pages of his weekly family magazine, *The Friendly Visitor*. At first the new connection grew and spread into neighboring states, claiming 2,000 members by 1825. But it was never more than a minority strand of Methodism and would eventually die out.[18]

Significantly, Methodism experienced no schism of similar dimensions to the conflict within American Baptist churches through the early nineteenth century, which culminated in the split at the Black Rock Convention in 1832, as "Primitive" Baptists sought to uphold the true historic faith against Arminian softening by corrupted modernizers; rooted in the rural South and West, the Primitives represented an atavistic reaction against "modernism" in their churches. Nor was there an American version of the English Primitive Methodist schism by which the victims of industrial and agrarian upheaval sought to sustain the revivalist enthusiasm of early Wesleyanism against "Buntingite" Toryism and early Victorian respectability. When British missionaries attempted to introduce Primitive Methodism into the United States, the effort failed, not the least because, as one Primitive Methodist minister ruefully and tellingly reported, "all sorts of labor (*except Primitive Methodist preaching*) is paid higher than in England."[19]

The truth is that American Methodism experienced no searing conflicts over its churches' embourgeoisement because, although the early Methodists were poor, they were far from hostile to enterprise and the capitalist ethic. The movement was indeed a radical, egalitarian counterculture, but it was reacting against genteel patriarchalism and aristocratic pretension, creating its own community against a world of deference and honor, and not against industrious effort and self-improvement.[20] It was a community whose *Discipline* encouraged its members to buy and sell to each other and to give employment to Methodists whenever possible.

From its early days, and emphatically from the turn of the century, the denomination comfortably embraced a number of socially prominent and economically successful members. In Maryland these included, for example, a number of leading Delmarva Peninsula planter families, as well as merchants, professionals, and successful craftsmen.[21] The Methodists embraced Thomas Worthington, United States senator from Ohio, a member of the Chillicothe Methodist Church, and a promoter of cotton, wool, and flax manufacturing in the town.[22] They numbered Philip Gatch, a pioneer Methodist evangelist who, after he had "located" from the itinerant ministry, settled as a slave-owning planter in Maryland and Virginia and then moved to Ohio. He served in Ohio's first state constitutional convention and encouraged the immigration of enterprising Southern Methodists who, like him, were disenchanted with slavery and alert to the opportunities for commercial agriculture in southern Ohio. ("We shall not want for trade in this Countrey," he declared in 1802.) One of those he tried to tempt was Edward Dromgoole, who never wholly succumbed to "Ohio fever" but who speculated in Ohio lands and managed them from Virginia through the Pelham family. Peter Pelham and his eight children had moved to Greene County in 1807; they saw no tension between their Methodist convictions (Pelham more than once acted as host to Asbury) and enthusiastically embracing the opportunities presented by commercial agriculture. They bought and improved lands, hired laborers, sold wheat and corn, established (with other leading Methodists) an interest in woolen and cotton manufacturing, opened a grocery store, and appear to have edited and published Xenia's first newspaper in 1810.[23]

Such socially and economically successful early Methodists, and those who observed them, saw no contradiction in declaring their skepticism about human riches while simultaneously striving for improvement of themselves and their families. In one breath Philip Gatch declared, "I generally look at a mans Spirit more than his wealth. This may fail but charity never failith, this includes the first and second command, which will abide forever." But at the same time Gatch reflected thoughtfully on commercial opportunities and made no attempt to conceal his pleasure when his daughters married "industrous" men. Though the itinerant minister John Sale might declare himself an enemy of worldly accumulation, he also lauded the fertile soil and good communications for trading in agricultural and other goods in early Ohio. Another Methodist itinerant, Bennett Maxey, urging Edward Dromgoole to leave Virginia for good, reflected with evident satisfaction on his four years in the "good land" of Ohio. You could, he reported, "do well by keeping a retail store in this country[;] three years ago our merchants began the wourld with little or nothing. This spring they opened . . . [their] stores with a large assortment. You may git farmers to cultivate your land." Dromgoole's profits might not be as gratifyingly large as they were under slavery, but, Maxey insisted, "I believe the . . . [far] greater parte that have come from Virginia have ad[d]ed fore fold to their property with industry and frugality. They live as well as heart could wish for."[24]

These examples suggest that Sellers's portrayal of Methodists' thought is both partial and misleading. Sellers emphasizes the movement's democratic egalitarianism and communalism, the values most congruent with his definition of subsistence culture. It is true that in rejecting the doctrines of election and Christ's limited atonement, Methodists did indeed proffer a radical, democratic, and inclusive alternative to the perceived restrictions and inequalities of Calvinist orthodoxy: Christ, they insisted, had died for all men and women; no one had been excluded from the offer of his love and the Spirit's grace. Here was a philosophy wholly in harmony with the antielitist, democratic methods of the early Republic. But these were not the only emphases in Methodist ideology. Asbury's itinerants bore a further message, one that underscored individual responsibility and ability. God's grace was freely given, as well as universally offered. What prevented its reception were human obstinacy and prevarication. These barriers were reinforced by the stultifying Calvinist doctrine that sinners had to wait passively for the operations of the Holy Spirit. But, Methodist preachers insisted, the initiative lay with sinners themselves. Although no one could experience the new birth without the intervention of the Holy Spirit, it was open to every sinner to take the steps that would put him or her in the way of salvation. At the core of Methodism, in other words, lay the empowerment of the individual.[25]

Determined to force Methodists into the mold of "antinomian" critics of Arminian heresy, Sellers is thus driven to ignore their primary defining characteristic: the trumpeting of Arminius's view of conversion as a process that involves human enterprise. In a purely theological sense, at the very least,

Methodists were thoroughgoing Arminians. They deemed the salvation of their own souls, the energetic pursuit of the conversion of others, and the prosecution of revivals as their paramount business, and to this end they had fashioned their highly centralized and effective connectional organization of circuits and stations, districts, and annual and general conferences. Faith in human ability and responsibility led them to seek out potential converts under the most daunting conditions and in all corners of the union, however remote, and to get there before their competitors; it underpinned the proliferation of camp meetings and their introduction of new decision-inducing techniques of evangelism. Revival-focused Methodists put their faith in itinerancy before pastoral care; in the repetition through pointed, colloquial preaching of the need for repentance; and in multiplying and protracting services until hearers achieved "liberty." Sinners were repeatedly told, "If you are damned, it is your own fault" and "All things are *now* ready." The preachers' quintessential revivalist tool, resolutely constructed on the foundations of human ability, was the "call to the altar" or to the "mourners' bench." By this means the "anxious," unconverted hearers separated themselves physically from the body of the congregations (who, even so, continued to look on), to open themselves to the instruction and prayers of those who would guide them to a liberating salvation. First adopted in a variety of frontier and urban settings at the end of the eighteenth century, the call to the altar was to become a standard element in Methodist spiritual engineering by the second decade of the new century.[26]

Methodism was both a perfect metaphor for the emergent capitalist market of the early Republic and the principal beneficiary of the free spiritual market that superseded the colonial system of energetic government support for religion. When state assistance to a single favored denomination was replaced not, as first seemed possible, by multiple establishment but by an entirely voluntary system of church support, the initiative passed to those whose denominational institutions proved most flexible and muscular. In 1780 the Methodists had perhaps 50 congregations, as compared with the Congregationalists' 750, the Presbyterians' nearly 500, and the Baptists' and Episcopalians' 400 each. Taking advantage of the end of legal constraints, though continuing to suffer considerable persecution in New England, the Methodists proved the most successful of the liberated groups, growing even more quickly than their principal competitors, the Baptists, the other denomination notably freed by voluntarism to proffer its spiritual wares on the open market. By 1820 the enterprising Methodists had so successfully "sold" their religious doctrines—and, in the significant language of their critics, "manufactured" revivals and converts—that they boasted 2,700 congregations, the same number as the Baptists, having easily overtaken the Presbyterians and Congregationalists, who counted 1,700 and 1,000, respectively. The Methodists prized every new convert and every indication of growth. The membership returns to the annual conferences became the balance sheets of the organization, and they were scrutinized for evidence of weakness or, more usually, stunning growth. In the "market revolution" in religion, the Methodists were the chief beneficiaries.[27]

The entrepreneurial instincts of Methodist ministers were not restricted to matters spiritual. As the market revolution advanced, many espoused the capitalist ethic. It was not unusual for poor itinerants, dependent on the hospitality of the laity, to lodge with rich laymen—often commercial farmers or manufacturers—or their wealthy widows. Traveling preachers carried in their saddlebags books for sale, books that were the production of the influential, competitive, and (ultimately) prosperous Methodist Book Concern, which had been established in 1789 as the first denominational publishing house in the United States.[28] Church leaders launched denominational newspapers and took such good advantage of their connectional system that the New York *Christian Advocate and Journal*, first published in 1826, quickly established itself as the most successful weekly newspaper in the world. Methodist ministers also entered the educational marketplace to win custom for their colleges; some institutions went to the wall, but many others flourished through the enterprise and determination of their administrators.

Despite Asbury's admonitions, the attractions of making a living from the market proved too powerful for many poorly paid Methodist itinerants. Reminiscing in 1818 after over forty years in the ministry, Freeborn Garrettson regarded the practice of preachers "locating" as the bane of early Methodism, a constant weakening of the itinerancy. Some left for marriage; others for relief from ill health; and others still, wearying of the arduous, penurious traveling life, for the attractions of business and material prosperity. For some, like John Hogan, all three influences combined. Hogan was an Irish-born orphan who had been raised under Methodist influence in Baltimore. Had he died young, in the late 1820s, he would have been remembered as a typical frontier preacher who spent several years itinerating on horseback around huge circuits in Indiana, Illinois, and Missouri. Eventually he was assigned to the St. Louis Circuit, a vast, wild, dangerous, and largely unsettled territory extending along the south bank of the Missouri River from St. Louis itself to Boonville. Traveling 15 miles a day on horseback from one log cabin settlement to another, reading as he rode (when not disturbed by Indians or wild beasts); holding two services daily for small congregations, sleeping in primitive accommodations and rarely having a room to himself, Hogan faced great hardships and privations, and eventually he suffered a serious breakdown in his health. In 1830, he located. The following month he married, and soon afterward he began a general merchandising business in Illinois with his brother-in-law. It quickly prospered. Moving to Alton, the largest town in southern Illinois, he established a successful wholesale grocery house and became president of the Alton branch of the State Bank of Illinois. Hogan lost his fortune in the panic of 1837–38, but in the 1840s he established a wholesale grocery partnership with his uncle in St. Louis, importing sugar and coffee. Extending his interests into insurance, banking, railroads, and manufacturing, "Honest John Hogan" was to become one of the wealthiest and best-known men in Missouri. He remained a Methodist local preacher to the end of his enterprising life.[29]

Ordinary members of Methodist congregations showed no less eagerness than locating preachers in pursuing the rewards of the market. Early nine-teenth-century itinerants commonly complained of members' laxness over Wesley's and Asbury's prescriptions about dress and material temptations. Yet the Methodists' seizing of market opportunities was a corollary of the very doctrines and values that ministers propounded and that helped make the church so attractive to Americans of the era: individualism, self-discipline, and self-improvement. Love feasts and revival services may have provided social networks, mutual support, and communalism, but Methodist doctrine also encouraged the atomistic and competitive elements integral to the early Republic's liberal capitalism. George Peck was skeptical about the supposedly selfless equality and brotherly love of early Methodism. This latter attribute "arose from the smallness of the social circle, and a sense of mutual depen-dence, as much as from real brotherly love," he wrote half a century later. "Methodists then were few, and were much persecuted, and they were natu-rally forced together. They were poor and mutually dependent, and they had motives for being kind to each other of a selfish nature which they have not at present." Methodists were not communitarians. The essential method for doing good, explained a writer in the *Christian Advocate* only a few months after Jackson entered the White House, was to "make every individual take care of himself." He posed the rhetorical question, "Can the good of the na-tion be secure in any other way than by first securing the good of those indi-viduals which compose the nation?" Each person had an obligation to make his or her individual spiritual pilgrimage. Each, according to the formal arti-cles of Methodism, had the right to the enjoyment of his or her private prop-erty. Not even the earliest and poorest Methodists favored the doctrine of community of goods.[30]

Methodism, like other evangelical movements of the era, also extolled and inculcated habits of self-discipline, industry, and temperance. With its con-cern for decency in personal behavior and encouragement of spiritual develop-ment, it both generated and helped slake a widespread "thirst for improve-ment" (as Freeborn Garrettson put it) among the young in particular. That improvement connoted both moral and material achievement. Sellers points to the correction that Methodism offered to "the patriarchal vices of drink-ing, swearing, gambling, and fighting" and its challenge—by giving women and slaves a more equal role—to narrow subsistence patriarchy. But he shies away from the conclusion that such corrections represented more a push toward the market's competitive individualism and material improvement than an endorsement of the fundamentals of premarket society. In his *Auto-biography*, a work that is in many ways a jeremiad for the passing of a lost era, the elderly circuit rider Peter Cartwright recalled the case of a miserly Illinois farmer who was saving all his money to buy land and, with $300 already hoarded, still kept his Methodist family in the most primitive of con-ditions. The preacher told the husband to refurbish his cabin with decent chairs, beds, eating implements, and so on and to provide his wife and daughters with smart clothing. "You ought to know," Cartwright told him, "that the Discipline of our Church makes it the duty of a circuit rider to recommend cleanliness and decency everywhere . . . and you ought to attend to [these things] for your own comfort, and the great comfort of your family." Though the farmer called Cartwright a proud preacher and told him that "he knew I was proud the moment he saw me with my broadcloth coat on," he did later refurbish his cabin and provide the women with new calico dresses. The instincts of each of the parties were beckoning them into one sector of the market or another, yet Sellers chooses to see the primary meaning of the story in the enforcement on the husband of communal (that is, subsistence) standards of behavior. But surely its main lesson is that the Methodists' concern for decency and self-improvement accelerated rather than resisted the advance of the capitalist market.[31]

In his recent study of rural religion in the early settlement and development of upstate New York, Curtis Johnson looks for evidence that might link Methodism and areas of subsistence agriculture. In Cortland County, which witnessed a rapid shift to commercial farming between 1810 and 1830 and which shared in the era's canal fever, he examines the church's strength in areas high in the production of homemade cloth (an indicator of subsistence economy). Far from flourishing in these communities, Methodism was much weaker than it was in areas of market penetration. In fact, "Methodism grew in strength as the market penetrated the county's farthest reaches," and by 1860 it was the largest denomination: "Arminian theology with its emphasis on individual free choice had conquered the Calvinist doctrine of election."[32]

By the 1820s, the decade of Jackson's insurgency and a period when the Methodist Episcopal Church almost doubled its membership from a quarter million to nearly half a million, it is clear that the movement had come to embrace many from the ranks of the socially well regarded. The more Methodists took pains to prevent "rant, extravagance, or 'strange fire'" in their revivals, the more they boasted converts who included "men of talents, science, and of the highest and most respectable standing."[33] At some camp meetings they had to protect themselves against "the baser sort of the community." Many joined the church "without laying off any of the garniture which characterizes the fashionable of the first grade." More refined revivalists like John Newland Maffitt found their natural constituency among "respectable" citizens who would protect them against "the baser sort."[34]

There are, then, strong grounds for believing that Methodism in the era of the Jacksonian revolution was already infused with a capitalist and market ethic. Even if, as Sellers contends, first-generation Methodism demonstrated a number of "antinomian" features—in its communalism, egalitarianism, and location in premarket areas—it was also "arminian" in its understanding of human ability, its stress on individual responsibility, and its stimulus to self-improvement and enterprise. It was certainly not *anti*market. By the 1820s the movement had entered its second generation. Though there were those who resisted its self-conscious pursuit of respectability, many others took heart from their own social betterment, from the movement's evident capacity to attract new members of wealth and status, and from its contribution to the discipline of the market and of the new industrial arena. Methodist reports of the religious revival in 1827 at Bozrahville, Connecticut, of which it was claimed that fewer than ten of the employees of the cotton factory were immune to religious reformation, happily extolled prayerful Methodism's capacity to fashion an attentive, industrious work force.[35] The denomination embraced many who conspicuously welcomed the new economic order. Its spokesmen saw the country's commercial growth (as well as its climate, natural resources, and republican institutions) as an earnest of the Almighty's grand design for the nation.[36]

In their capacity for embracing modernity and economic improvement, the Methodists showed themselves members of a church with a national, not local or merely congregational, vision. Methodist connectionalism—involving the countrywide transfer of information through denominational publications and networks like the Sunday School Union, as well as the itinerating ministry— worked to inform and broaden the horizons of ordinary members. "Until after the period in which I became a subscriber to your useful and valuable paper," explained a reader of the *Christian Advocate and Journal*, "my views were entirely local." Sellers treats Methodism as part of the magical spirituality of a parochial and fatalist countryside against the cosmopolitan and activist market." Yet few agencies in the early Republic could boast Methodism's capacity for national integration and the breakdown of localism.[37]

It is only in recent years that writing on religion in the early Republic has ceased to serve a "Calvinist synthesis"—that is, the perception that Calvinism is the shaping force in American religious culture and that other expressions of Christianity, as well as non-Christian religions, are either marginal or warrant exploration principally in relation to Calvinist experience. Sellers's evident sympathy for the egalitarianism and communalism of "antinomian" culture, and his placing early Methodism at its heart, suggests his own break with that Calvinist-centered tradition; so, too, does his recognition of "antinomian" Methodist influence on the revolutionary Finney. Yet he does not wholly escape that tradition, for it is to Calvinists of the Moderate Light that he attributes the reconciliation of "the anxious antinomian majority to a capitalist world."[38] He devotes considerable space to the conflict within Calvinism, in New England and its diaspora, over the response to the market but only a sentence or so to the ultimate embourgeoisement of Methodism. He develops a model of middle-class clerics and entrepreneurs who combined to impose capitalist values and "collective repression" on reluctant antinomians who sought to remain loyal to rural ways; he misses the possibility that the Methodist lives of ordinary folk may have provided them with their own channel of self-improvement. He regards the conflict in Finney's burned-over district as one between Presbyterians and antinomian rebels; but the differences between the fast-growing Methodists and the "Presbygationalists" had more to do with missionary zeal in the spiritual marketplace, with conflict over style and comportment in the competition for souls, and with the residual theological conflict between Arminianism and Calvinism.

The triumph of Arminianism (in the conventional doctrinal sense, not Sellers's Jacksonian variant) over Calvinism was the crucial ideological reorientation brought about by the Second Great Awakening.[39] As the principal purveyors of Arminian doctrine, the Methodists engaged in bitter conflict with their evangelical competitors. No reader of the diaries and reminiscences of the popular preachers of this period can avoid their interdenominational polemics. Methodists saw themselves engaged in a battle not just against sin but also against false doctrine, principally Calvinism, and their principal targets among Calvinist adversaries were the Baptists. Far from being commonly engaged in an "antinomian" battle against the market, Arminian Methodists and Calvinist Baptists were antagonistically locked into one of the fiercest popular conflicts of the first half of the century as each strove to become the preeminent denomination of the common people. Peter Cartwright's struggle with various stripes of Baptists provides a leitmotif for his recollections of early Methodism in Kentucky, Tennessee, and Illinois. In a similar vein, Allen Wiley's memoirs are a celebration of the Methodists' achievement in becoming the principal denomination in the Whitewater Valley and other regions of Indiana where the Baptists had earlier predominated (a story made all the more pointed by Wiley's having been raised a Baptist). Similar conflicts marked upstate New York.[40]

Nowhere was sectarian antagonism more bitter, however, than in southern Appalachia. In Tennessee, Methodist growth and proficiency in revivals provided a target for Calvinists from the early 1800s to the late antebellum years: Both Baptists and Presbyterians regularly abused Methodist organization, theology, and morality, especially in the pages of Frederick Augustus Ross's *Calvinistic Magazine* and James R. Graves's *Tennessee Baptist*. Interminable theological wrangles over complete immersion, the baptism of infants, and Calvinistic predestination jostled side by side with salacious charges and countercharges of debauchery, drunkenness, and sexual seduction. Methodist allegations that Baptists were ignoramuses opposed to formal learning confronted Baptist claims that Wesley's troops were unscrupulous recruiters who would make any promise—even rebaptize their adult members by immersion—to secure their church's supremacy. Venomous polemics reached their climax in the 1850s when Graves published a compilation of his newspaper articles, which had denounced Methodism as an autocratic, overcentralized, and cryptopapist organization, under the title of *The Great Iron Wheel, or, Republicanism Backwards and Christianity Reversed*. This elicited a reply in kind from "Parson" William Gannaway Brownlow. In *The Great Iron Wheel Examined*, the combative Methodist preacher and editor mixed invective, personal insult, and innuendo in a bellicose riposte that confirmed his status as southern Appalachia's most redoubtable polemicist. Reflecting on these sectarian contentions, Brownlow—an Arminian pot bubbling alongside Graves's Calvinist kettle—concluded that "there is, perhaps, more of the spirit that prevailed in Geneva [when Servetus was burned] . . . in East Tennessee, than in any other place in this Union."[41]

These conflicts between the two principal popular denominations undoubtedly expressed more than competing certainties over baptismal practices. In part they represented a conflict between the Methodists' energetic connectionalism and the Baptists localism, in part a conflict between Arminian self-advancement and Calvinist determinism. In part, too, they may have been related to differing attitudes toward the market and economic development. Brownlow and his fellow eastern Tennessee Methodists were much more strongly identified than Baptists with Whiggish enterprise in moral and economic affairs. Brownlow, Samuel Patton, and (in western North Carolina) David R. McAnally typified Methodist leaders who regarded their part of the southern highlands as "the garden spot of the Union, and the El Dorado of America," which needed only investment and internal improvements to link it to a wider market and realize its economic potential. As Whig-Methodist boosters they connected religious, social, and economic improvement. "The interests of Education, Agriculture, and Commerce," one of them explained, "are more nearly allied to the prosperity of Christ's Kingdom than most men, perhaps, are willing to admit."[42]

Conversely, Baptists in the same region, particularly but not exclusively those "hard shell and iron sided" believers who made up the Primitive Baptists, seem more often than not to have been Jacksonians. High Calvinists, the Primitives (or "Antimission" Baptists) were strenuously opposed to Arminian "effort" and denounced theological education, temperance and missionary societies, "new measure" revivalism, "steam religion," and a "money-hunting" Protestant priesthood; they had no enthusiasm, as one elder explained, for plans "for the improvement of the moral, intellectual, and physical condition of mankind." James R. Graves and the "Missionary" Baptists shared many of the Primitives' perspectives on the Methodists, lamenting the heterodoxy of their preaching ("You can get religion and lose it at pleasure") and the threat to local autonomy and congregational independence from their centralizing ecclesiastical machinery. There was a history of Tennessee Baptists' rebuking Methodist preachers for their finery and warning the people to beware those who "wore black broadcloth coats, silk jackets . . . fair-topped boots, and a watch in their pockets; [and] that rode fine fat horses." *The Great Iron Wheel* can be seen as a Baptist-Democratic manifesto for republicanism, decentralization, and provincialism in the face of a movement that Baptists saw as despotic, open to cosmopolitan corruption, and too energetically nationalist. These ideological and cultural antipathies ran deep. They produced in Anderson County in eastern Tennessee what the Methodist circuit rider Frank Richardson judged "the bitterest denominational prejudice I have ever known anywhere." At the county seat, Clinton, the two communities "had no dealings with each other whatever": Each had its own churches, schools, taverns, stores, blacksmith shops, and even ferries across the river. "Most of the Methodists, were Whigs," Richardson noted, "and most of the Baptists were Democrats."[43]

The Methodist Vote

If, then, the Methodists cannot be easily classified as "antinomian" opponents of the capitalist market, and indeed in some incarnations have rather to be seen as its allies, how is one to explain their loyalty to Jackson and his party? Accepting that Methodists in 1828 were mainly Jacksonians in politics, we do not have to suppose that those who voted for Old Hickory as Methodists did so because they saw him as their best defense against the market's advance. As I have argued at greater length elsewhere, evangelicals' motivation in voting was highly complex and varied, and their party preferences depended considerably on context.[44] But there is no doubt that many saw in Jackson's party, as they had earlier seen in Jefferson's, their best defense against illegitimate religious power—that is, against the influence of "formalist" denominations, including those that had previously benefited from the statutory alliance of church and state and that sought to impose legal controls on moral and religious behavior. The ranks of the Methodists—and the Baptists—were made up very largely of those who emphasized the primacy of personal piety; believed churches' main responsibility was to save souls; saw social change as coming through the Holy Spirit, working in the hearts of individuals; and tended to distrust government schemes for social reform. Politically they sought the best means of sustaining the laissez-faire principles by which churches fought for members on equal terms and sustained themselves by their own voluntary efforts.

Methodists' experience at the hands of Calvinist orthodoxy had a profound effect on their partisan allegiance. In New England, in the early years of the young Republic, they were forced to pay taxes to sustain the Congregationalist Standing Order; poor ministers faced heavy fines for conducting marriage ceremonies; hostile mobs intimidated worshippers, broke up services, destroyed meeting places, and attacked as "incarnate demons" itinerant "intruders into the land of steady habits." In areas of Yankee settlement outside New England, though they suffered no legal disabilities, Methodists faced social discrimination from higher status Calvinists: In the Western Reserve, for example, an informal establishment of "Presbygationalists" strove to hobble them by closing off preaching places. In consequence, it was hardly surprising that most Methodists became enthusiastic Jeffersonians, not because they cherished Jefferson's personal religious beliefs, of course, but because his party provided a principled defense of religious pluralism against Federalist church establishments.

The motives that made Methodists Democratic-Republicans under the first party system—in particular their determination to defend Arminian truth against Calvinist arrogance and bigotry—similarly drew many into Jacksonian Democracy, especially in the party's early years.[45] Political managers were no strangers to the truth expressed by a writer in the *Farmer Herald*: "To subserve the interests of party, different sects must be made to hate, and vilify each other."[46] The Democrats offered a home to those who, encouraged by Jacksonian publicists in the 1828 election, believed that John Quincy

Adams's National Republicans and subsequently the early Whig party represented the Calvinist establishment in new clothing. Methodists' anxieties over
political favoritism to formalist churches revealed themselves especially in
reactions during the 1820s and 1830s to campaigns for Sabbatarian and temperance legislation. Presbyterian and Congregationalist demands for a state-
supported Sabbath and for the prohibition of the liquor trade seemed to many
Methodists, as to Baptists and others who also had suffered at the hands of a
religious establishment, to be part of a momentum that would bring the young
Republic's delicate structure of civil and religious liberty crashing to the
ground. Methodists' strong Democratic loyalties persisted in many areas into
the late antebellum period, sustained in part by these earlier antagonisms.[47]
William X. Ninde recalled that even into the 1850s, Democratic politicians
continued to exploit the residual tensions between Congregationalists and
Methodists in Connecticut and "went in heart and soul to help the Methodists. It was a sort of 'you tickle me, and I'll tickle you' system, a kind of see-
saw arrangement. When the Whigs and the Congregationalists went down the
Democrats and the Methodists went up and *vice versa*."[48]

Yet by no means all Methodists were Jacksonians, even in the early days
of the second party system, and as the years passed the Democrats' hold on
the denomination grew increasingly less assured. "Lord deliver us from
Whiggery!" prayed a Tennessee Methodist preacher at a camp meeting; "God
forbid!" came the response. It is impossible, given the well-established problems of identifying the church affiliations of aggregated anonymous voters,
to be precise about the number or proportion of anti-Democratic Methodists.
But there were plenty who shared the perspective, if not the exhibitionism,
of the "violent anti-Jackson preacher" of the Baltimore Conference who visited
the White House with colleagues in 1831 and took the opportunity to pray
that the general might be converted, "which he did so loud that he could be
heard at the President's gate." Some, like William Winans, followed Henry
Clay from Antifederalism through National Republicanism to Whiggery.
Others took a route that led through Antimasonry. Paul Goodman's analysis
of that movement in New England suggests substantial Methodist support for
Antimasonry, morally and electorally, especially in Maine and Vermont,
where Methodists contributed a number of prominent leaders. Subsequently
a majority of Antimasonry's supporters, including Methodists, amalgamated
with the new Whig party. An examination of Methodists' published and
manuscript diaries and journals makes it clear that, by the 1840s, many
denominational leaders regarded Clay's party as a truer heir of Jeffersonian
democracy than the radically anti-Bank Locofocos. So, too, do some of the
sparse survivals of individual voting records: one analysis indicates that 74
percent of Methodists in Green County, Illinois, voted Whig; another points
to a similar orientation in parts of Rhode Island.[49] The Indiana Conference
was said to be mainly of that outlook by 1840. James Dixon, a British Wesleyan
representative at the General Conference of the Methodist Episcopal Church
in 1848, was clear that the vast majority of Methodists that he met on his travels
were Whigs.[50]

There can be little doubt, as Brownlow's example has already suggested, that Whiggery's attraction lay to some degree in its posture as the party of economic and social improvement. But it is also clear that a major element of its appeal was its standing as a party of Christian—or, more specifically, Protestant—integrity. Inheriting the Jeffersonian mantle of defenders of religious pluralism and freedom of conscience, Jacksonian Democrats found themselves charged with moral myopia in a society increasingly threatened, as it seemed, by paganism, rationalism, and degraded theology. Whigs played mercilessly on Jacksonians' readiness to shelter freethinkers and heterodox Christians. Thus William Crane, a Methodist minister and devoted free trader and internationalist, chloroformed his economic conscience and voted Whig through the 1830s, despite "their cardinal doctrine" of a protective tariff, because the Whigs offered evangelical Protestant Americans a more secure morality than their free-trade opponents. The Jacksonians, he believed, were irredeemably flawed since "they despised no man for his sins." Crane and other Methodists froze at the names of Fanny Wright, Abner Kneeland, Robert Dale Owen, and other Locofoco heirs of Tom Paine. Significantly, Crane had been sympathetic to Antimasonry, a movement that had successfully played on fears of a spreading conspiracy against orthodoxy and that acted for some as a route into Whiggery.[51]

Most threatening of all the groups embraced by the Democrats, it seemed, were the burgeoning communities of Irish and German Catholics. The smoldering embers of the Methodists' historic antipathy toward Romanists burst into open flame through the 1830s and 1840s, fanned by the energetic propaganda of skilled publicists. Few were more vitriolic than Parson Brownlow, who charged Catholic immigrants and the papal hierarchy with near unanimous support of the corrupt Locofocos and denounced Van Buren as a Romish sycophant. More temperate in language, but equally clear about the political logic of Catholic leaders' attempts to secure state funding for their schools in New York City and elsewhere, was the *Christian Advocate and Journal*: "The school question is now made a political one . . . with the Romanists claiming the aid of the Democrats, and the Whigs asking the help of Protestants. . . . The Protestants have not done this . . . the Romanists are the aggressors." The Methodists were also aggrieved by the Jacksonians' complicity in the Irish Catholic efforts to secure the repeal of the Anglo-Irish Union of 1800.[52]

To counter this convergence of Methodism and evangelically oriented Whiggery, the Democrats stressed their opponents' "Federalist" roots and their threat to the separation of church and state, but with decreasing effect. In the early years of the century, the fear of federalism and state-supported Calvinism had been sufficient to bind within the Republican party both Methodism and, as Alfred Brunson recalled, "the infidel portion of the community, though these classes were antipodes in all things pertaining to religion."[53] By the 1830s, however, some Methodists were ready to support a party that also embraced the very Calvinists they had once had cause to fear. Many Methodists were beginning to question their faith in moral suasion alone as a means of preserving a Protestant social order and came to see the legitimacy of and need for legal measures, whether to prohibit the selling of liquor or to protect camp meetings from outside disturbance. Those Methodists who were afraid that such measures represented a threat to the separation of church and state and a betrayal of their historic credo were reassured by church spokesmen, who sought to distinguish between an improper sectarian establishment and an entirely proper subjection of legislation to Christian influence. As Benjamin Tefft explained, "No Enlightened Christian wishes to see Church and State united in government; but if men [are] to do ill to the glory of God, they must be constantly governed by the moral law of God; and hence religious principle must be made the basis of political action."[54]

That important body of Methodists who traveled into Whiggery from Jeffersonian Republicanism and Jacksonian Democracy were moving from being cultural outsiders to insiders; their journey was associated with the increasing social status of Methodists and the Arminianization of Calvinism, which made political coalition with non-Methodists even more possible. There is, however, little evidence to suggest that it had to do with latter-day conversion to the merits of the market, for they had long been in tune with the individualist, entrepreneurial values of capitalism. Sellers is critical of liberal historians who argue for an early nineteenth-century capitalist consensus, and he may well be right in relating political divisions in Jacksonian America principally to the responses to the market.[55] But Methodists provide a more feeble support than his argument supposes. They do not fit comfortably into his "antinomian"—"arminian" dichotomy. Their experience suggests that capitalist values were shared across the partisan divide. But this does not make those political divisions artificial or insubstantial. Methodists' partisan allegiances were neither accidental nor incidental. They were to a considerable degree shaped by meaningful ideological conflict between competing religious groups. The concerns underlying Methodists' political allegiance in the early national period were shaped as much by the religious as by the economic marketplace.

Notes

1. In two influential works William G. McLoughlin discusses the ideological shift effected by the Second Great Awakening and relates it to deep cultural change. He sees each of America's awakenings "as a period of fundamental social and intellectual reorientation of the American belief-value system, behavior patterns, and institutional structure." In the early nineteenth century, American Calvinism had to be "revitalized" as society moved toward a new cultural era, emphasizing independence and self-reliance. William G. McLoughlin, Jr., *Modern Revivalism: Charles Grandison Finney to Billy Graham* (New York: Ronald Press, 1959), 3–121; and McLoughlin, *Revivals, Awakenings, and Reform. An Essay on Religion and Social Change in America, 1607–1977* (Chicago: University of Chicago Press, 1978), quotation 10, 98–140.

2. Representative of the genre are Paul E. Johnson, *A Shopkeeper's Millennium: Society and Revivals in Rochester, New York, 1790–1865* (New York: Hill & Wang, 1978); Mary P. Ryan, *Cradle of the Middle Class: The Family in Oneida County,*

New York, 1790–1865 (Cambridge: Cambridge University Press, 1981); Randolph A. Roth, *The Democratic Dilemma: Religion, Reform, and the Social Order in the Connecticut River Valley of Vermont, 1791–1850* (Cambridge: Cambridge University Press, 1987); and David G. Hackett, *The Rude Hand of Innovation: Religion and Social Order in Albany, New York, 1652–1836* (New York: Oxford University Press, 1991). All draw inspiration from a classic work: Whitney R. Cross, *The Burned-Over District: The Social and Intellectual History of Enthusiastic Religion in Western New York* (Ithaca, N.Y.: Cornell University Press, 1950). Other valuable local and regional studies include John B. Boles, *The Great Revival, 1787–1805: The Origins of the Southern Evangelical Mind* (Lexington: University Press of Kentucky, 1972); and Terry D. Bilhartz, *Urban Religion and the Second Great Awakening: Church and Society in Early National Baltimore* (Rutherford, N.J.: Fairleigh Dickinson University Press, 1986).

3. Curtis D. Johnson, *Islands of Holiness: Rural Religion in Upstate New York, 1790–1860* (Ithaca, N.Y.: Cornell University Press, 1989), relates the growth of evangelical churches in Cortland County to the experience of rapid economic change, but he attaches primary importance to ideology in fostering revivalistic religion. Hackett, *Rude Hand of Innovation*, 164, also emphasizes how ideological conflict "directly influenced the course of social change."

4. Richard Carwardine, "The Second Great Awakening in Comparative Perspective: Revivals and Culture in the United States and Britain," in *Modern Christian Revivals*, ed. Edith L. Blumhofer and Randall Balmer (Urbana: University of Illinois Press, 1993), 84–100.

5. Charles Sellers, *The Market Revolution: Jacksonian America, 1815–1846* (New York: Oxford University Press, 1991), 5–6, 9–33.

6. Ibid., 137–38, 157–61, 164–65, 178, 299–300.

7. Ibid., 202–36.

8. Ibid., 158, 161.

9. Particularly valuable in understanding the status and worldview of early Methodists are Nathan O. Hatch, *The Democratization of American Christianity* (New Haven, Conn.: Yale University Press, 1989); Donald G. Mathews, *Religion in the Old South* (Chicago: University of Chicago Press, 1977), esp. 28–37; Russell E. Richey, *Early American Methodism* (Bloomington: Indiana University Press, 1991); and William Henry Williams, *The Garden of American Methodism: The Delmarva Peninsula, 1769–1820* (Wilmington, Del: Scholarly Resources, 1984). See also George Claude Baker, Jr., *An Introduction to the History of Early New England Methodism 1789–1839* (Durham, N.C.: Duke University Press, 1941); and Emory S. Bucke, ed., *The History of American Methodism*, 3 vols. (New York: Abingdon, 1964).

10. John Sale to Edward Dromgoole, February 20, 1807, in William Warren Sweet, *Religion on the American Frontier 1783–1840, Vol. 4, The Methodists* (New York: Cooper Square, [1946] 1964), 160.

11. Nicholas Snethen to George Roberts, December 31, 1801, Methodist Manuscript Connection, Methodist Center, Drew University (hereafter MCDU).

12. Freeborn Garrettson to Bishop George, August 1, 1825, MCDU. See also the New York *Christian Advocate and Journal* (hereafter *CAJ*), December 7, 1827.

13. J. Tackaberry, "Day Book" and "Journal," MCDU, 1: 8–9.

14. *CAJ*, January 25, 1828. See also, for example, issues for March 14, 1828; January 13, February 25, and April 15, 1831; [?] Smith to G. Coles, May 12, 1842, MCDU.

15. Teresa Anne Murphy, *Ten Hours' Labor. Religion, Reform, and Gender in Early New England* (Ithaca, N.Y.: Cornell University Press, 1992); Jama Lazerow, *Religion and the Working Class in Antebellum America* (Washington, D.C.: Smith-

sonian Institution Press, 1995; and William R. Sutton, *Journeymen for Jesus: Artisans Confront Capitalism in Jacksonian Baltimore* (University Park: Pennsylvania State University Press, 1998). See also Ronald Schultz, "The Small-Producer Tradition and the Moral Origins of Artisan Radicalism in Philadelphia, 1720–1810," *Past and Present* 127 (1990): 84–108. Murphy is especially sensitive to the complexities and ambiguities of the evangelical response to industrialization, and she explores both the disruptive and disciplining element of Methodist revivals: "Religion was . . . a strongly contested terrain in which working people and employers had very different interpretations and goals." See Murphy, *Ten Hours' Labor*, 82. By contrast, the socially controlling, self-disciplining, and restrictive effects on mechanics and factory workers of evangelicalism in general and Methodism in particular is the principal, though not the sole, burden of Paul Johnson, *Shopkeepers' Millennium*; Bruce Laurie, *Artisans into Workers: Labor in Nineteenth-Century America* (New York: Hill & Wang, 1989); Laurie, *Working People of Philadelphia, 1800–1850* (Philadelphia: Temple University Press, 1980); and Anthony F. C. Wallace, *Rockdale: The Growth of an American Village in the Early Industrial Revolution* (New York: Knopf, 1978). Sean Wilentz, like Murphy, also sees the "gamut of impulses" deriving from Methodism "all tied to the tensions between submissiveness and egalitarianism that lay at Methodism's core." Ultimately, though, he sees Methodism becoming "more closely identified with efforts to enforce an industrious morality of self-discipline." See Wilentz, *Chants Democratic. New York City and the Rise of the American Working Class, 1788–1850* (New York: Oxford University Press, 1984), 80–81. Cf. Barbara M. Tucker, *Samuel Slater and the Origins of the American Textile Industry, 1760–1860* (Ithaca, N.Y.: Cornell University Press, 1984), 163–85.

For a parallel debate on the influence of Methodism on English social relations during the industrial revolution, see Edward P. Thompson, *The Making of the English Working Class* (London: Gollancz, 1963); see, in response, Thomas W. Laqueur, *Religion and Respectability: Sunday Schools and Working-Class Culture, 1780–1850* (New Haven, Conn.: Yale University Press, 1976); and David Hempton, *Methodism and Politics in British Society, 1750–1850* (London: Hutchinson, 1984). Sellers writes, "Urban stress converted the antinomian ecstasy of Methodist conversion into the arminian effort of capitalist discipline, and a 'changed heart' was most clearly evidenced by the capitalist virtues. Spiritual intensity functioned to motivate and sustain the personal transformations required for survival under intensifying market relations." See his *Market Revolution*, 285.

16. Beverly Waugh to George Coles, August 25, 1856, MCDU; George Peck, *The Past and the Present. A Semi-Centennial Sermon: Preached before the Oneida and Wyoming Conferences at Ithaca, N.Y., April 19, 1866* (New York: Carlton & Porter, 1866), 10, 24–27, 32; *CAJ*, May 12, 1827; June 12 and July 31, 1829; April 8, 1831; January 5, 1848; May 15, June 12, and August 7, 1856; January 1, 1857.

17. Benjamin F. Tefft, *Methodism Successful, and the Internal Causes of Its Success* (New York: Derby & Jackson, 1860); *CAJ*, July 27 and September 15, 1841; June 14 and July 5, 1843; October 25, 1848; February 21 and March 13, 1856; June 11, 1857; and Richard Wheatley, *The Life and Letters of Mrs. Phoebe Palmer* (New York: W. C. Palmer, Jr., 1876), 600.

18. Sutton, *Journeymen for Jesus*, 69–100; Bucke, *History of American Methodism*, 1:629. Methodism suffered divisions over the church's proper response to Freemasonry in the late 1820s and the early 1830s. Insofar as the Antimasonic movement can be seen (as Paul Goodman argues) as a passionate reaction against the transition to a new industrial society and consumerism, its Methodist supporters might seem to be

antimodernists and Methodist Masons part of an emergent bourgeoisie. But that conclusion has to be tempered by the fact that many New England Methodists were drawn to Freemasonry primarily because of its devotion to religious tolerance and that, as a predominantly Calvinist movement, Antimasonry was not a natural refuge for Arminian Methodists.

19. Bucke, *History of American Methodism*, 1:633.

20. For Methodists and patriarchy, see Richey, *Early American Methodism*, 25–26, 56–57. Donald Mathews argues that early Southern Methodism "was a means through which a rising 'new' class sought authentication outside the archaic social hierarchy." Methodists tried to replace class distinctions based on worldly honor, but they could not escape the social system in which they were moving to better farms and improving their status. See Mathews, *Religion in the Old South*, 37–38.

21. Williams, *Garden of American Methodism*, 99–105.

22. Peter Pelham to Edward Dromgoole, Jr., June 21, 1811, in Sweet, *Religion on the American Frontier*, 4:196–97. See Rutsen Suckley to J. B. Wakeley, May 7, 1866, MCDU, for the friendship of Asbury, Coke, and Garrettson with Suckley's Methodist father, the American agent and managing partner of the Sheffield-based manufacturers, Newbould and Holy.

23. Peter Pelham to Edward Dromgoole, June 20 and July 27, 1807; April 16, 1810; June 21, 1811; in Sweet, *Religion on the American Frontier*, 4:163–67, 172–74, 187–88, 196–97. The family of Frederick Bonner, friend of Gatch and Asbury, presents a similar Greene County mix of staunch Methodism, land acquisition, manufacturing, and openness to the market. Frederick Bonner to Edward Dromgoole ["Drumgole"], July 19, 1807, in Sweet, *Religion on the American Frontier*, 4:170–71.

24. Philip Gatch to Edward Dromgoole, June 1, 1805; John Sale to Edward Dromgoole, February 20, 1807; and Bennett Maxey to Edward Dromgoole, July 27, 1807, in Sweet, *Religion on the American Frontier*, 4:157, 159–60, 174–75.

25. For Methodist doctrine and its implications for human ability, see, for example, Hatch, *Democratization of American Christianity*, 171–72. George M. Thomas discusses the tension between Calvinists' predestinarianism and their daily experience of the market; Methodists, by contrast, enjoyed a doctrine of human ability and control that was wholly compatible with their daily economic and political lives. See Thomas, *Revivalism and Cultural Change: Christianity, Nation Building, and the Market in the Nineteenth-Century United States* (Chicago: University of Chicago Press, 1989), 7–8. Hackett, in *Rude Hand of Innovation*, 90–95, 98–99, also relates the conflict between Calvinist and Methodist to the changing socioeconomic order of the early Republic and sees Methodist ideology appealing to Albany's rising poor through its "new insistence on individual autonomy."

26. *CAJ*, November 18, 1831; Richard Carwardine, "The Second Great Awakening in the Urban Centers: An Examination of Methodism and the New Measures," *Journal of American History* 59 (1972): 327–40.

27. Jon Butler, *Awash in a Sea of Faith: Christianizing the American People* (Cambridge, Mass.: Harvard University Press, 1990), 268–82.

28. When the New York Book Concern burned down in 1836, the loss of building and stock was so great that the New York insurance companies almost collapsed. Bucke, *History of American Methodism*, 1:579.

29. Freeborn Garrettson to [?], December 22, 1818; June 15, 1824, MCDU; Sophia Hogan Boogher, *Recollections of John Hogan by His Daughter* (St. Louis, Mo.: Mound City, 1927), 8–29, 51.

30. Peck, *Past and the Present*, 32; *CAJ*, July 24,1829; A. A. Jimeson, *Notes on*

the Twenty-five Articles of Religion, as Received and Taught by Methodists in the United States (Cincinnati, Ohio: Applegate, 1855), 362–66. Cf. Phoebe Palmer, *Faith and Its Effects, or, Fragments from My Portfolio* (London: n.p., 1856), 192–97.

31. Freeborn Garrettson to [?], December 22, 1818, MCDU; Sellers, *Market Revolution*, 155, 160; *Autobiography of Peter Cartwright. The Backwoods Preacher*, ed. W. P. Strickland (New York: Methodist Book Concern, n.d.), 251–54.

32. Curtis Johnson, *Islands of Holiness*, 82–85.

33. *CAJ*, February 3, 1827 (on the Ithaca revival of 1826–27), October 10, 1828 (on the Hempstead, Long Island, revival). See also, for example, *CAJ*, April 8, 1831, which states that the Troy revival brought in "men in their prime . . . high in talent, office, influence and wealth."

34. *CAJ*, October 7, 1826; July 31, 1829; J. N. Maffitt to G. Coles, n.d., MCDU.

35. *CAJ*, April 7, 1827.

36. Methodists' enthusiasm for the new economic order did not make them blind to the greed and exploitation of the capitalist market, but in general the Methodist critics sought to Christianize or sanitize the new order, not resist it. The efforts of Methodists in both the United States and Britain to devise Christian rules for commercial practice deserve analysis in their own right and are necessarily beyond the scope of this chapter. Attention to stewardship, voluntary giving, and schemes of private, systematic benevolence was an ever present element of midcentury Methodists' discourse. For quintessential examples, see Jimeson, *Notes on the Twenty-five Articles*, 204–12 ("Good Works"), 362–70 ("Of Christian Men's Goods"); Andrew Carroll, "Proper Use of the Mammon of Unrighteousness," in *The Ohio Conference Offering, or, Sermons and Sketches of Sermons, on Familiar and Practical Subjects, from the Living and the Dead*, ed. Maxwell P. Gaddis (Cincinnati, Ohio: Methodist Book Concern, 1854), 95–109; Calvin Kingsley, "Practical Benevolence," in *Original Sermons by Ministers of the Pittsburgh, Erie, and Western Virginia Conferences of the Methodist Episcopal Church*, ed. William Hunter (Pittsburgh: Geo. Parkin, 1850); *Memorial of E. P. Williams, Late Member of the Oneida Conference of the M. E. Church: Being a Choice Collection of the Sermons and Sketches of Sermons, with a Brief Biographical Sketch* (Syracuse, N.Y.: Daily Journal Print, 1859), 54–78 ("Christian Benevolence"). *The Successful Merchant* (London: Hamilton, Adams, 1852), William Arthur's well-regarded prescription for the blending of commercial and Christian values, enjoyed an extensive transatlantic circulation among Methodists during the 1850s. See, for example, *CAJ*, March 13, 1856.

37. *CAJ*, September 23, 1826; February. 17, 1827; August 31, 1827; Sellers, *Market Revolution*, 30–31.

38. Sellers, *Market Revolution*, 204.

39. McLoughlin, *Revivals, Awakenings, and Reform*, 113–40. For a recent discussion, see Curtis Johnson, *Islands of Holiness*, 33–52, which shows that the decades of Cortland County's rapid economic development were also years when Arminians won "an astonishing victory over Calvinist opponents" and when Calvinism itself was Arminianized. But "the popularity of Arminian ideas . . . was more a secondary consequence than a primary cause of new attitudes [of individualism, self-determination, and free will] among the general population" (51).

40. Cartwright, *Autobiography*, 64–72, 107–10, 118–22, 133–38, 147–51, 215–19, 226–28; Elizabeth K. Nottingham, *Methodism and the Frontier: Indiana Proving Ground* (New York: Columbia University Press, 1941), 41–55; Curtis Johnson, *Islands of Holiness*, 43–52.

41. Richard Carwardine, "Religious Revival and Political Renewal in Antebellum America," in *Revival and Religion Since 1700: Essays for John Walsh*, ed. Jane Garnett and Colin Mathew (London: Hambledon, 1993), 145–46.

42. Ibid., 138–51.

43. Frank Richardson, *From Sunrise to Sunset: Reminiscence* (Bristol, Tenn: King Print, 1910), 107–8. Yet it is hard to press all Baptist-Methodist conflict into the mold described here, given that so many Baptists in various parts of the Union were pursuing and enjoying the benefits of the market, too.

44. Richard J. Carwardine, *Evangelicals and Politics in Antebellum America* (New Haven, Conn.: Yale University Press, 1993), 97–132. For a judicious and persuasive analysis of voters' characteristics in Jacksonian America based on lists of identifiable voters in New England, see Paul Goodman, "Politics in Jacksonian America," *Journal of the Early Republic* 6 (1986): 23–58. Goodman stresses the importance of context in voting behavior and sees the important connections between class and religion without reducing denominational conflict to economics alone.

45. For Methodists' strongly Democratic preferences in early Jacksonian New Hampshire, Maine, Vermont, and Massachusetts, see Paul Goodman, *Towards a Christian Republic. Antimasonry and the Great Tradition in New England: 1826–1836* (New York: Oxford University Press, 1988), 109–19, 130, 153–54.

46. Quoted ibid., 130.

47. William E. Gienapp, *The Origins of the Republican Party* (New York: Oxford University Press, 1987), 432–34, concludes that although Methodist ministers in the free states may have been vocally Republican by 1856, lay Methodists "seem not to have backed the Republican cause to any decisive degree." Some rallied to Fillmore, but many remained loyal to the Democrats, especially in western states. Even in 1860, large numbers of Methodists continued to adhere to the Democratic party, and of all the evangelical bodies in the North, Methodists were the least strong for Lincoln. Gienapp, "Who Voted for Lincoln?" in *Abraham Lincoln and the American Political Tradition,* ed. John L. Thomas (Amherst: University of Massachusetts Press, 1986), 75.

48. Mary L. Ninde, *William Xavier Ninde. A Memorial* (New York: Eaton & Mains, 1902), 50–51.

49. Carwardine, *Evangelicals and Politics*, 125–26; Goodman, *Towards a Christian Republic*, 116, 130; Ronald P. Formisano, The *Transformation of Political Culture: Massachusetts Parties, 1760s-1840s* (New York: Oxford University Press 1983), 219. Formisano cites John M. Rozett's study of Green County in his comments on Robert Kelley, "Ideology and Political Culture from Jefferson to Nixon," *American Historical Review* 82 (1977): 573. Goodman, in "Politics in Jacksonian America" (55), offers data from East Greenwich and Providence, Rhode Island, that suggest Whig support among Methodist church members of 47% and 75% in the two towns.

50. James Dixon, *Personal Narrative of a Tour through a Part of the United States and Canada, with Notices of the History of Methodism in America* (New York: Lane & Scott, 1849), 62–63.

51. William W. Crane, *Autobiography and Miscellaneous Writings* (Syracuse, N.Y.: A. W. Hall, 1891), 85–86; Carwardine, *Evangelicals and Politics*, 108–9.

52. William G. Brownlow, A *Political Register Setting Forth the Principles of the Whig and Locofoco Parties in the United States* (Jonesborough, Tenn.: Jonesborough Whig, 1844), 77, 109–11, 113–16; *CAJ*, April 27 and November 18, 1842; Cincinnati *Western Christian Advocate*, November 12, 1841; July 8, 1842.

53. Alfred Brunson, *A Western Pioneer, or, Incidents in the Life and Times of Rev. Alfred Brunson*, 2 vols. (Cincinnati, Ohio: Walden & Stowe, 1880), 1:43, 285–86.

54. Benjamin F. Tefft, *The Republican Influences of Christianity. A Discourse* (Bangor, Me.: n.p., 1841), 10–11.

55. Sellers, *Market Revolution*, 268.

5

E. P. Thompson and Methodism

David Hempton and John Walsh

E. P. Thompson's general picture of religion and the working classes in early industrial England arose from his controversial portrait of English Methodists. The dramatic brilliance of his influential book, *The Making of the English Working Class*, in which that portrait appears, has generated enormous interest on both sides of the Atlantic since its publication in 1963. There is scarcely a serious work of historical scholarship on the Methodist tradition in the North Atlantic world that does not cite Thompson, either as exemplar or pariah.[1] Although Thompson was hardly an uncritical admirer of Methodism in Wesley's day, he turned most of his firepower on the generation that emerged after Wesley's death in 1791. His picture of second-generation Methodists becoming, wittingly or unwittingly, the facilitators of dehumanizing industrialization and political repression has become such an important reference point for all those interested in the modern social history of religion that an examination of Thompson's own history with Methodism is long overdue. Thompson's striking account of evangelical Protestantism, money, and industrial capitalism arose not from disinterested observation but, as this chapter will show, from a personal history steeped in the very tradition he came to challenge.

A friendly *Times Literary Supplement* reviewer of Edward Thompson described his portrait of Methodism as "horrific." It showed, he said, "the spiritual squalor of the narrow and out-worn circle of theological ideas, fatal to culture, happiness and self-respect . . . Matthew Arnold never painted the "dissidence of Dissent" in gloomier colours."[2] But this was in 1935. The book was not *The Making of the English Working Class* but *Introducing the*

Arnisons, one of several autobiographical novels by Edward Thompson senior, E. P. Thompson's father. Edward John Thompson was a cradle Noncon- formist, son of the Wesleyan manse, a one-time Methodist minister and ex- Bengal missionary who left the church and ended his days as a lecturer in Bengali at Oxford and a research fellow of Oriel College. Abandoning his early Methodism, he did not exchange belief for total skepticism, but he seems to have moved fairly easily from a quasi-mystical, Johannine, *logos* Chris- tianity toward a broad, undogmatic theism that owed much to the eastern reli- gions he had encountered in his Indian years.[3] The catena of quotations in his anthology *O World Invisible* (1931) suggested an eternal Spirit, reaching out to humankind not only through Christianity but also through other great world faiths—Buddhism, the Vedas, and Sufism.[4] In India he described himself as "a mystic in my spare moments."[5] In later life Thompson seems to have felt no need to anchor his religious belief to any church. He disagreed with John Wesley's maxim that "the Bible knows nothing of a solitary religion"; for him, religion rose out of human solitude. His fictional John Arnison spoke for his author when he reflected, "God did not come close to him, except in an artistic or semiartistic fashion, which he distrusted, for to the roots of life he was a Puritan of the Puritans, and his imagination was of the kind which is most at home in desert places, where no veil hangs between God and man."[6] In his later, postmissionary decades Edward John Thompson's life began markedly to foreshadow that of his younger son; he became well known as a prolific writer, poet, novelist, journalist, and historian of India. An anti- imperialist, in the last phase of his life he became a campaigner for Indian independence, a familiar figure on progressive platforms.[7]

At several way stations on his literary career, the senior Edward Thompson reflected critically on the chapel culture of his youth. In his autobiographical novels there are several passages about Methodism that presage themes in his son's *Making of the English Working Class*. He resented its failure to nurture the fierce intellectual traditions of Milton's Puritanism—"the real Protestantism." He was irked by its otherworldly fear of political activism; saddened at the way in which it "worked in suppressed lives and starved minds that carried a load of inhibitions which, because they were sublimated into religious belief and emotion, led to no disturbance outside themselves."[8] He was ashamed of the timidity of a denomination that had once "sacked and expelled any who showed the least tendency to Radicalism."[9] At the same time, Thompson's retrospect on his Nonconformist inheritance was highly nuanced. He recognized the countervailing virtues that it also nurtured; un- selfishness, fidelity to the call of conscience, and the impressive "brotherhood of service" that bonded its ministry.[10] "English Nonconformity," he wrote, "kept alive poetry (though it would not have called it this, had it recognized it) and courage and an infinite flow of kindliness from man to man. Patience, valour, sympathy, these have never lacked in the English poor. Their springs were largely in religion—a religion narrow, uninstructed, ugly, but known to God, and with majesties of power and imagination strangely rising from its darkness."[11] The second Arnison novel, very much a Bildungsroman, ends with the hero being caught up in the ferment of the pre-1914 years, exulting in the hope that the old Puritan spirit had been rekindled and was forcing its way once again into public life. Young Arnison encounters the Social Gospel as it was expounded by Nonconformist preachers like Dr. Clifford. At the end of the novel we see him moving hesitantly toward an Edwardian socialism, which seemed to exemplify "that pity and love which his mind had caught from Christ."[12] It was a destination that the author himself never quite reached.[13]

The Thompson household on Boar's Hill, Oxford, was, in E. P. Thompson's words, "supportive, liberal, anti-imperialist, quick with ideas and poetry and international visitors"—Nehru himself instructed the young Edward on how to hold a cricket bat.[14] The senior Edward Thompson retained enough atavistic attachment to Methodism to dispatch his younger son, Edward Palmer Thompson—known to the family as Palmer—to his old school, Kingswood, founded by Wesley for his preachers' sons. When the younger Thompson arrived there in 1936 from the Dragon School in Oxford, sons of the Methodist manse still outnumbered those of laymen by some two to one, giving the school a highly distinctive flavor. But the Methodism of Edward Palmer Thompson's schooldays was not identical with that of his father's youth. Under the headship of A. B. Sackett, Kingswood had relaxed the tone of late Victorian pietism and, though strongly Christian in its general outlook, was more liberal about matters of faith. Freedom of religious debate was encouraged. Religion was expressed through culture, as well as doctrine: In its chapel services one was almost as likely to hear a reading from Gerard Manley Hopkins as one from the Old Testament. It pleased the headmaster to include occasional heretics and unbelievers on the Chapel Committee, which organized the services: Here Edward Thompson junior filled a countercultural role as a self-professed "Walt Whitman pantheist" (perhaps one can see some paternal influence here?), while his friend Arnold Rattenbury was coopted as the school atheist. Together they arranged services on Friendship, Peace, and Nature: Their readings for these included William Morris's *Dream of John Ball*, as well as Richard Jefferies, Wilfrid Owen, and Rex Warner.

But if classical pietism was in some respects recessional at Kingswood, nonconformity was not. It was an argumentative, literary place, encouraging poets and artists and tolerating rebels. Boys who were themselves the sons of preachers were not averse to preaching at each other. Although many of those who left Kingswood were confirmed in an updated version of the Methodism of their homes, others did not find it difficult to slough off home piety and move on from Methodism to Marxism. Arnold Rattenbury recalls of his Kingswood circle that "we moved, as the decade and its adult poetry were also moving, from pacifism towards anti-fascism and then, since negatives please no-one and boys will be men, to socialism in the classic sense, supposing a necessary revolution." In the early war years Thompson and Rattenbury applied to join the Communist party, and during the school's wartime exile at Uppingham they distributed the *Daily Worker*.[15]

Thompson certainly emerged from John Wesley's school no devout Methodist. He wrote in a wartime letter to a school friend (soon to be ordained), "I still don't like your God."[16] His direct exposure to Methodism had been fairly brief, and he had encountered it in a cultivated and relaxed form. He did not share the nostalgia that suffused his father's memory-laden vision of his Methodist childhood. Yet he could not entirely escape Methodism. It was writ large on his cultural family tree. He had not only an ex-missionary father but also two missionary grandfathers: His paternal grandfather, John Moses Thompson, was a Methodist missionary in South India, and his mother, Theodosia, was the daughter of one American Presbyterian missionary in Syria and the granddaughter of another. In a household so articulate and closely bonded as that of the Thompsons on Boar's Hill, it is inconceivable that E. P. Thompson did not hear or read his father's reflections on his Nonconformist inheritance. And Methodism in Victorian form was still embarrassingly close at hand in the person of his eccentric uncle, Alfred, something of a skeleton in the family cupboard, who hovered near the Thompson household in a state of near destitution, lay preaching in local chapels and sending on unsolicited copies of the *Methodist Recorder*.[17] After serving as a tank commander in Italy, Thompson read for the English Tripos at Corpus Christi College, Cambridge, and then went off to be an adult education lecturer in the Yorkshire mill town of Halifax. There he encountered the legacy of the industrial revolution face to face and saw a lot of those "blackening chapels" (his phrase) of the North, whose bygone community life he was then beginning to reconstruct in *The Making of the English Working Class*.

Trace elements of parental Nonconformity remained in Thompson, as they did in many other radicals incubated in Dissent—like Christopher Hill, also of impeccable Methodist ancestry—for whom socialism offered an alternative agenda for social and personal salvation. Those who heard him speak may have discerned pulpit echoes in his platform oratory: the prophetic call to national righteousness, the fervent appeal to the individual conscience, and the desire to stand up and "Witness against the Beast" (the title he chose for his last book). "I was brought up a dissident," he claimed proudly.[18] Comparing his youthful self with that of his brother, Frank, he wrote that his "own political cultural drive was already . . . low-brow, moralizing—perhaps even Methodistical."[19] Thompson had a good deal in common with Victorian evangelicals. He possessed an exceedingly clear-cut view of the distinction between right and wrong, together with the confident conviction that he could locate them in society. Although approaching history through Marxist categories of class formation and changes in the modes of production, he nonetheless insisted on the need to take full account of the cultural forms—especially the moral forms—in which social experience is expressed. Writing in 1976, he declared himself "transfixed by the problem of the degeneration of the theoretical vocabulary of mainstream, orthodox Marxism—the impoverishment of its sensibility, the primacy of categories that denied the effective existence in history or the present of the moral consciousness."[20] An intense sense of need for personal responsibility informed his lifelong hatred of defeatism and apathy (one of his books is entitled *Out of Apathy*). Evil should not be condoned by inaction. When he quoted Marx's dictum that "to leave error unrefuted is to encourage intellectual immorality," he was invoking a principle that his evangelical forebears might well have endorsed.[21] As a schoolboy he was remembered for the frequency and vehemence of his exclamation, "I wouldn't put up with that if I were you."[22] Thompson saw history as shaped by agency and moral choice, by activists, campaigners, speakers, platforms: It was not merely predetermined by impersonal "evolution" or the inexorable march of economic forces. His socialism of cultural and moral sensibility marked him off from the high theoreticians, be they economistic Stalinists or structuralist Althusserians.[23]

Yet if he assimilated some of the moral values of evangelicalism, Edward Palmer Thompson viewed early nineteenth-century Methodism from a more distant vantage point than his father, who had been born into it and belonged to it, with mingled exasperation and affection, through his formative years. The younger Thompson's own religious opinions were never clearly on public view. Christian imagery surfaced in his writing, as, for example, in his cycle of Christmas poems on the Nativity theme, where the Massacre of the Innocents became a paradigm of the aggression of modern "principalities and powers," trampling down the weak and helpless.[24] In 1978 he concluded that for all their quietism, the Christian churches had helped preserve the values of love and community and that "it was within the Christian myth that symbols could be found unpolluted by the language of power."[25] But in his 1980 postscript to *The Making of the English Working Class*, he says firmly, "I am not a Christian."[26] Readers of his book on Blake, published in 1993, were left to make what they wished of his jocular remark that he was a "Muggletonian Marxist."[27]

Edward Palmer Thompson seems to have remembered his father's strictures on Methodism while forgetting the affectionate qualifications that balance them in the Arnison novels. His critique is altogether harsher. No doubt this is partly because it is refracted through Marxist theory. But there is probably more to it than that. Methodism is treated with a ferocity not visible in the writing of his fellow Marxist Eric Hobsbawm, whose perspective is far more detached. Why was this? Was he trying to shock the respectability (perhaps Methodist-originated) of his Yorkshire adult education students, among whom his lot was cast and for which he felt a quasi-Bohemian impatience?[28] Perhaps the root lay deeper. One senses in Thompson some kind of continuing interior engagement with religion that runs back to his early days. Was he in subterranean debate with his father's mysticism? Or the Marxist agnosticism of his brother, Frank? Or the religious sensibility of his mother, a strong and beneficent force in the family, who remained more overtly Christian than her husband? At all events, something deeply complicated, ambivalent, and personal lies buried beneath the rhetorical violence of the celebrated chapter 11 of *The Making of the English Working Class*, entitled "The Transforming Power of the Cross." It throbs with the kind of submerged emotion that smolders in chapter 15 of Gibbon's *Decline and Fall*.

"Puritanism—Dissent—Nonconformity: the decline collapses into a surrender."[29] So begins the controversial chapter. With the possible exception of Elie Halévy's work, Thompson's relatively short account has been the single most important influence on the writing of the history of Methodism in the early industrial revolution.[30] Its theoretical and descriptive richness has inspired—and provoked—a burst of creative reinterpretation that has helped to take Methodist history out of the realm of conventional "church history" and bring it firmly into the sights of the social historians. Part of the remarkable impact of the book stems from the imaginative power of Thompson's language. It is the sarcasm, the color, the passion, the eloquence, the overstatement, the metaphoric and symbolic affluence, and the foreshortening of argument in the interests of coherence and force that make his work on popular religion so gripping. His phrases have livened up many a tired page of undergraduate prose: "psychic masturbation," "the chiliasm of despair," "religious terrorism" (taken from Lecky), "the box-like, blackening chapels" (echoing Blake), and "Sabbath orgasms of feeling." But there is more to it than that. What really makes Thompson stand out from the run of denominational hagiographers and social commentators on Methodism is his relentless quest for the heart of Methodist experience and culture during the transformation of industrialization:

> I was anxious to find out what the experience of Methodism—in particular between 1780 and 1820—was about—why working people who had been deserting, or were resistant to, the more rational dissenting churches, should accept this passionate Lutheranism. Too much writing on Methodism commences with the assumption that we all know what Methodism was, and gets on with discussing its growth rates or its organizational structures. But we cannot deduce the quality of the Methodist experience from this kind of evidence. . . . Are we studying a genuine spiritual experience, or is it only to be understood as a rendering of, or displacement of, other mental energies?[31]

One of his last books, on Blake, provides a clue to the historical angle of attack he employed on the Methodists. When he playfully declared himself a "Muggletonian Marxist," he was not merely aligning himself with a radical antinomian tradition that stretched from the Commonwealth to the metropolitan sects of the later eighteenth century but also associating himself with a particular analytical method. The method is made explicit in the old Muggletonian rhyme, "Since by contraries all things are made clear, without contraries nothing can appear." Thompson saw this as the "dialectic which came to influence Blake's whole stance—his historical, moral and utopian thought."[32] For Thompson, this dialectical principle would probably have possessed some Hegelian, Marxist resonance.

His delineation of Methodism is largely built on the model of "the two contraries," on the dynamics of oscillation. Sometimes this is made explicit, as in his theory of the "chiliasm of despair," in which the hopes of political radicals, when shattered, were transferred to religion, finding expression in emotional revivals and millennial expectations: "We may suppose something like an oscillation, with religious revivalism at the negative, and radical politics (tinged with revolutionary millenarianism) at the positive pole."[33] As a point of analytic departure, the approach by way of contrarieties makes a good deal of sense of Methodist history, in which ambiguity is ubiquitous. Thompson is at his most persuasive when he wrestles with the "many tensions at the heart of Wesleyanism." He sees them everywhere: in the spiritual egalitarianism of the Methodist message and its authoritarian ecclesiastical structure; in the religion of the heart and the suppression of spontaneity; in a piety allegedly founded on love, which yet "feared love's effective expression, either as sexual love or in any social form which might irritate relations with Authority"; in the libertarianism of free grace, which was sabotaged by the works-righteousness to which converts were driven by the fear of backsliding; in the "moral civil war" between the chapel and the pub, the wicked and the redeemed; and in the tension between the kingdom without and the kingdom within, the kingdom here and now and the kingdom hereafter.[34]

Such tensions, Thompson freely admits, could be creative and energizing, but it is the burden of his argument that the reverse was more usually the case. Methodism siphoned off energy into unproductive channels. It was an enemy to joy, cheerfulness, and spontaneity. It inculcated self-reproach and melancholy and taught its converts to seek solace from their morbid guilt feelings in ceaseless work. For the Methodist, "to labour and to sorrow was to find pleasure, and masochism was love."[35] Sexual energy was repressed and diverted into a sterile, pietistic emotionalism—it was "ritualised psychic masturbation." "The idea of a passionate Methodist lover in these years is ludicrous," he exclaimed (an assertion not easily proven).[36] In a famous passage, he linked the Methodist message of godly industriousness and self-control to the time-work discipline required by industrialization, as thousands of artisans and field workers, accustomed to the irregular, task-oriented rhythms of labor in traditional societies, were fed into a new system dominated by the time-orientation of the factory clock and the production schedule. It was the historic role of evangelical religion to internalize the new work discipline among the poor by transmuting it into religious duty. This transformation was wrought in the crucible experience of the evangelical conversion, that "psychic ordeal in which the character-structure of the rebellious pre-industrial labourer or artisan was violently recast into that of the submissive industrial worker." Methodism turned the laborer into "his own slave driver."[37] Here was a major reason that an exploitative industrial capitalism did not meet with more resistance.

Thompson's chapter the "Transforming Power of the Cross" is dominated by the "satanic" presence of the authoritarian Wesleyan leader Jabez Bunting, brooding over Methodism in its critical years. The apostate son of a Manchester Methodist "Jacobin," Bunting had been educated by Unitarians in some of his formative years and so in Thompson's eyes blended in himself two religious traditions that he strongly disliked: cold Socinian rationalism and callow Methodist emotionalism. Bunting comes across as a kind of Nonconformist Stalin—perhaps no coincidence, for Thompson was gestating his book soon after Khrushchev's exposure of Stalin in 1956, which had astounded him. Like Stalin, Bunting packed vital committees with henchmen, set up inquisitorial tribunals, and purged the Connection of radicals. Thompson saw him as a leader who set himself against every popular cause in Methodism, from the teaching of writing in Sunday Schools to the sanctioning of female preaching, and from Luddism to artisan radicalism. Bunting pushed the campaign for Sabbatarianism. He was a scarcely concealed sacerdotalist who wished to ape the Church of England and turn Wesley's lay preachers into parsons. "In Bunting and his fellows," wrote Thompson, "we seem to see a deformity of the sensibility complementary to the deformities of the factory children whose labour they condoned."[38] To highlight Bunting's full awfulness, Thompson set his portrait beside cameos of other Methodist leaders. He compares him with John Wesley, equally authoritarian, but still a "great-hearted warhorse," a man who stood up at the market cross to be pelted and spent his life searching out "Christ's poor."[39] He contrasts Bunting with the founding fathers of Primitive Methodism, Bourne and Clowes, who were truly of the people, and with the passionate Methodist Chartist Ben Rushton, in whom the spirit of the New Model Army lived again.[40]

Thompson's strictures carry weight, for Bunting was a man who had a host of enemies in his own Connection. Many Methodist historians have found him an embarrassment and blamed him for the internal disruptions that almost shattered Wesleyan Methodism in midcentury (though seldom noting that Bunting's full ascendancy did not begin until 1820, by which time some of the damage had already been done).[41] There is no doubt about the timidity and conservatism of many Conference leaders who expelled Luddites and issued sonorous pastoral letters that avowed the loyalism of the Connection to those in authority. But these public pronouncements were deemed necessary by leaders who had to negotiate with governments at a time of crisis and—as in 1811 with Lord Sidmouth's bill—fend off hostile legislation to curb Methodist preaching. And they could conceal some surprising personal opinions. Adam Clarke, Methodist patriarch and thrice President of the Conference, assured Sidmouth that the Methodist preachers had "persuaded the whole land powerfully, and, thank God effectively, proclaiming 'Fear God and honour the King.'" Yet though Clarke held a deeply religious view of magistracy, he was no high Tory but a man who saw much good in the French Revolution, supported peace with Napoleonic France, favored parliamentary reform, and assured a visitor that "the Duke of Wellington is a despotic man and so are all the kings that have ever reigned."[42] Moreover, the latent effects of religious movements are often very different from those sanctioned by its official rhetoric. A case could be made that Methodism's evangelistic success did as much as the radical insurrectionary tradition to undermine the structures of the old order in church and state. Certainly many contemporaries thought so. The Toleration Act of 1812 (which repealed the Five Mile and Conventicle Acts) and the momentous "constitutional revolution" of 1828–32, which entailed the repeal of the Test Acts and the passage of Catholic Emancipation and the Reform Act, were arguably brought about as much by intensified religious pluralism (Irish Catholics and English evangelical Dissenters) as by pressures from working-class radicalism, despite the importance of the latter in the long run.[43]

A case of sorts can also be made out for Bunting's role within the Wesleyan Connection, in which he and his acolytes faced problems that Thompson made no attempt to understand. After the death of the aged and enfeebled Wesley, they struggled to govern a sprawling Connection, which at first was almost entirely without direction. Between one preachers' Conference and the next, Methodism was seen to be "a rope of sand," totally lacking overall direction which would prevent it from dissolution.[44] Bunting's authority in the Wesleyan hierarchy did not so much reflect the dominance of one strong personality as epitomize structural shifts in Methodism, as it moved awkwardly from the loose-knit, missionary movement of Wesley's day to become the settled denomination of the Victorian era. It was largely produced by a need for consolidation. Though the Conference authorities could be brutally high-handed, the raison d'être of their authoritarianism lay not so much in their capitulation to conservative ideology as in their desire to stabilize Methodism, consolidate its gains, and give it coherence and enduring ministerial structures. The dangerous dependence of the Buntingite bureaucracy in some northern cities on the munificence of propertied elites ("our leading friends") was due in part to the sharp financial crisis that overtook Methodism in the post-Napoleonic period as it tried to cope with deflation and falling subscriptions. Debt rather than Toryism was a key factor here. How to appeal to a lower class audience and still keep expanding, without courting bankruptcy, was a formidable challenge to the Buntingite generation.[45]

Some of Thompson's originality comes from his concern to examine popular evangelicalism, not by way of its denominational aggregates, but as it was socially experienced in its local habitats. In a celebrated article he rebuked historians so mesmerized by structural anthropology that they neglected vital "specificities of context." Yet in his own study of Methodism he often failed to take account of the huge diversity of local chapel cultures and the great variation in the level of commitment shown by its adherents. Methodism could have sharply different inflections, even within the same circuit. For example, at the Octagon Chapel in Chester, which by 1800 had begun to attract members of the civic elite, a visitor who had strayed in from some nearby country chapel shocked the congregation by shouting from the gallery when the preacher mentioned the Devil, "Eh, punch him in't guts Lord."[46] Thompson tends to homogenize Methodism. Like many eighteenth-century critics who saw the Connection as something like the Jesuit order, he assumes it to be an efficient machine with an authoritarian chain of command that ran straight from the apex to the extremities, from Conference to the card-carrying society members (like some Communist parties of his time, perhaps?).

But despite Wesley and Bunting, the reality was often very different. Away from the splendid new chapels in the larger towns, where the preachers often resided, the class leaders and the local preachers were what counted. There were far too many chapels in a "circuit" for the full-time ministers to control their activities very effectively. On any given Sunday the great majority of Methodist pulpits were occupied by local part-timers, usually working men. In the small out-townships and industrial villages, local cultures determined the ethos of Methodism as much as the thunderbolts of the Conference or pressure from the full-time preachers appointed by it, who were transitory beings, only staying a year of two, traveling around their circuits before being posted elsewhere. There was constant give and take between center and periphery, connectionalism and localism. At times it ended in victory for officialdom (which could be pyrrhic) in the form of expulsions. At times— especially among the irrepressible Cornish, "the mob of Methodism" in Bunting's curdled view—it did not.[47]

Since Thompson was teaching for the Workers Educational Association in Halifax when writing *The Making of the English Working Class*, it is hardly surprising that his primary focus would be on Methodism in a cluster of north- ern industrial cities. No doubt this provided useful insights, but it also helped distort his angle of vision. Thompson had a deep love of the countryside (at Kingswood the headmaster had talked him out of his plan to become a farmer and study estate management, and at his home in Wick Episcopi his garden was much in his thoughts), but he chose not to bring rural Methodism to life, even though he was well able to do so, and the space has been filled by others— notably by James Obelkevich in his fine study of Lincolnshire religion.[48] Thompson allows briefly that village Methodism "assumed a more class- conscious form" than urban Methodism and that "the chapel in the village was inevitably an affront to the vicar and the squire, and a center in which the laborer gained confidence and self-respect."[49] Here we get a fleeting glimpse of a more irreverent, nondeferential Methodism, whose laborers cocked a snook at Anglican paternalism and, in the view of their masters, improvidently wasted their time and energy by walking miles to hear preachers and sitting up late at night to sing hymns of Zion. But this is explained away quickly as a matter of "reactive dialectic." He recognizes briefly that the chapel com- munity could offer "mutual aid, neighborliness and solidarity" for those whose lives were disrupted by economic change, but he interprets these as standard plebeian values "injected" into Methodist culture from outside, not as endogenous qualities nurtured by religion. Deferential and conservative Methodism is seen as an authentic expression of evangelical theology, whereas unsubmissive and undeferential Methodism is not.

Perhaps Thompson is at his most one-sided in his depiction of the cultural effects of evangelicalism, no doubt because he saw it at odds with some of the "customs in common" that bound together preindustrial community life and which he later analyzed in a series of innovative essays. He certainly had a case. From the start, Methodist opposition to the "cakes-and-ale" side of popular culture provoked litanies of complaint from paternalist Tory squires and parsons and the "mobs" who harassed preachers and broke into chapels. Here, as elsewhere, Thompson lent his voice to a long polemical tradition, sustained by William Cobbett, Leigh Hunt, and William Lecky (on whom he drew), which is worthy of respect and is too often overlaid by romantic de- nominational pieties.[50]

Methodism was often deliberately countercultural. The moral code of its serious-minded preachers collided with many traditional customs, pastimes, and recreations (especially if they desecrated the Sabbath); with fairs and wakes; and with rural sports like football, wrestling, cudgel playing and cockfighting. The blanket condemnation of plebeian convivialities by early evangelical leaders caused some resistance in the ranks: It was not easy to wean Welsh Calvinistic Methodists from the neithior, or wedding feast, or insulate English Methodists from the annual parish "wake."[51] Historians of popular culture have celebrated these social events as important, life-enhancing affirmations of community and opportunities for the therapeutic discharge of tension.[52] But they often had their darker side, too, as saturnalian occasions for drunken brawling and brutality—not least brutality to animals, in the cockfighting and bullbaiting that evangelical humanitarians so strongly deplored. Festivities that could begin with children dancing on the green could end in wife beatings and an upsurge of unwanted pregnancies.[53] Thompson himself concedes that such customs were "not all harmless or quaint."[54] As John Rule has observed, in this context Methodism must be considered a counterattractive, as well as a counteractive, force, for it provided its own alternative, "improving," "respectable" recreations, which were assimilated into the calendars of local society.[55] Viewed in a secular light, chapel-going itself, with its opportunities for sociability, choral singing, music, and the occasional excitement of a revival, was in some sense "entertainment."[56] Eloquent preachers were accorded "star" status, especially in Wales, where revivalists like John Elias or Christmas Evans drew vast pilgrim crowds from a wide hinterland, and the preaching seasons of popular evangelicalism became the principal holidays for many people, being "ranked among the festivals of the nation." A host of humbler preachers like the blacksmith Sammy Hick or Billy Dawson were local celebrities.[57] Once chapel occasions became accepted and indigenous they became important components within the varied repertoire of local cultures and rapidly acquired the patina of tradition. Cultural energy radiated out from the Sunday Schools and choirs, from anniversary concerts, picnics, and processions with banners. The social role of Methodism here was not one of mere destruction. The inscription over the chapel door did not only read THOU SHALT NOT. Methodism at this point ran closely parallel to the movement for "rational recreation" that took hold among trade unionists, plebeian radicals, and the "labor aristocracy," who often shared the dislike of drunken wakes and brutal sports, which they as saw as childish and degrading and obstacles to the progress of the working man.[58]

Any emphasis on religion as unadorned social control needs to be balanced by attention to the powerful pressures from below for self-restraint, "improvement," self-culture, and the acquisition of education. If John Wesley had been mobbed in some places, he had been welcomed in others as the bringer of light and life: The rough colliers of Rainton had chided him affectionately for "being too long a coming."[59] There was immense plebeian demand for Sunday Schools and for their religious, as well as their secular, education.[60] The pressure for teetotalism came initially from the bottom of the movement, not from the top: Indeed, in 1841 the Wesleyan Conference actually banned temperance orators from chapel premises as potential extremists.[61] Much the same might be said about revivalism. Thompson's suggestion that revivals were used to canalize energy into safe channels needs to take account of the "folk" element in revivals and the opposition to "noisy" revivalism (especially in Cornwall) from connectional leaders like Bunting, fearful of its disruptive force and its potential for undermining pastoral authority.[62] The "opium of the people" theme needs closer attention. A good deal of evidence has been assembled to suggest that popular evangelicalism had a greater appeal to the aspiring artisan classes in search of decency and self-respect than to defeated workers who were swinging back from heady political activism to the chiliasm of despair.[63]

Thompson is not unaware of all this, and he concedes that some tight-knit communities were able to make Methodism "their own" and "soften its forbidding outlines." But this is subordinate to his ringing assertion that Methodism was a religion not of the people but for the people.[64] His stress is on the supply side, not on the demand side. It is a top-down picture of Methodist activity. This leads him to produce startlingly divergent stereotypes of the working man. In most of *Making of the Working Class*, the worker is a sturdy "free-born Englishman," shrewd, self-reliant, not easily fooled, and possessed of "an eye for Brechtian values—the fatalism, the irony in the face of establishment homilies, the capacity for self-preservation." Yet when Methodism appears, he is suddenly transformed into an abject, credulous being, easily cowed by "religious terrorism."[65] Sometimes one needs to pinch oneself to remember that Methodism was, after all, operating in an open market and not sustained by a confessional state. It was a purely voluntary association, freely supported and freely abandoned. There is much evidence that the working classes often took what they wanted from Methodism and were perfectly capable of appropriating Methodist forms and institutions for their own use and on something like their own terms. Sunday Schools, though funded by churches, remained, as Thomas Laqueur has shown, organizationally and in style of operation largely in the control of their consumers.[66] All this could hardly be otherwise, for Methodism depended overwhelmingly for its effective leadership on the goodwill of a huge army of unpaid, local lay officers as its class leaders, band leaders, trustees, stewards, and local preachers. It relied utterly on voluntary contributions for its survival. As a result, it could not avoid being an adaptive, as well as an innovative, force, accommodating itself to plebeian life in myriad cultural contexts.

This symbiotic interpretation would fit in with recent historiography on popular culture and missiology, which has moved away from rigid "two-tier" models to an emphasis on "cultural brokerage," which stresses the process of reactive exchange and negotiation between different social groups. To be fair, Thompson himself half accepted such a view in some of his later work. In his review of Keith Thomas's *Religion and the Decline of Magic*, he suggested that one reason for the success of eighteenth-century Wesleyanism lay in its counter-Enlightenment appeal to the "superstitious" poor by way of Wesley's own powerful supernaturalism, his affirmation of "bibliomancy, old wives' medical remedies, the casting of lots, the belief in diabolical possession and in exorcisms by prayer, in the hand of providence, [and] in the punishment (by lightning stroke or epilepsy or cholera) of ill-livers and reprobates." Methodism here is seen to have a close "fit" with the indigenous superstition of tinners, colliers, and fishermen, who were dependent on hazard in their everyday lives.[67] Such a congruity, it has been suggested, perhaps helps to explain why it caught on so quickly in "Celtic" areas like Cornwall, Wales, and the Isle of Man, where pre-Christian religion survived at its strongest.[68] It is characteristic of Thompson to locate his version of Methodism's successful cultural brokerage in Wesley's own superstition and "enthusiasm" rather than in the sobriety of Bunting and his supporters. Yet it remains a stubborn fact that Buntingite Methodism, notwithstanding the shocks of repeated expulsions and secessions, continued to grow rapidly until at least the 1840s, when the system came under insupportable strain and consolidation tipped over into bureaucratization.

To some extent Thompson fell victim to the bias of his Methodist sources. Much of the immediacy of his writing derives from his occasional use of biographies and conversion narratives, but he does not make due allowance for the exemplary nature of these accounts, which were often heavily didactic in purpose. He seems to envisage becoming a Methodist as a traumatic event, necessarily involving a sudden and complete change of life and the rupture of traditional cultural ties; it was the "psychic ordeal" in which the preindustrial laborer or artisan was "violently recast" into his new mold. To be sure, this immediatist paradigm found some encouragement in Wesley's theology. One must not deny the living reality of the conversion experience in a huge number of lives, but many conversions were far from sudden or cataclysmic and were accomplished by quiet recruitment along networks of friendship and kinship. There were not a few Methodist aspirants who sought "Christian perfection" by the path of intraworldly asceticism: No doubt devout chapelgoers and boisterous ale-house frequenters did provide contrasting behavioral models in many communities. Nonetheless, the actualities of denominational religious life were often far more nuanced than Thompson's sharp dichotomy might suggest. Throughout his treatment of Methodism runs a "displacement of energy" paradigm, which assumes it to be a totalitarian institution, preempting time, talent, and money from other activities by the intensity of its demands. But the divide between "rough" and "respectable" was seldom so sharp. The great majority of those who attended the Methodist chapels were never card-carrying "members," who submitted to the discipline of the society, but "hearers," more or less fellow-traveling adherents, for whom commitment was less rigorous, for whom the services were often one of several competing attractions, and whose social life fell somewhere between the polarities of pub and chapel. The proportion of "hearers" to "members" was variously estimated but was probably something like three to one.[69] As David Luker has observed, the mass of adherents "brought secular concerns . . . into the chapel and took religion out into the community: the lines between alternative cultures tended to be blurred, the links between them very real."[70]

The idea that Methodism might be more fruitfully interpreted as some kind of cultural exchange rather than as a matter of repression and displacement can be applied to Thompson's treatment of work discipline. The quest for self-control, industriousness, and "improvement," as well as a keen sense of the dignity of labor, was characteristic of many plebeian radicals, as well as Methodists, and was a common feature of the artisan culture that was a social seedbed for both categories. Moreover, few preachers can have been more eloquent on the work ethic than Thompson's radical hero, William Cobbett, author of a vigorous "sermon" on the evils of sloth.[71] It is not in doubt that evangelical religion often encouraged industriousness and financial prudence (why should it not?): Indeed, much of the attraction of the Methodist society lay in its offer of a more purposeful, disciplined way of life. But there is very little empirical evidence to suggest that many adherents acquired an obsessive and slavish devotion to labor. Thompson's view of Methodism as propagating a "pitiless ideology of work," which turned thousands into their own slave drivers, is manifestly extreme. Was some of its vehemence derived from the parallels he perceived with the place of work discipline in Stalinist ideology, which he deplored and explicitly linked to that of the Methodists?[72]

Here, as elsewhere, Thompson's argument is driven more by theory (Weber and Fromm) than by observation. His striking quotations from Andrew Ure about the need for the "moral machinery" of time-work discipline to complement the material machinery of the factories have polemical force, but they are not (as many incautious readers have assumed them to be) by a Methodist writer but rather by a Scots industrial scientist of Utilitarian leanings, making his pronouncements from what Thompson himself concedes are "transcendental heights."[73] In the period covered by *Making of the Working Class*, Methodism appealed far more to artisans than to factory hands.[74] It is not clear that they stood in need of much "slave-driving." Many of those drawn into the chapels—like the northeastern miners, Cornish tinners, and Craven textile workers—although operating within the context of an increasingly demanding capitalist system, still maintained some control over their own work patterns.[75] It is ironic that throughout the eighteenth century few charges were more constantly leveled against the Methodists than that they destroyed the incentive to labor: Methodism was perceived as "the bane of industry." In Wesley's time many moralists and economic writers anticipated Thompson's belief that evangelicalism made a total claim on the energies of its converts, but in their view it displaced energy from work, not from radical protest. By its psychological demands and the multiplicity of its services—some of them late at night—it was widely held to be draining time and strength away from productive labor.[76]

The narrowness of Thompson's approach also hindered him from grappling very closely with the social geography of popular evangelicalism. The movement spread rapidly in areas that he does not examine. In the 1821 census, the county in which the Wesleyans had the highest proportion of members to population (one to nineteen) was Cornwall. In the industrial West Riding of Yorkshire, which concerned Thompson the most, it was indeed high (one to twenty-one), but in rural Lincolnshire it was almost equally high with (one to twenty-three) and may well have had more cultural influence.[77] And popular evangelicalism (mostly in Calvinistic form) can seldom have taken much deeper root than in rural North Wales. Methodism also spread prolifically in the United States, in very different environments, to become by the middle of the nineteenth century the largest single denomination in the country. The strong impression is given in *Making of the Working Class* that the typical Methodist was a male factory worker, but this is belied by statistical evidence. Skilled craftsmen made up around 57.5% of its membership in the last half of the eighteenth century, a proportion that does not appear to have changed significantly before 1830. The movement also had a majority of women (some 57.7% according to the most recent estimate).[78] Given the publication date of Thompson's book (1963), the almost complete absence of women from his Methodism is understandable enough—though he makes a notable exception for "the greatest prophetess of all, Joanna Southcott"— but one wonders what he would have made of recent work on the strongly libertarian language of women in early American Methodism, "coaxed into speech" by its "subversive spirituality"?[79] Deborah Valenze has shown how English Methodism opened opportunities not only for the small (and sadly dwindling) band of women preachers but also for the myriad women who ministered by their own firesides at "cottage meetings," making Methodism a religion of the hearth, as well as of the chapel.[80]

The point here is not to castigate Thompson for failing to possess the knowledge and insights of a later generation of historians, not a few of whom, it must be remembered, were inspired by him. It is to suggest that there are great acreages of Methodist spirituality, appealing to humble people in all sorts of contexts in the North Atlantic world during an age of upheaval, that spoke of liberty, release, and fellowship and cannot be explained away simply as psychic displacement.[81] The "unanimity of conceptualization," which is at once the great strength and weakness of Thompson's evocation of Methodism, did not allow him to accept the fact that the Liberty Tree and the Methodist's Cross (his opposing symbols) could ever stand in the same garden.[82] This is why he virtually ignored the popular antislavery movement that flourished mightily among Methodist artisans, who were, as Seymour Drescher has shown, the most dedicated of all emancipationist groups (possibly nine out of ten members signed the great emancipation petition of 1833).[83] Thompson's desire for theoretical coherence helps him ignore the fact that many of the class tensions he detected between Methodism and its surrounding working-class constituencies (over the control of Sunday Schools, for example) were in reality as bitterly fought out within Methodism as they ever were between the religious and the irreligious or the middle and the working classes.[84]

On rereading *Making of the Working Class*, one is struck to see how Thompson anticipated some of the major lines of critical counterattack that might be launched against him. He acknowledged the claim of Wearmouth and others that Methodism trained working-class political leaders in the art of mobilization, sent out orators from the pulpit to the platform, and acted as an organizational model for radical societies.[85] He conceded that Methodism could offer "genuine fellowship" and self-respect to lowly people. He perceived very clearly the contradictory tendencies in a movement that could host both the pastoral authoritarianism of the Buntingites and the "democratic and intellectual elements," which produced the many irrepressible secessionists who left the parent body. For all the rhetorical forcefulness of his prose, Thompson had caution enough to admit the conjectural and provisional nature of some of his arguments (though this concession seldom restrained the confidence of his authorial tone).[86] After suggesting that Methodist hymns were replete with sexual symbolism, for example, he allows that the topic is "due for . . . more expert attention." Similarly, the idea of an oscillation between revivalism and radicalism is "a hypothesis."[87] But of course the attraction of the book lies not in its qualifications but in its bold assertions. This is why it will probably go on being read as a kind of vehement, radical, historical prose-poem long after more balanced or insipid versions of Methodist history have been forgotten.

Would Thompson have written quite so intensely about Methodist religion in his later years? Readers of one of his last books, on Blake, will see that he was capable of a sympathy for spirituality, which hardly surfaced in *Making of the Working Class*, though that is because Blake was a poet and artist, not a theologian, and because Thompson perceived him to be drawing on antinomian, pantheistic, quasi-mystical, Commonwealth traditions that epitomized freedom from the constrictions of law. Do we glimpse in *Witness against the Beast* the "Muggletonian Marxist" E. P. Thompson, saluting, ironically or seriously, consciously or unconsciously, Palmer Thompson, the young poet who half a century before had acknowledged himself a "Walt Whitman pantheist" at Kingswood School? Maybe. But there are no signs that he came to admire the Methodism in whose long shadow he had been reared or to recognize that it, too, was concerned, as much as William Blake, with building a world governed by mercy, pity, peace, and love.

For the purpose of studying evangelical religion in the North Atlantic world in the historical period for which *The Making of the English Working Class* has supplied such a powerful paradigm, Thompson's vision remains critically important. Certainly he pointed American historians down some blind alleys, but he also unsettled some cozy convictions. Passionate commitment to his own certainties about Methodists and the market prevented him from seeing the many instances in which Methodists resisted the newer industrial order. More important, that same passion blinded him to ways in which Methodists embraced the new economic order *for Methodist reasons*, or where, again for Methodist reasons, they structured their lives with little thought at all for material concerns. The weaknesses of a great historical work are always instructive. In Thompson's case those weaknesses point to the preeminent necessity of understanding the *religion* of the religious movements that experienced the market and industrial revolutions firsthand. They also illustrate the truism that a connection exists between the interpretation of Protestants and the market economies of the early nineteenth century and the attitudes of historians toward money and religion in the late twentieth century. E. P. Thompson's analysis of the profound interrelatedness of religious, social, and economic change may have been highly colored, even grotesque, at times, but the challenge for those who come after him is not to deny the depth of the relationship between religion and the economic order but rather to integrate them in a way that does not lead to the kind of "condescension of posterity" toward humble people against which Thompson struggled all his life. Whatever he would have made of this chapter, it is hard to resist the conclusion that he would have warmly welcomed the idea of bringing together Protestantism and political culture, work and money, mission and markets. The mischievous controversialist in him would have been delighted to discover that the hornet's nest he stirred up in 1963 has not yet subsided.

Notes

1. See, for example, Nathan Hatch, *The Democratization of American Christianity* (New Haven, Conn.: Yale University Press, 1989); David Hempton, *The Religion of the People: Methodism and Popular Religion c. 1750–1900* (London: Routledge, 1996); and Dee E. Andrews, *The Methodists and Revolutionary America, 1760–1800: The Shaping of an Evangelical Culture* (Princeton, N.J.: Princeton University Press, 2000).

2. *Times Literary Supplement*, October 3, 1935, 610.

3. For E. J. Thompson and his family, see Bryan D. Palmer, *E. P. Thompson: Objections and Oppositions* (London: Verso, 1994), chap. 1.

4. E. J. Thompson, *O World Invisible* (London: Benn, 1931). The title is taken from the poem by Francis Thompson.

5. E. P. Thompson, *Alien Homage* (Delhi: Oxford University Press, 1993), 16.

6. E. J. Thompson, *John Arnison* (London: Macmillan, 1939), 114.

7. L. G. Wickham Legg and E. T. Williams, eds., *Dictionary of National Biography 1941–1950* (Oxford: Oxford University Press, 1959). The entry is by H. M. Margoliouth, an Oxford friend.

8. *John Arnison*, viii–x.

9. Ibid., 136.

10. Ibid., 293.

11. E. J. Thompson, *Introducing the Arnisons* (London: Macmillan, 1935), 97.

12. E. J. Thompson, *John Arnison*, 330–5.

13. E. J. Thompson seems to have remained a kind of radical liberal. See Palmer, *E. P. Thompson*, chap. 1.

14. E. P. Thompson, *Beyond the Frontier* (Woodbridge, Eng.: Merlin, 1997), 47.

15. We are grateful here for information from Arnold Rattenbury.

16. Information was kindly supplied by the Rev. Ernest Goodridge.

17. E. J. and Theodosia Thompson papers, Ms Eng 2699, Bodleian Library, Oxford.

18. Interview with Thompson in "E. P. Thompson, a Life of Dissent," Channel 4 (Britain), screened September 18, 1993. For comment on the theme of E. P. Thompson as "secular Protestant," see Penelope Corfield, "E. P. Thompson, the Historian: An Appreciation," *New Left Review* 201 (1993): 16–17. F. M. Leventhal notes of the

Marxist historian Brailsford: "He was bred in the Nonconformist religious tradition and despite his rebellion against that upbringing, its style was ingrained in him. His moral fervour, his capacity for indignation, the certainty of his convictions, were an inheritance from his Methodist background." Leventhal, *The Last Dissenter. H. N. Brailsford and His World* (Oxford: Clarendon, 1985), 2. See Raphael Samuel, "British Marxist Historians, 1880–1980," *New Left Review* 120 (March–April, 1980): 42–55, for a succinct discussion of this theme. For example, in the recollection of Christopher Hill, the transition from Nonconformity to Marxism in the 1930s was frequent enough for it to form an ironical form of introduction in Communist society: "I'm a Methodist, what's your heresy, I think we used to say."

19. E. P. Thompson, *Beyond the Frontier*, 52.

20. Henry Abelove et al., *Visions of History* (New York: Pantheon, 1983), 21.

21. E. P. Thompson, *The Poverty of Theory and Other Essays* (London: Merlin, 1978).

22. Information was kindly supplied by the late G. W. Kingsnorth.

23. See Harvey Kaye, *The British Marxist Historians* (Basingstoke, Eng.: Macmillan, 1995).

24. E. P. Thompson, *Infant and Emperor* (London: Merlin, 1983).

25. E. P. Thompson, *Poverty of Theory*, 241

26. E. P. Thompson, *The Making of the English Working Class*, (Harmondsworth: Penquin, 1968), postscript, 918.

27. E. P. Thompson, *Witness against the Beast. William Blake and the Moral Law* (Cambridge: Cambridge University Press, 1993), xxi.

28. We owe this suggestion to the late Raphael Samuel. Did Thompson project onto a bygone Methodism not only the conformism but also the Puritanism of Communist life as he met it in a northern mill town? For an evocation of Communist society at this time, emphasizing its close similarities with sectarian Protestant subcultures, see Raphael Samuel's autobiographical "The Lost World of British Communism," *New Left Review* 154 (November–December 1985) and 156 (March–April 1986). This contrasts with the more open, argumentative, Bohemian, "happily idiosyncratic" Communist coteries of the metropolitan theatrical and literary world, described by Arnold Rattenbury in his piece on Edward Thompson's brother, Frank, in *The London Review of Books* 19, no. 9 (May 8, 1999).

29. E. P. Thompson, *Making of the Working Class*, 385.

30. Elie Halévy, *A History of the English People in the Nineteenth Century*, 4 vols. (London: Benn, 1949–51); and Halévy, *The Birth of Methodism in England*, trans. and ed. Bernard Semmel (Chicago: University of Chicago Press, 1971). See, too, J. L. and Barbara Hammond, in *The Town Labourer 1760–1832* (London: Longman, 1917), and *The Bleak Age* (London: Longman, 1934).

31. E. P. Thompson, *Making of the Working Class*, 917–18.

32. E. P. Thompson, *Witness against the Beast*, 44, 72. Muggletonians were followers of Lodowick Muggleton, who in the tumultuous days of Cromwell promoted an anti-Trinitarian form of religious radicalism.

33. E. P. Thompson, *Making of the Working Class*, 429.

34. Ibid., 430–37.

35. Ibid., 409.

36. Ibid., 405. Thompson may have borrowed the phrase "psychic masturbation" from the English subtitles of Ingmar Bergman's *Wild Strawberries*, shot in 1957. Since Henry Abelove's innovative *The Evangelist of Desire* (Stanford, Calif.: Stanford University Press, 1990), work has begun on the question of Methodist sexuality. In a

recent case study, John Tosh concludes that "a belief in the sacred nature of marriage was not inherently hostile to the enjoyment of sex." Tosh, "Methodist Domesticity and Middle Class Masculinity in Nineteenth Century England," in *Gender and Christian Religion, Studies in Church History*, ed. R. Swanson (Woodbridge, Eng.: Boydell, 1998), 34:23–45. John Wesley's relations with women were indeed complex and sometimes fraught, but Thompson's suggestion that Charles Wesley's hymns suggest a "repressed" sexuality needs to take account of his extremely happy marriage and his eight children. As for the Moravian Brethren, Thompson's idea that their hymns revealed a "perverse" and repressed sexual imagery is odd since the Moravians were attacked for their unusual openness about sexuality and their celebration of sexual love as a God-given blessing. For the Moravians' remarkable frankness on these issues and their pioneering views on sex education and marriage guidance, see Colin Podmore, *The Moravian Church in England 1728–1760* (Oxford: Clarendon, 1998), 129–32.

37. E. P. Thompson, *Making of the Working Class*, 390–98.

38. Ibid., 390. The Methodist Connection was the official name of the denomination in England, while the Conference was the gathering of ministers in the Connection.

39. Ibid., 389.

40. Ibid., 440.

41. See Hempton, *Religion of the People*, 91–108; W. R. Ward, *Religion and Society in England 1790–1850* (London: Batsford, 1972); *The Early Correspondence of Jabez Bunting 1820–1829, Camden 4th series*, ed. W. R. Ward, (London: Royal Historical Society, 1972), 4–20.

42. A. Clarke to Sidmouth, May 14, 1811, Sidmouth papers, Devon County Record Office, Exeter, C.1811/OE; J. Everett, *Adam Clarke Portrayed*, 2 vols. (London: Hamilton, Adams, 1843), I:323, 329; II:243; W. Pollard, "Conversations with the Revd. Adam Clarke, 1828," Methodist Archives Mss, John Rylands University Library of Manchester.

43. Historians with different ideological perspectives have come to this conclusion from different routes. See Ward, *Religion and Society*; D. N. Hempton, *Methodism and Politics in British Society 1750–1850* (London: Hutchinson, 1984); J. C. D. Clark, *English Society 1688–1832* (Cambridge: Cambridge University Press, 1985); D. W. Lovegrove, *Established Church, Sectarian People: Itinerancy and the Transformation of English Dissent 1780–1830* (Cambridge: Cambridge University Press, 1985); T. Bartlett, *The Fall and Rise of the Irish Nation: The Catholic Question 1690–1830* (Dublin: Gill & Macmillan, 1992); A. D. Gilbert, "Methodism, Dissent and Political Stability in Early Industrial England," *Journal of Religious History* 10, no. 4 (1978–79): 381–99.

44. Bernard Semmel, *The Methodist Revolution* (New York: Heinemann, 1973), 117.

45. Ward, *Religion and Society*, 97–104.

46. Charles Atmore, *The Methodist Memorial* (London: Hamilton, Adams, 1871), 8.

47. See, for example, Mark Smith, *Religion in Industrial Society: Oldham and Saddleworth 1740–1865* (Oxford: Clarendon, 1994); Robert Colls, *The Collier's Rant* (London: Croom Helm, 1977); B. Rees, *Chapels in the Valley* (Upton, Wirral: Ffynnon, 1975); David Luker, "Revivalism in Theory and Practice: The Case of Cornish Methodism," *Journal of Ecclesiastical History* 37, no. 4 (1986): 603–19. For "the mob of Methodism," see J. H. Drew, *The Life, Character, and Literary Labours of Samuel Drew* (London: Longman, 1834), 491.

48. James Obelkevich, *Religion and Rural Society: South Lindsey 1825–1875* (Oxford: Clarendon, 1976). For the later impact of Primitive Methodism on agricultural trade unionism, see Nigel Scotland, *Methodism and the Revolt of the Field* (Gloucester: A. Sutton, 1981); and Alun Howkins, *Poor Labouring Men: Rural Radicalism in Norfolk, 1872–1923* (London: Routledge, 1985).

49. E. P. Thompson, *Making of the Working Class*, 437.

50. For an account of similar popular resentment against the early Methodists in America, see Christine Leigh Heyrman, *Southern Cross: The Beginning of the Bible Belt* (New York: Knopf, 1997).

51. E. M. White, "'The World, the Flesh and the Devil', and the Early Methodist Societies of South West Wales," *Trans. of the Honourable Society of Cywmmrodorion*, 3 (1996): 49. See the *Minutes of the Methodist Conferences*, 19 vols. (1862): I:171, for a complaint that the spiritual growth of members was "exceedingly hurt" by their continued attendance at wakes and fairs.

52. See, for example, R. W. Malcolmson, *Popular Recreations in English Society, 1700–1850* (Cambridge: Cambridge University Press, 1973), 75–88.

53. See J. Colwell, *Sketches of Village Methodism* (London: Elliot Stock, 1877), 131–32; S. P. Menefee, *Wives for Sale* (Oxford: Blackwell, 1981), 40. For the saturnalian aspects of the wake, see D. A. Reid, "Interpreting the Festival Calendar," in *Popular Culture and Custom in Nineteenth Century England*, ed. R. D. Storch (London: Croom Helm, 1982), 125–30. For the seamy side of the "harvest frolic," see *The Autobiography of Joseph Arch,* ed. J. G. O'Leary (London: Macgibbon & Kee, 1966), 88–89.

54. Thompson, *Making of the Working Class*, 451.

55. John Rule, "Methodism, Popular Beliefs and Village Culture in Cornwall," in Storch, *Popular Culture*, 53.

56. For chapel music and choral singing, see Roger Elbourne, *Music and Tradition in Early Industrial Lancashire 1780–1840* (Woodbridge, Eng.: D. S. Brewer, 1980). For Methodist social life as "entertainment," see Obelkevich, *Religion and Rural Society*, 212–23.

57. D. M. Evans, *Christmas Evans: A Memoir* (London: Heaton, 1863), 163–73, 205; E. Morgan, *A Memoir of John Elias* (Liverpool: J. Jones, 1844); Owen Jones, *Some of the Great Preachers of Wales* (London: Passmore & Alabaster, 1885), 252–59; J. Everett, *The Village Blacksmith: A Memoir of Samuel Hick* (London: Hamilton, Adams, 1832) and *Memoirs of William Dawson* (London: Hamilton, Adams, 1842). For Sunday School culture, see T. W. Laqueur, *Religion and Respectability. Sunday Schools and Working Class Culture 1780–1850* (New Haven, Conn.: Yale University Press, 1976), 175–79, 227ff.

58. See P. Bailey, *Leisure and Class in Victorian England: Rational Recreation and the Contest for Control* (London: Methune, 1978); R. Holt, *Sport and the British* (Oxford: Clarendon, 1989), 136–48.

59. *The Journal of the Rev. J. Wesley*, ed. N. Curnock, 8 vols. (London: Epworth, 1938), 3:288.

60. Laqueur, *Religion and Respectability*, 61–62. For a later period see Sarah C. Williams, *Religious Belief and Popular Culture in Southwark c. 1880–1939* (Oxford: Oxford University Press, 1999), 126–39.

61. See the discussion of this theme in L. Billington, "Popular Religion and Social Reform: A Study of Revivalism and Teetotalism 1830–1850," *Journal of Religious History* 10, no. 3 (1978–79): 266–93. For the debate in Wesleyanism (in which

Cornwall was facetiously described as being in a state of "fermentation" on the subject), see Benjamin Gregory, *Sidelights on the Conflicts of Methodism* (London: Cassel, 1898), 318.

62. Ward, *Early Correspondence of Bunting*, 11–12; D. Luker, "Cornish Methodism, Revivalism and Popular Belief c. 1780–1870" (D.Phil. diss., Oxford University, Oxford, 1987), 305ff.

63. See A. D. Gilbert, *Religion and Society in Industrial England* (London: Longman, 1976), 83–85; Albion Urdank, *Religion and Society in a Cotswold Vale* (Berkeley: University of California Press, 1990), 52.

64. Thompson, *Making of the Working Class*, 417

65. Ibid., 63.

66. Laqueur, *Religion and Respectability*, 188–89.

67. E. P. Thompson, "Anthropology and the Discipline of Historical Context," *Midland History* 1, no. 3 (1972): 41–55.

68. Rule, "Methodism, Popular Beliefs," 61–67; Robert Currie, *Methodism Divided. A Study in the Sociology of Ecumenism* (London: Faber, 1968), 21–22.

69. The ratio of "hearers" to "members" varied considerably. Estimates ranged between two to one and six to one. The average has been estimated as about three to one; see C. D. Field, "The Social Composition of English Methodism to 1830: A Membership Analysis," *Bulletin of the John Rylands University Library of Manchester* 76, no. 1 (1994): 153–54.

70. Luker, "Cornish Methodism," 378.

71. William Cobbett, *Twelve Sermons* (London: J. M. Cobbett, 1823), Sermon 6, "The Sluggard."

72. E. P. Thompson, *Poverty of Theory*, 296–97.

73. E. P. Thompson, *Making of the Working Class*, 395–98.

74. For similar patterns of artisan Methodism in the United States see Andrew, *Methodists and Revolutionary America*, 155–83; and William R. Sutton, *Journeymen for Jesus: Evangelical Artisans Confront Capitalism in Jacksonian Baltimore* (University Park: Pennsylvania State University Press, 1998).

75. See P. Rycroft, "Church, Chapel and Community in Craven, 1764–1851" (D.Phil. Diss., Oxford University, Oxford, 1988), 38–39, 267; W. J. Rowe, *Cornwall in the Age of the Industrial Revolution* (Liverpool: University Press, 1953), 26–28. On the relative autonomy of northern miners, see Robert Colls, *The Pitmen of the Northern Coalfield: Work, Culture and Protest, 1790–1850* (Manchester: Manchester University Press, 1987), 29–33.

76. See John Walsh, "'The Bane of Industry?' Popular Evangelicalism and Work in the Eighteenth Century," in *Work, Rest and Play: Religious Aspects of the Use of Time, Studies in Church History*, 36 (forthcoming).

77. *Wesleyan Methodist Magazine* 42 (1824): 378–79.

78. Figures taken from Field, "Social Composition of English Methodism," 153–69. See Andrews, *Methodists in Revolutionary America*, app. A, for the percentage of female Methodist society members at the turn of the eighteenth and nineteenth centuries in New York (64.7%), Philadelphia (64.1%), and Baltimore (60.1%).

79. D. H. Lobody, "'That Language Might Be Given Me': Women's Experience in Early Methodism," in *Perspectives in American Methodism*, ed., R. E. Richey, K. E. Rowe, and J. M. Schmidt (Nashville, Tenn.: Kingswood, 1993). See the critique of Thompson in Joan W. Scott, *Gender and the Politics of History* (New York: Columbia University Press, 1988), 71–81. It is perhaps a contradiction in Thompson that he

shows sympathy to one form of quasi-irrationality—Blake's mysticism—as some kind of defensive strategy against the onslaught of establishmentarian rationalism, while attacking Methodism, including its prophetesses, for being "irrational."

80. See D. M. Valenze, *Prophetic Sons and Daughters. Female Preaching and Popular Religion in Early Industrial England* (Princeton, N.J.: Princeton University Press, 1985).

81. D. N. Hempton, "Motives, Methods and Margins in Methodism's Age of Expansion," *Proceedings of the Wesley Historical Society* 49, pt. 6 (October 1994): 189–207.

82. See Suzanne Desan, "Crowds, Community and Ritual in the Work of E. P. Thompson and Natalie Davis," in *The New Cultural History,* ed. Lynn Hunt (Berkeley: University of California Press, 1989), 58.

83. Seymour Drescher, *Capitalism and Antislavery* (New York: Oxford University Press, 1987), 118–31.

84. Hempton, *Religion of the People*, 171–78.

85. Robert F. Wearmouth, *Methodism and the Working-Class Movements of England 1800–1850* (London: Epworth, 1937); and Wearmouth, *Some Working-Class Movements of the Nineteenth Century* (London: Epworth, 1948).

86. E. P. Thompson, *Making of the Working Class*, 372. See F. S. Piggin, "Religion and the Industrial Revolution. An Analysis of E. P. Thompson's Interpretation of Methodism," *University of Wollongong Historical Journal* 2, no. 1 (March 1976).

87. E. P. Thompson, *Making of the Working Class*, 388.

II

FINANCE AND
THE EXPANSION OF
AMERICAN PROTESTANTISM

6

A Tale of Preachers and Beggars: Methodism and Money in the Great Age of Transatlantic Expansion, 1780–1830

David Hempton

And numbers are all we have to boast of, for our money matters are little improved.
　　　　　　　　　—Miles Martindale to Jabez Bunting, July 9, 1816

When it was objected "that we spoke for hire"; it was answered, "No—it was only for a passing support."
　　　　　　　　　—Francis Asbury, September 22, 1783

Two very different kinds of facts and figures can be brought to bear on Methodism's great era of expansion throughout the English-speaking world and beyond in the generation after the American and French Revolutions.[1] The first are membership statistics, which reveal spectacular gains across the North Atlantic world. In England members of Methodist societies expanded from 55,705 in 1790 to 285,530 in 1830. In the same period, Irish Methodist membership almost doubled and Scottish Methodist membership tripled. Even more dramatically, Methodist membership in Wales increased by a factor of twenty.[2] But the most dramatic growth of all occurred in America, which had fewer than 1,000 members in 1770 and over 250,000 just fifty years later. By 1850 the Methodist share of the religious market in the United States had increased to 34% of the national total.[3] In addition, Methodism was in the process of becoming a genuinely international religious movement in the wake of the foundation of the Methodist Missionary Society in England in 1813 and in America in 1820.[4] By any standard these figures represent a quite remarkable religious revolution, comparable in scope to the Reformation itself and one that is still comparatively underresearched.

A second set of facts and figures, those more familiar to the accountant than the demographer, suggest a much gloomier picture, for throughout this period Methodism was racked by serious financial problems, which both reflected and exacerbated deep-seated structural deficiencies in the Methodist organization. Religious movements, as with big businesses, can, up to a point, thrive on deficit finance in periods of expansion, but in religious organizations the great disadvantage is that there is generally no access to liquid capital beyond that which is voluntarily granted by their members or raised by their supporters. Hence, financial problems in churches, as with families, are never confined to the account books and are rarely about money alone. Financial stresses and strains within Methodism exposed manifold layers of other conflicts endemic in the wider society. Money was not only a sufficiently important commodity to be worth fighting over but also served as a scorecard for even more important contests. Both for those who raised the money and for those who spent it, balance sheets reveal as much about personal and religious priorities as they do about profit and loss. Where money is, the heart is also.

John Wesley's early experimentation with the primitive church's practice of the community of goods did not long survive its encounter with basic human acquisitiveness, which, disappointingly for him, proved to be a stronger force in Methodist societies than the social expression of perfect love.[5] Although Wesley abandoned the experiment, he never lost his deep-seated distrust of the corrupting power of money on the spiritual life.[6] While he remained in charge, however, money was regarded not only as a threat to spiritual progress but also as an opportunity to fund ministry, poor relief, and fellowship in an expanding Methodist connection. His example of the rigorous keeping of accounts was passed on to his followers in a quite remarkable way. The records of Methodist churches in all parts of the world are full of neat rows of figures constructed by circuit stewards, chapel committees, and mission secretaries.[7] Behind the accounts lay the principle of accountability, to God and to men, for Wesley recognized that his movement was vulnerable to all kinds of accusations, from sexual immorality to material exploitation.[8] Although Wesley's Methodism was by no means free from allegations of financial corruption, the fate of all new religious movements, it was largely free from corruption itself. Not only did Wesley set a powerful personal example of frugality and financial discipline, but also it was well known that the Methodist itinerant ministry was scarcely a bed of roses for those who devoted their lives to it.

After Wesley's death a combination of unseemly jockeying for power and influence among the senior preachers and the crippling economic depressions of the 1790s revived older allegations of financial malpractice, only this time from within the ranks of Methodists themselves.[9] The literature arising from the first secession from the parent Connection, led by Alexander Kilham in 1797, is positively riddled with allegations of preachers' lording it over a starving people.[10] The essence of Kilham's allegations was that a combination of secrecy, extravagance, and chicanery had undermined the people's confidence in the Methodist preachers, whom he accused of behaving like "begging friars; and whining canting Jesuits." In this way the campaign for lay rights within Methodism was fueled, at least in part, by a lack of confidence in the financial probity of the Methodist leadership. Nor was that particularly surprising. By devolving responsibility for Methodist fund-raising and accounting to the laity in the localities and by insisting that preachers should be both itinerant and take their orders from above, the Wesleyan system was a recipe for financial discord. What largely held conflict in check was the belief that itinerant preaching was a holy calling and the fact that in the early stages of the movement preachers were expected to forswear the emotional and material comforts of the settled life. The ideal, therefore, was that spiritually enlivened preachers would have their needs met by a grateful laity, and in addition, deficiencies in one part of the country would be made good by surpluses in another. All was to be held together by bonds of affection. What put the system under stress and strain in the early nineteenth century, however, was the unpleasant reality that Methodist fund-raising always fell short of Methodist aspirations. Equally problematic was the fact that the original ideal of a celibate, self-sacrificing, and ascetic brotherhood of preachers proved to be as unrealizable among the second generation of Methodists as had Wesley's earlier experiment with the community of goods among the first generation. Where money was concerned, Christian perfection, as the Calvinists had always contended, was easier to talk about than to achieve in practice.

Squabbles about money may be bad news for religious movements, but they supply excellent sources for religious historians, for they shed light on deeper structural tensions within religious communities. The aim of this chapter then is to use money as a device to explore the tensions within Methodism in two different locations, England and New England, in the generation after Wesley's death. A comparative approach exposes both the similarities and the dissimilarities of the two Methodist traditions and also points up some significant differences in their social and cultural contexts.

Methodism in England

The second decade of the nineteenth century was the most crucial period in the financial history of English Methodism, for it was in that decade that a combination of the rising social aspirations of Methodist preachers and the postwar price deflation exposed serious structural problems in the Methodist organization. A convenient window into these positions is provided by the correspondence of Jabez Bunting, the key Methodist networker of his generation. And a convenient place to start to unravel the social dynamics of the relationship between Methodism and money is with a letter written to Jabez Bunting from a northern English city in the midst of the economic depression at the end of the Napoleonic Wars:

> You throw out some hints as if this Collection [the annual July congregational collection] was to become *permanent*—not adverting to the number of Collections already established among us, which I can prove to be not less than 155 every year. Was there ever such a begging system in existence before? Almost *every other day* we have our hands in the pockets of our people. You have often maintained in Conference that 450 members are sufficient for the support of a married and single preacher. Theoretically this may do on paper; but the uniform practice invalidates the system and I can prove that neither 450, nor 500 members either do or can pay their way, generally speaking. You have 306 circuits and out of these only 76 are clear; and a great number of those who have upwards of 500 members are paupers.
>
> You speak of *prudent retrenchment of expenditure*—where will you begin? You have 181,710 members −736 preachers, 400 wives = 1137 persons besides children to provide for: so that you have 159 members to each person, which at one penny per week &c. is £66.5.0. Here then is the *core* of your embarrassments, and I maintain that you ought to have either 50,000 more members, or 50 preachers and 50 wives less: but as the latter is impracticable, and to realise the former is a work of time; we must remain in *statu quo*, till the *decrease* of preachers and the *increase* of members come to a sort of par.[11]

Put crudely, the expansion of Methodist members, though dramatic, was not fast enough to keep up with the financial demands of its increasing number of married preachers, their wives, and their children.[12] Moreover, according to Bunting's correspondent, the problems were exacerbated by the propensity of poor circuits to invite eminent preachers from distant circuits on all-expenses-paid preaching trips. A gravy train, not unlike the modern lecturing circuit, was eroding Methodist capital. Still worse, according to the letter writer, was the impact of new demands to finance home and foreign missions:

> Another cause of our debts has been the *Home Missions*, a system that ought never to have been tolerated among us, and which originated to cover *lazy* preachers. We were all missionaries once—*Thirty years* ago I used to preach *ten times a week*, and travel many miles, I did so fifteen years ago. I do so now, even the *last week*—And if you will keep your circuits large enough, let them contain a sufficient number of members, say 600 or upwards, and let your preachers labour as we used to do, and as I do yet, and you will not have so many demands, the people will be more joyful to see you, and you will close the mouths of some idle talkers to whom our begging system gives but too much occasion.[13]

To continue with a business metaphor, the expansion of the Methodist enterprise at home and abroad depended on a careful balancing of its income and expenditure, circuit by circuit, at a time when rising overheads, in the shape of increased numbers of married preachers with families, stretched its resources. Two possible solutions were either to increase its membership faster than its unit preaching costs or to recruit richer members. Bunting's correspondent was pessimistic about the possibility of achieving either of these options, especially in a period of rising social tension and incipient class conflict:

> I suppose you have heard of our revival in this circuit, as it has been trumpetted around the nation: if your Cornish and other revivals are as mechanically produced and carried on as this has been they will neither be any credit nor lasting good to the connexion. When Nelson came here we had not a single seat to let in our chapel, and though we are building a new chapel, and have added 1,100 new members in the City and Circuit, yet last Sunday evening I had to read the advertisement of 6 whole pews and many single seats for which there is no application, nor one seat in the new chapel ever enquired after, while the Calvinist Chapel tho' not as forward as ours, have all their seats engaged! The fact is, the noise and nonsense has produced a number of the lowest order, and driven many respectable people from our chapel.[14]

Financial exigencies, styles of evangelism, and contested expressions of social class and culture were inextricably linked with one another in the cash crisis of early nineteenth-century Methodism.

The nub of the problem was that the financial outgoings of the Connection had increased dramatically at a time of severe economic stress and strain. According to Reginald Ward, the Methodist Connection in those years had come to "resemble a modern cut-price motor-car insurance company tempted or deluded by a large cash flow into prodigal disregard for its future liabilities."[15] Financial demands were legion: Money was required to fund home missions and convert the Irish and the Welsh,[16] to build new chapels and extend old ones,[17] to retire chapel debts and fund interest payments,[18] to build schools and supply educational materials,[19] to bail out poor country circuits and distant villages,[20] to meet the expenses of the itinerant preachers and their increasing numbers of dependents,[21] to supply the needs of infirm preachers and societies in distressed districts, and to pay for foreign missions.[22] In essence, the burgeoning costs of chapels, preachers, and missions virtually rendered obsolete the old pattern of Methodist fund-raising, based on voluntary class subscriptions. Moreover, as Methodism became a powerful new force in the world, the expectations of its itinerant foot soldiers also began to rise.[23] Itinerancy, as a system of ministry, was remarkably cheap and effective in its pioneering mode but became less so as circuits became more established. A new generation of preachers began to discriminate between good and bad hospitality and between "easy" and "hard" circuits. Bunting, who increasingly took on responsibility for administering the system, had his ears bent by a bewildering range of requests for more comfortable lodgings, better educational facilities for children, and accommodation of a suitable standard for newly married preachers.[24] One preacher even had the audacity to make known his new bride's requests and preferences.[25] Besides the ordinary costs of itinerancy, attendance at Conferences, expensive removals every three years (a costly consequence of the circuit system), and, where necessary, maintenance expenses for horses all had to be paid for.[26] Where was the money to come from?

There was general agreement among the senior preachers in the Connection that "a permanent plan of finance" was better than the "temporary expedients" of special collections, which both irritated the laity and failed to supply a recurrent income.[27] In a response to yet another request for a new subscription, Bunting outlined the traditional Methodist approach to fund-raising:

> Could they [Methodists] be brought uniformly to contribute 1d per week, & 1s. per quarter, those *deficiencies*, to which you allude, would have no existence. Without Yearly Collection, & the profits of the books, we could then bid defiance to *want*. . . . In many circuits of the South, indeed, & in some of the North, the present rules are more than *kept*, they are considerably exceeded. Now how can we press the people of such Circuits to a new & *permanent* subscription, in order to make up for the indolence & covetousness of their neighbours?[28]

In theory the 1 pence a week was to help supply the "weekly wants of the preachers, first in the place or society where it is made, and then in the Circuit at large," whereas the quarterly moneys were to meet the more general expenses of the itinerant system.[29] The delicious simplicity of the old Methodist plan, combining as it did piety and payment, commitment and contributions, and savings and sanctification, had one fatal flaw—it simply could not be made to work.[30] Devised as a system for bringing together societies and preachers in bonds of affection and mutual dependency, the old Methodist plan had also the capacity to generate financial conflicts that undermined the goodwill on which the whole system depended. Once the lubricant of adequate revenue dried up, the moving parts simply ground together and generated heat and friction. The preachers blamed circuit stewards and class leaders for poor performance of duties, whereas the laity, especially in periods of economic hardship, blamed the preachers for unrealistic expectations. Preachers who were laboring mostly in poorer circuits accused those in wealthier circuits of a lack of commitment to genuine equality in the sharing out of connectional resources.[31] More substantially, the system was cumbersome to administer and set up tensions between poorer and richer Methodists. In some districts the latter preferred to settle up quarterly at the renewal of class tickets, often in excess of what was nominally required, whereas the former, who had also lost the weekly habit, could not make good the deficiency at the end of the quarter.[32] Moreover, many Methodist chapel-goers were "hearers" only, not society members, and were therefore under no obligation to pay their dues to receive their class tickets. Some no doubt kept clear of membership to escape not only the discipline but also the costs and the fear of eviction for debt. What is clear from the private correspondence of the preachers is that the old Methodist plan was not working properly and that it was difficult for *them* to solve it since they were both its beneficiaries and its custodians. Either the system needed to be cranked up more assiduously to maximize contributions from class members or the number of preachers and their families would have to be scaled down to bring profit and loss into due balance. Although some of the preachers retained confidence both in the spiritual potential of the old system and in its capacity to deliver the goods, others favored retrenchment or fresh initiatives for tapping into new wealth.

Meeting the day-to-day running costs of the Connection would have stretched Methodist resources even without the added burdens of chapel building and foreign missions. Although the great, new urban chapels were ferociously expensive, there were at least established ways and means of raising the capital.[33] Subscriptions from "our leading friends" (a recurring Methodist phrase), special collections (openings, anniversaries, and interest repayments), pew rents, and borrowed money all played their part.[34] Debts were sometimes crippling, especially during the postwar price deflation, when interest payments became artificially inflated, but the system chugged along reasonably well with the help of new wealth and frequent collections. Indeed, in some cases chapel fund-raising was so successful that there were calls to divert chapel surpluses into the financing of the preachers, a plan that would have brought Methodism closer to the dissenting model of congregational independence.[35] Bunting, predictably, held the line against suggestions that would have undermined the integrity of the Pastoral Office and insisted on chapel surpluses being deployed either on new chapels or on proper accommodation for circuit preachers. As with English monarchs in the seventeenth century, Methodist preachers wanted adequate supply, but they did not wish to be placed under the authority or to be circumscribed by the expectations of the suppliers.

Whatever one makes of the evangelistic strategy of building impressive urban chapels as a focal point of Methodist ministry in towns, cities, and regions, there is no denying the sheer scale of the financial achievement. Foreign missions, however, were a different story. With no property to act as collateral and with no pew rents to offer a regular income, the financing of missions posed particular problems. By the time this issue was seriously addressed in the second decade of the nineteenth century, there were already allegations circulating within Methodism that money raised for foreign missions had been surreptitiously and disgracefully diverted into the domestic coffers.[36] Moreover, there was widespread agreement that the "missionary intelligence" published by the Connection was so late, inadequate, and depressing that it was virtually useless as a primer for missionary giving. Yet interest in foreign missions was rising as fast as the Methodist membership.

At the end of the Napoleonic Wars, therefore, an intriguing combination of financial pessimism and missionary optimism left the Connection in dire need of fresh initiatives. Whereas Methodist chapels were leaking money to collectors from the London Missionary Society, its own fund-raising was largely dependent on the personal networks established by the mercurial Dr. Coke. Clearly, revenue for the support of overseas missions, as with all other Methodist endeavors, needed to be placed on a more secure footing, but how was this to be done? In the midst of a cheerless rant about the deficiencies of metropolitan Methodist organization and the baleful consequences of chapel debts, Joseph Entwistle told Bunting that "we have already so many irons in the fire, that we shall do little in the 1d. a week way."[37] Bunting, as ever, was ready with his remedy:

> I do not believe that missionary exertions will lessen our home-resources. Zeal, once kindled, will burn in every direction at once. The pains taken in Missionary Meetings to evince the value & necessity of a Christian Ministry & of Christian Ordinances will shew our people *their own privileges*, as well as the wants of others. They will estimate the Gospel more highly, & of consequence be more ready to support it. Besides on the new plan, a large proportion of the Mission Fund will be raised among persons not in Society, & even among persons who are not so much as stated Hearers. This will set us more at liberty in our applications to our own members for our own particular collections.[38]

In this way Bunting was prepared to raise capital from outside the company to expand the business, without adding to the burdens of the existing shareholders. Nor is the business metaphor entirely fanciful, for the great public meetings on behalf of foreign missions were addressed by the best available talent and were advertised through press notices and "posting bills."[39] Large public meetings backed by the formation of auxiliary missionary societies throughout the country delivered unimaginable financial rewards, but both in principle and in practice Methodist finances were now no longer the private business of a pious sect.[40] Whatever was lost in terms of connectional intimacy in this transition from the political economy of a religious sect to that of a national church was, according to Bunting, more than made up by the sheer financial success of the new strategy:

> The Halifax report is this day come to hand. . . . From that report it appears that the annual public collections for missions in that district in 1813 amounted to £154.19.5., and the extraordinary Missionary collection to £145.8.7. whereas the moneys raised in 1814 by the new system of public meetings & Auxiliary Societies were £1292.14.8. This *fact* is worth more than a score of theories.
> There is nothing like public meetings! We have had one at Albion St., and another at the old chapel for the purpose of shewing the necessity of the proposed Chapel in Meadow Lane. . . . More is offered daily; so that we expect to raise, in this easy & delightful way, £1600 or £1800. . . . The mission business is no hindrance, but a great help. The poorer classes have now learned by experience the privilege of giving. They know the consequence & efficiency conferred on them by their number, & they seem resolved to maintain by actively assisting every good work, the dignity to which they have been raised.[41]

Bunting's sanguine assessment of the "privilege of giving," however, is not borne out by most of the correspondence that arrived on his desk in the first quarter of the nineteenth century. Itinerant preaching and the connectional system, which historians have justifiably lauded as instrumental in the remarkable expansion of Methodism throughout the world in the early nineteenth century, was an accountant's nightmare. Because preachers were paid for out of voluntary subscriptions to societies and because they were paid allowances rather than a salary, the whole enterprise depended on frugality, goodwill, and bonds of affection. The system was at its most cost-efficient when it was supporting single preachers with no dependents, but by 1820 the ratio of married to single preachers was in excess of three to one.[42] Moreover, many of the older preachers were in no doubt that the younger brigade were less frugal and less willing to accept hardship than their predecessors. One of Bunting's correspondents stated that the finances of the Connection lay "at the mercy of inexperienced, if not vain & extravagant young housekeepers."[43] He also estimated that by 1819 married preachers cost the Connection just over £100 per annum: "The young officer in the Army or Navy— The Young Surgeon & Attorney—The Young Curate in the Church—the junior Clerk in the Bank or Counting House, never dream of being in these circumstances which they have reason to hope for after 20 or 30 years' hard labour."

Even if all preachers had been infused with moral excellence and material restraint, the connectional system was not an easy one for which to make a budget. There were sharp differences in wealth within and between circuits. Membership did not expand and decline in direct proportion to the expenses of the preachers. Although guidelines were laid down for voluntary subscriptions, there were well-understood limits to the amount of moral coercion that could be employed on volunteers. In addition, trade cycles, economic depressions, war demobilization, and price and wage deflation could wreck even the most careful of plans.[44] Not only was the administration of the system as dependent on the voluntary principle as the contributions to be administered, but also the demands on it were everywhere increasing at an exponential rate both at home and abroad. The Conference tried to regulate the system, but there were as many opinions about what to do as there were demands on the connectional purse.[45] There was also evidence of both moral and structural decay within the Methodist organization itself as some unscrupulous districts and circuits carelessly passed on their burden of debt to Conference.

At its best connectionalism was an ecclesiastical system of mutual care and support; at its worst it was a ramshackle model of financial evasion and irresponsibility.[46] Connectionalism had the capacity of converting failures at the periphery into crises at the center and vice versa. The nub of the matter, however, was that Methodism had embarked on a roller coaster ride of national and international expansion that simply outgrew the old method of a pence-a-week class subscriptions. As expensive new chapels were built in the great, new industrial cities and as the Methodist message was taken to Africa, India, Europe, and the Caribbean, the balance of financial power in the Connection moved steadily away from class members toward men of substance, wealthy suburban chapels, and the general public. Methodism in England grew its business not only by expanding its markets and recruiting new consumers but also by expanding its support base beyond its dedicated core of members. The increase in Methodist membership was thus accompanied by a formidable mobilization of fresh resources. Between pietism and filthy lucre, it seemed there was tension but no ultimate incompatibility. By 1820, to judge from the correspondence of the preachers, Methodist finances, as with real wages in the nation as a whole, were in a better position than they had been a decade earlier.[47] Membership had increased rapidly, dozens of new chapels had been built, and foreign missions had been given a major injection of cash. The ordinary members of Methodist societies in the towns and country had played their part in all of this, but the wealthier suburban chapels in the growing towns and cities made the decisive difference.[48] And power certainly followed wealth, for the alliance between well-connected preachers and propertied laymen, for better and for worse, was set to lead the connection into the Victorian era.

Methodism in New England

The system of raising money for the support of the Methodist cause in America was broadly similar to that in Britain, but there were also some significant differences. On both sides of the Atlantic the main engines of financial support were the class and quarterly meetings, which had the added advantage of devolving financial responsibility to committed members and not just to hearers. In the early stages of Methodist expansion, wherever it took root, the chief expense was the itinerant ministry, which was vital to the whole connectional enterprise. Local preachers may have borne the responsibility for the day-to-day ministry of Methodism in its manifold peripheries, but it was the itinerancy that more than anything else forged a national movement.[49] The early Methodist itinerancy in America has been described as "a brotherhood of poverty," with all—regardless of social status, length of service, productivity, or talent—expected to share both the same salary level and the same privations of ministry in service of the gospel and the people.[50] The episodic references to money in Asbury's *Journal* make it clear that he regarded the itinerant ministry as a sacrifice undertaken for the sake of the gospel. He had the same deep-seated fear of the corrupting power of money as Wesley himself and repeatedly refused the gifts that his status in the connection attracted. In his vision of the brotherhood of preachers, fraternity and purity were dependent on and guaranteed by equality and poverty. In a joint episcopal message with Dr. Coke in 1798, Asbury stated:

> Those who read this section attentively will see the impossibility of our ministers becoming rich by the gospel, except in grace. And here there is no difference between bishops, elders, deacons, or preachers, except in their travelling expenses, and consequently in the greater labours of one than the other. . . . And we may add the impossibility of our enriching ourselves by our ministry, is another great preservation of its purity. The lovers of this world will not long continue travelling preachers.[51]

With equality as the principle and frugality as the practice, the 1796 Annual Conference fixed on an annual salary for an itinerant preacher of no more than $64 per annum, a figure that was raised to $80 in 1800 and then to $100 in 1816.[52] Such increases were regarded not as rewards for the increased social status of itinerant preachers but as the bare minimum required for bringing an already inadequate salary into line with inflation.[53] It has been pointed out that the salaries of Methodist preachers were about a fifth of those paid to Congregational ministers and were probably less than those earned by many unskilled laborers, but the salaries themselves do not tell the whole story.[54] Not only were the salaries rarely paid in full, but they were also supplemented by a complicated set of allowances to take account of expenses, dependents and enterprise. While on the road, preachers were theoretically entitled to expenses for traveling, food and fuel (after 1816), though preachers' journals and autobiographies testify to the fact that this was a haphazard process and obviously depended to some extent on the wealth of the circuits to which they were assigned.[55] Moreover, some preachers were more prepared for privation than others. A similar problem arose over allowances for dependents. As in Britain, most of the early itinerants were unmarried, a practice approved by Asbury and blessed with cheapness. Yet without a formal declaration of celibacy as a necessary precondition of acceptance for the Methodist itinerancy, it was inevitable that pressures to marry and have children were bound to increase. Surveys from different parts of the country suggest that the great majority of Methodist itinerants in the period 1780–1830 eventually married and located.[56] For example, of the bumper crop of 37 men who were received into the New England Conference in 1822, only 15 either died in the ministry or became superannuated preachers; the rest located, discontinued their ministry, or were expelled for undisclosed offenses.[57]

Although some Methodist itinerants remained celibate and others delayed marriage, provision had to be made for married preachers and their families lest all the prime talent drifted out of the itinerancy altogether. Accordingly, the General Conferences of 1796 and 1800 established a chartered fund to make provision for married, superannuated, and worn-out preachers and their wives, widows, and children. The fund was to be supplied by preachers' subscriptions; circuit surpluses; quarterly, annual, and conference collections; and the profits from book sales. Under the conditions of the fund, a sum not exceeding $16 was allowed for the children of preachers, and quite remarkably, the allowance of a married preacher was deemed to be double that of a single preacher. In addition preachers were allowed to keep their fees for conducting marriages and were paid a commission of between 15% and 25% per cent for selling books.[58] The phenomenal success of Methodist book sales, regarded by many as the jewel in the crown of the early Methodist mission, was owing, at least in part, to the entrepreneurship of an impoverished itinerant ministry.

Salesmanship and enterprise alone could not save the Methodist system from duress. In fact the propensity of itinerant preachers to marry and have children imposed the same kinds of stresses and strains on American Methodism as those that bedeviled the English Connection. The system had worked tolerably, if erratically, under Asbury's egalitarian and ascetic regime. After his death in 1816, however, growing concerns about the extent of itinerant locations, partial locations (which created even more difficult pastoral problems for circuit superintendents), and the low quality of new recruits forced the General Conference of 1816 to address the matter with some urgency. Resolutions were duly passed to increase preachers' allowances, to provide houses for them, to pay for their food and fuel, and to improve their education.[59] Ironically, therefore, the consequences of the underfunding of the itinerancy—location and low-quality recruitment—led to the implementation of policies that over time would lead to the demise of the itinerancy itself. It seemed that the Methodist system of itinerant ministry, devised by Wesley for his societies in England and further adapted by Asbury for use in America, could only continue to serve the poor as long as it was characterized by celibacy and as long as it depended mostly on hospitality rather than cash allowances. The only available alternatives were either to demand more money from Methodist societies, thereby risking expressions of anticlericalism and demoralizing allegations of corruption, or to devise structures, which would have the effect of contributing inexorably to the decline of itinerancy itself.

The resolutions of the General Conference of 1816, whatever their long-term effects on the Methodist system, did not immediately solve the problems for which they were designed. The impeccably presented accounts of the New England circuits during a remarkable period of membership growth in the 1820s illustrate the extent of the problem. Preachers were paid expenses for travel, table, board, rent, and fuel and were then due their quarterage or annual allowance, which was fixed at $100 for single preachers, $200 for married preachers, $24 for children between 7 and 14 and $16 dollars for children under 7. For the year ending June 1820 the sum required to meet the quarterage was $16,487 and the deficiency was calculated to be $11,062.[60] In short, less than half the quarterage of the itinerant preachers was actually met by the circuits they served, and the sums received from central funds, including money from the book concern, the chartered fund, and special collections, amounted to less than 10% of the deficiency.[61] The reason for such a huge deficit was the fact that by the 1820s, married preachers in New England outnumbered unmarried ones by a ratio of two to one, and the average number of children per family was around two. Successive reports in the 1820s and 1830s showed no signs of a decline in the annual deficit, despite the growth of members and despite impassioned pastoral addresses from circuit superintendents.[62] Two striking facts emerge from these data. The first is the remarkably generous allowances, in principle at least, for wives and children. Married preachers were entitled to double the quarterage of their single co-itinerants, and allowances for children were not subject to a sliding scale. Second, Methodist expansion seemed to thrive on deficit finance and the relative poverty of its preachers. Despite the propensity of married preachers to locate, the Methodists appeared not to experience unsustainable problems in recruiting for the itinerancy, though there were periodic expressions of concern about the quality of some of the recruits. Within the Methodist community, itinerant preaching was regarded as a high-status, low-pay enterprise,

sustained more by the prayers and hospitality of its constituents than by financial or material rewards.

The relative poverty of the itinerants must have made life uncomfortable
for their bodies, but it did their spiritual reputation no harm whatsoever. The
evidence from the diaries and reminiscences of the preachers themselves
suggests that their lofty conception of the itinerant ministry far outweighed
the financial hardships they undoubtedly experienced. Some complained to
circuit stewards about low provisions and others sent their wives to work or
tried to supplement their income by teaching school or exploiting the black
economy, but most had their expenses, if not their quarterage, met by the local
Methodist community. One old itinerant warhorse who labored in New
England for fifty-six years declared at the end of his ministry that there was
only one year "in which my expenses exceeded my receipts."[63] Conversely,
most itinerants frankly admitted that they were rarely paid their full "disci-
plinary estimate" and were never far from the poverty line.[64] Those who stayed
in the ministry long enough to review their lives from the standpoint of a ripe
old age looked on the privations of their early ministry as part of the great
adventure of "gospel ministry," but poverty, as with other forms of suffering,
is a far nobler thing in retrospect than in prospect.

Although there are suggestive comments both in Conference minutes and
in preachers' journals that some were more adept at securing expenses and
presents than others, in general terms the financing of the itinerancy in America
occasioned much less controversy than in England.[65] Just what chiefly explains the difference is that the American itinerants had more confidence in
the egalitarianism of the system of appointments and rewards than their counterparts in England, for which Asbury, albeit operating in a more democratic
and populist culture, deserves much of the credit. Second, the American
Methodist laity accepted more easily than their British counterparts that the
life of a Methodist preacher was one of such affliction and poverty that none
would choose it out of mere financial self-interest or personal ambition. Third,
the general absence of established churches in the United States led to the
establishment of the voluntary principle as the only viable model of church
support and the relative absence of bitter class conflict compared with early
industrial England. Fourth, there is also evidence to suggest that there were
subtle differences in attitudes to money and religion between the two cultures.
On the whole, Americans were more transparent in their acknowledgment of
money as a suitable subject of public religious discourse than their British
counterparts and were, at least within Methodism, less inclined to believe in
the malpractice of private clerical castes.[66] Indeed, Methodist expansion in
the United States was itself facilitated by the chorus of anticlerical and anti-
Calvinist sentiment, often expressed in verse, which lambasted the careerist
financial aspirations of educated clergy.[67] Nevertheless, among Methodists
there seems to be no American equivalent of the sheer volume of suspicion,
criticism, and unwillingness to contribute that infected a sizable proportion
of the Methodist community in England in the period 1780–1830.

In the longer term, however, the Methodist itinerancy in the United States, as in Britain, could not escape the consequences of its own evangelistic success nor fail to benefit from the upward social mobility of its support base. A few snapshots from anniversary conventions in New England indicate clearly enough the changes that took place in the culture and financing of Methodist preaching over the course of the nineteenth century. The first is taken from the Methodist Centenary Convention, held in Boston in the somber aftermath of the Civil War. The Convention was essentially a celebration of Methodist expansion in the New England states from a membership of 5,829 in 1800 (1 in 211 of the population) to 103,961 (1 in 31) on the eve of the Civil War.[68] It was also a celebration of increased financial muscle and enhanced social status. Pride and pleasure blended with nostalgia and millennial optimism in speeches highlighting the beneficial symbiosis of Methodism and American values.[69] Although there were ritual appeals to primitive simplicity and renewed zeal, the overwhelming endorsement of the convention was for a more educated, more cultivated, and more influential ministry. "The wealth of the Church, like water," suggested one delegate, "ought to be poured out, so that the intellectual culture shall be found even in the humblest ones in the service of God."[70] In the wealth of statistical information presented to the convention on the state of Methodism in antebellum New England, three features are worthy of comment. First, Methodism's circuit system had all but collapsed;[71] second, "pastors" in wealthy states were being paid almost twice as much as those in poorer states; and third, the capacity of the church to generate wealth from its members had grown dramatically in the middle decades of the nineteenth century.[72] Despite greater material affluence, or more likely because of it, Methodism in the eastern states on the eve of the Civil War had virtually abandoned the old ideal of an itinerant and egalitarian ministry supported by local hospitality.

In yet another statistical survey of New England Methodism, this time carried out on the eve of the World War I, it was reported to the convention that the salaries of Methodist preachers were still too low for the role required of them. "The minister of today," the convention was told,

> must be a community leader, broad minded, progressive, and aggressive. The age insists that he be a good preacher, an efficient pastor, a true leader of men. For these lofty requirements the minister must be well equipped. He must dress acceptably, for he is a leader; he must attend many assemblies, for service and the deepening of his spiritual life; he must travel for vision and culture; he must read extensively for intellectual impetus and resourcefulness. The high requirements on the part of the public make necessary heavy expenditures for adequate equipment.[73]

Traveling for vision and culture is not exactly how Asbury's preachers conceived of their itinerant ministry a century before. A conception of ministry based more on the cultivated expectations of settled congregations than on missionary expansion had triumphed within American Methodism. It was not that the history of the early itinerancy was repudiated by later generations;

rather it was simply confined to the realm of nostalgic reminiscences and wrapped in condescension.[74] Early itinerants were regarded as those not endowed with much learning who nevertheless made good their deficiency by speaking with "tongues of fire."[75] In this way the early privations of Methodist itinerants were confined to a different dispensation and then sacralized by memory.

Their own memories tell a different story, however. Beginning in the 1880s the New England Methodist Historical Society supplied a forum for aging Methodist preachers to recount their experiences of the "heroic age" of Methodism. What emerges from these revealing, often moving, sometimes bizarre, and always romanticized accounts is an entirely different conception of ministry from that of the later nineteenth century.[76] Even allowing for the notorious capacity of aging Methodist "croakers" to favor their own pioneering efforts over those of their more "worldly" successors, the contrast between the qualities admired by successive generations of nineteenth-century Methodist preachers remains striking.[77] The attributes admired most by the older preachers included a robust constitution; a capacity for endurance; a commanding personal appearance; a voice sufficiently powerful for preaching at camp meetings; an ability to "affect" an audience; a capacity for weeping; an ability to produce religious revivals and spiritual fruit; an earnest and candid disposition; plain-speaking; a good-hearted, generous, and sociable personality; evangelical unction; a passion for reforming causes (temperance, teetotalism, and antislavery); practical and scriptural preaching; frugality and exactness; demonstrated experience of special providential guidance; a fixed gaze on doing good in the here and now with the hope of eternal rewards in the hereafter; a meek, gentle, and childlike spirit; a melodious voice for singing; a keen wit; a capacity for explosive shouts of praise and enthusiasm; a desire for learning and education; a carelessness about material comforts; an ability to confound the rich, the learned, and the powerful; a willingness to celebrate the contribution of "mothers in Israel"; and above all, the experience of a good death (peaceful, thankful, faithful, and anticipatory). In contrast, the qualities most admired by the end of the century were learning and self-improvement, the ability to hold a large and intelligent congregation, a preaching ministry informed by careful preparation, facility of expression, clarity of thought and freshness of style, a status that could not be disparaged by other denominations; a capacity to keep pace with the best thought of the age, a reputation for philanthropy, and evidence of educational and intellectual attainment. Over time, a change in job specification, and the personal attributes required to fulfill it also led to a change in status and rewards.

Although the funding of the itinerant system of ministry was the chief cause of financial disquiet within transatlantic Methodism in the period 1780–1830, there were, of course, other demands on the connectional purse. During the same period, for example, even larger sums of money were raised to build chapels. The financing of chapel building was not free from controversy, as the heated conference debates on pew renting made clear,[78] but the problems were less severe and more confinable.[79] Although the connectional system facilitated some transfer of funds for chapel building from wealthier to poorer areas,[80] in general terms the sums raised to build chapels were given by the laity in a particular locality for the benefit of the laity in that locality.[81] Although the raising of such large sums of money inevitably delivered more power into the hands of those who volunteered the largest subscriptions, those who raised the money benefited directly from their investment. By contrast, raising money for the itinerancy, and also for the international missionary movement, forced into the open big questions about the nature of ministry, the relationship between committed members and the general public, and the conflict between God and mammon in the expansion of the Methodist movement. Buildings were tangible assets; preachers and missionaries were speculative investments.

Tangible assets nevertheless became ever more tangible over the course of the nineteenth century. Wherever in the world Methodism took root, a powerful combination of perceived ecclesiastical inferiority and a surge of confessional pride resulted in the building of enormous neoclassical and gothic "cathedrals." From the cities of northern England to those on the eastern seaboard of the United States, Methodists built big, not only to accommodate burgeoning numbers, but also to demonstrate that they sought an ecclesiastical rapprochement with the values of order, stability and respectability. Precisely the same pattern has been detected by students of second- and third-generation Canadian and Australian Methodism.[82] From Leeds to Boston and from Toronto to Adelaide, huge sums of money were raised to invest in buildings that declared the social arrival of the once despised. Indeed, it was Methodism's genius that throughout the English-speaking world it was able to act for so long both as a countercultural movement of populist revivalism and as an enforcer of social stability and sobriety, though not always in the same place at the same time. It was Methodism's misfortune, at least in terms of its capacity to recruit, that it could not oscillate between these poles forever.

Conclusion

As every undergraduate knows, John Wesley had a particular genius for organization. The system he put in place for funding the evangelistic expansion of the Methodist movement has been regarded as stunningly successful. Its essence was that the ministry of traveling preachers was made dependent on the hospitality and regular subscriptions of a grateful laity. Whereas other churches in the eighteenth century depended for their resources either on ecclesiastical taxation or autonomous congregational giving, the Methodist pattern combined voluntarism with connectionalism. Moreover, itinerancy militated against the comfortable conformity so much dreaded by Wesley in England and Asbury in America. The regular giving of the laity in small amounts in class or quarterly meetings encouraged them to feel that they had a stake in the movement and helped keep the controlling power of money away from the rich and the powerful. The fact that money was mostly raised and spent within the same geographical locality further aided transparency and accountability. Meticulous methods of accounting trained the laity in prudence, and careful delineation of expenses trained the preachers to expect nothing in this life but the bare necessities. If never fully egalitarian in practice, the system had an egalitarian logic, based on the premise that money was merely the servant of mission and not a commodity to be desired for the benefit of the individual.

For the system to work well, however, three conditions had to be established. First, supply and demand needed to be kept in some sort of rough harmony. Second, there was a need for trust, both among preachers and between preachers and people. Third, it was vital that money raised by voluntary donations in particular locations was spent on the agreed religious objectives of those making the contributions. In practice it proved impossible to meet these conditions either in the British Isles or in North America in Methodism's great age of expansion. In both locations, the desire of preachers to marry and raise families imposed dramatic new demands on resources, as did the building of new chapels and the financing of world missions. That much was shared by the Methodist traditions on both sides of the Atlantic, but there were also important dissimilarities. In America, a combination of the self-sacrificial leadership of Asbury and the tendency of married itinerants, without too much reproach or bitterness, to locate and become local preachers promoted greater connectional harmony than in Britain. There, allegations of financial corruption and abuse of power often accompanied the complaints of the laity against the leading preachers after Wesley's death. Class conflict was therefore generated from both outside and inside the Methodist community in Britain in the early industrial revolution.[83] American Methodism was not entirely free from such tensions,[84] but it was to be another form of property, human slavery, that created the real trouble for Asbury's successors. John Wesley, as his most recent biographers have made clear,[85] was a flawed leader who left many unresolved tensions for his followers to cope with, but his strong warnings about the evils of money and slavery were both prophetic and largely ignored by many of his followers. Unsurprisingly, his economic triptych of gaining, saving, and giving all that one could proved to be incapable of mass realization.[86] The ultimately contradictory principles of serious acquisitiveness and careless generosity were played out in Methodist minds and societies wherever the movement took root.[87]

The history of Methodism in a transatlantic context reveals at least four different stages in the financing of a popular religious movement. The first, an attempt to replicate the early church's experiments with the community of goods, collapsed just as completely in early English Methodism as it appears to have done in early Christianity. The second, and under eighteenth-century conditions an equally radical approach, was to combine voluntarism and egalitarianism in the financing of an itinerant ministry by a grateful and supportive laity. The egalitarian component of that approach survived longer in the United States than in Britain, but it ultimately disappeared in both places over

time. A third approach, which was in nature broadly paternalistic and bureaucratic, was to exploit more structured forms of raising revenue in which richer
donors came to have more power and influence than the poorer ones. A fourth
stage, namely, the virtual collapse of Methodist finances as a result of a declining membership and an unresponsive wider public has been the unfortunate fate of many Methodist congregations in modern Britain. Even allowing
for the reductionism implicit in such a typology, it is clear that each successive phase of fund-raising involved not only a different approach to financing Methodist operations but also a very different social experience of religion for those giving and receiving the money. In religious organizations,
money is not simply a necessary and neutral commodity for getting things
done but rather carries with it a symbolic revelation of the values for which it
was collected.

Notes

1. I am grateful to John Walsh, Mark Noll, Richard Carwardine, and John Wigger
for their comments on an earlier draft of this chapter. Although a great deal has been
written about the political, economic, and social factors that facilitated the growth of
Methodism in the British Isles and North America in this period, almost nothing of
substance has been written about how that expansion was financed and organized.
One reason for that is that the financing of a connectional system is not as easily
understood as that of a conventional established or dissenting church, and another is
the lack of available sources beyond the rather drab columns of figures reproduced in
Conference minutes and circuit account books. For some of the more influential gen-
eral interpretations of this remarkable period of Methodist growth, see Edward P.
Thompson, *The Making of the English Working Class* (Harmondsworth, Eng.:
Penguin, 1968); William R. Ward, *Religion and Society in England 1790–1850* (Lon-
don: Batsford, 1972); Alan D. Gilbert, *Religion and Society in Industrial England:
Church, Chapel and Social Change, 1740–1914* (London: Longman, 1976); Richard
Carwardine, *Trans-Atlantic Revivalism: Popular Evangelicalism in Britain and
America, 1790–1865* (Westport, Conn.: Greenwood, 1978); Nathan O. Hatch, *The
Democratization of American Christianity* (New Haven, Conn.: Yale University Press,
1989); Michael R. Watts, *The Dissenters, Vol. 2: The Expansion of Evangelical Non-
conformity* (Oxford: Clarendon, 1995); David Hempton, *The Religion of the People:
Methodism and Popular Religion c. 1750–1900* (London: Routledge, 1996); and, most
recently, John H. Wigger, *Taking Heaven by Storm: Methodism and the Rise of Popu-
lar Christianity in America* (New York: Oxford University Press, 1998).
    2. Robert Currie, Alan Gilbert, and Lee Horsley, *Churches and Churchgoers:
Patterns of Church Growth in the British Isles since 1700* (Oxford: Clarendon, 1970).
    3. Wigger, *Taking Heaven by Storm*, 3–7; John Wigger, "Taking Heaven by
Storm: Enthusiasm and Early American Methodism, 1770–1820," *Journal of the Early
Republic* 14 (Summer 1994): 167–94.
    4. Bernard Semmel, *The Methodist Revolution* (London: Heinemann, 1974), 152–
66. See also Frederic Stuart Piggin, "Halevy Revisited: The Origins of the Wesleyan
Methodist Missionary Society: An Examination of Semmel's Thesis," *The Journal
of Imperial and Commonwealth History* 9 (1980): 17–37; Roger H. Martin, "Missionary Competition Between Evangelical Dissenters and Wesleyan Methodists in the

Early Nineteenth Century: A Footnote to the Founding of the Methodist Missionary Society," *Proceedings of the Wesley Historical Society* 42 (1979): 81–86; Roger H. Martin, *Evangelicals United: Ecumenical Stirrings in Pre-Victorian Britain, 1795–1830* (Metuchen, N.J.: Scarecrow Press, 1983); and Wade Crawford Barclay, *History of Methodist Missions*, vol. I (New York: Board of Missions and Church Extension of the Methodist Church, 1949).

5. John Walsh, "John Wesley and the Community of Goods," in *Protestant Evangelicalism: Britain, Ireland, Germany and America, c. 1750–1950, Studies in Church History*, Subsidia 7 (Oxford: Blackwell, 1990). Walsh, "Methodism and the Common People," in *People's History and Socialist Theory*, ed. Raphael Samuel (London: Routledge, 1981), 354–62. For a wider discussion of these issues see Hempton, *Religion of the People*, 85–87.

6. See, for example, *The Works of John Wesley*, ed. Frank Baker, *Letters II 1740–1755* (Oxford: Clarendon Press, 1982), 26:544–45.

7. For a splendid example, which can stand for the rest, see the meticulous account books relating to the first Methodist society and church (Methodist Ally Church) in Boston. Boston University, School of Theology, Methodist Mss (hereafter BU STh Mss), Stewards' books 1792–95 and 1795–97, which contain daily entries of income (male and female class, quarterly, and Sunday collections) and expenditure (preachers' room and board, quarterage, and candles).

8. Allegations of financial malpractice are almost as frequent as those dealing with sexual impropriety. See Oliver A. Beckerlegge, "The Lavington Correspondence," *Proceedings of the Wesley Historical Society* 42 (1980): 101–11, 139–49, 167–80.

9. David Hempton, *Methodism and Politics in British Society 1750–1850* (London: Hutchinson, 1984), 67–73.

10. Alexander Kilham, *The Progress of Liberty, Amongst the People called Methodists* (Alnwick: n.p., 1795); *A Candid Examination of the London Methodistical Bull* (Alnwick: n.p., 1796); *A Short Account of the Trial of Alexander Kilham, Methodist Preacher, at a Special District Meeting Held at Nexcastle* (Alnwick: J. Catnach, 1796); *An Appeal to the Methodists of the Alnwick Circuit* (Alnwick: n.p., 1796); *An Account of the Trial of Alexander Kilham, Methodist Preacher, before the General Conference in London* (Nottingham: n.p., 1796). For the official defense against these allegations see Jonathan Crowther, *Christian Order: or Liberty without Tyranny and Every Man in his Proper Place* (Bristol: R. Edwards, 1796); Thomas Hanby, *An Explanation of Mr. Kilham's Statement of the Preachers' Allowance* (Nottingham: n.p., 1796).

11. The John Rylands University Library of Manchester, Methodist Archives Research Centre, Jabez Bunting manuscripts (hereafter MARC Bunting Mss), Miles Martindale to Jabez Bunting, July 9, 1816.

12. For the clearest delineation of the issues involved in funding preachers and their dependents, see MARC, Bunting Mss, Robert Miller to Jabez Bunting, April 1, 1819.

13. MARC, Bunting Mss, Miles Martindale to Jabez Bunting, July 9, 1816.

14. Ibid.

15. Ward, *Religion and Society*, 97.

16. MARC, Bunting Mss, Thomas Allan to Jabez Bunting, July 28, 1810.

17. MARC, Bunting Mss, Matthew Lamb to Jabez Bunting, June 13, 1812.

18. MARC, Bunting Mss, Joseph Entwistle to Jabez Bunting, December 15, 1814; Edward Hare to Jabez Bunting, January 9, 1815.

19. MARC, Bunting Mss, John Gaulter to Jabez Bunting, April 6, 1809; Joseph

Entwistle to Jabez Bunting, December 18, 1812; Joseph Butterworth to Jabez Bunting, November 9, 1814.

20. Deciding on the appropriate number of preachers to meet the needs of a circuit caused much debate. The issue was that poorer and more distant circuits needed extra help but could not raise sufficient money to pay for it. These internal transfers of resources from wealthier to poorer circuits could, depending on circumstances, produce both magnanimity and great resentment. See MARC, Bunting Mss, Theophilus Lessey Sr. to Jabez Bunting, June 7, 1811.

21. This was a particularly thorny issue. Since itinerant preachers were the chief beneficiaries of class and society collections, it was difficult for them publicly to criticize either the level or the administration of the giving. See MARC, Bunting Mss, E. Hare to Jabez Bunting, November 12, 1810; John Barber to Jabez Bunting, April 24, 1812.

22. MARC, Bunting Mss, Thomas Coke to Jabez Bunting, April 21, 1812.

23. It was difficult for connectional leaders at the time, and difficult for historians since, to estimate with any degree of accuracy the average cost of maintaining single and married itinerant preachers. When asked by the Property Commissioners in 1810 for an official estimate, Bunting replied that "as the ministers in that Connexion have no regular salaries, but are maintained by different small allowances made to them by their Societies, which vary considerably according to circumstances, there is some difficulty in determining our proper annual income. But I am informed that some years ago, when the whole case was laid before the Commissioners, & left to their decision, they were pleased to fix on the sum of Sixty Pounds per annum as the sum for which the Methodist ministers of this town should be chargeable to the duty of professions." A decade later, some careful calculations by Robert Miller estimated the cost at £103 per annum. To what extent this represents a real increase in the early decades of the nineteenth century or is merely the result of different calculations for different audiences is difficult to determine with any precision, but the sheer volume of correspondence that complained of increased allowances for preachers indicates that something was happening. See MARC, Bunting Mss, Robert Miller to Jabez Bunting, April 1, 1819.

24. MARC, Bunting Mss, E. Hare to Jabez Bunting, November 12, 1812; W. Midgley to Jabez Bunting, May 22, 1812; Walter Griffith to Jabez Bunting, February 4, 1814.

25. MARC, Bunting Mss, William Williams to Jabez Bunting, February 17, 1809.

26. United Methodist Church Archives, Lake Junaluska, Jabez Bunting to James Wood, late August or early September 1811. See also MARC, Bunting Mss, Thomas Hutton to Jabez Bunting, March 17, 1815; W. Worth to Jabez Bunting, January 23, 1818.

27. MARC, Bunting Mss, Jabez Bunting to Joseph Dutton, December 27, 1813.

28. Ibid.

29. MARC, Bunting Mss, Edward Hare to Jabez Bunting, May 30, 1816.

30. Complaints were legion about the administrative frailties of Methodist giving and fund-raising. See MARC, Bunting Mss, Edward Hare to Jabez Bunting, November 12, 1810; Thomas Hutton to Jabez Bunting, February 4, 1811; John Barber to Jabez Bunting, April 24, 1812; Joseph Entwistle to Jabez Bunting, December, 18, 1812; Joseph Cusworth to Jabez Bunting, April 14, 1815; George Morley to Jabez Bunting, October 28, 1815; Edward Hare to Jabez Bunting, May 30, 1816; Jonathan Roberts to Jabez Bunting, July 17, 1818.

31. W. R. Ward, ed., *The Early Correspondence of Jabez Bunting 1820–1829,*

Camden 4th Series (London: Royal Historical Society, 1972). See letter from John Mercer to Jabez Bunting, March 14, 1820, detailing the privations of laboring in the debt-ridden societies of the Isle of Man (from MARC, Bunting Mss).

32. MARC, Bunting Mss, Joseph Cusworth to Jabez Bunting, April 14, 1815.

33. Edward Royle, *Queen Street Chapel and Mission Huddersfield* (Huddersfield: Local History Society, 1994). This huge chapel built in 1819 to accommodate some 2,000 people, cost about £10,000. Similar-sized chapels were constructed in other northern towns in this period.

34. MARC, Bunting Mss, R. Smith to Jabez Bunting, March 30, 1809; Jabez Bunting to James Wood, January 7, 1815; Theophilus Lessey Sr. to Jabez Bunting, May 6, 1815.

35. MARC, Bunting Mss, Jabez Bunting to Isaac Clayton, July 14, 1815. Bunting stated, "I do not think it desirable that any *large or considerable* portion of the supplies necessary for the maintenance of our ministry should come from the chapels. This, I know, is the dissenting plan. But the longer I live, the less I am disposed to assimilate our plan to theirs. *Our* plan is to support our ministry by the voluntary subscriptions of our *societies*. This I think is almost infinitely preferable."

36. MARC, Bunting Mss, Thomas Allan to Jabez Bunting, July 28, 1810.

37. MARC, Bunting Mss, Joseph Entwistle to Jabez Bunting, December 15, 1814.

38. MARC, Bunting Mss, Jabez Bunting to Samuel Taylor, March 10, 1814

39. MARC, Bunting Mss, Jabez Bunting to T. S. Swale, November 19, 1814.

40. The key years in this transformation of missionary funding were 1814 and 1815. For accounts of what was going on up and down the country, see MARC, Bunting Mss, Walter Griffith to Jabez Bunting, February 4, 1814; John Furness to Jabez Bunting, February 28, 1814; Jabez Bunting to Samuel Taylor, March 10, 1814; Jabez Bunting to James Everett, March 29, 1814; George Morley to Jabez Bunting, September 22, 1814; Joseph Butterworth to Jabez Bunting, September 27, 1814; John Beaumont to Jabez Bunting, October 18, 1814; Thomas Kelk to Jabez Bunting, October 25, 1814; George Morley to Jabez Bunting, October 29, 1814; Jabez Bunting to T. S. Swale, November 19, 1814; Thomas Kelk to Jabez Bunting, December 14, 1814; Joseph Entwistle to Jabez Bunting, December 15, 1814; George Morley to Jabez Bunting, December 30, 1814; Jabez Bunting to James Wood, January 7, 1815; Edward Hare to Jabez Bunting, January 9, 1815; Richard Watson to Jabez Bunting, January 17, 1815; Robert Newton to Jabez Bunting, January 18, 1815; Jabez Bunting to George Morley, January 25, 1815; Thomas Hutton to Jabez Bunting, March 17, 1815; J. Shipman to Jabez Bunting, April 22, 1815; John Barber to Jabez Bunting, April 27, 1815; Jabez Bunting to George Morley, June 24, 1815; Walter Griffith to Jabez Bunting, September 21, 1815.

41. MARC, Bunting Mss, Jabez Bunting to George Morley, January 25, 1815.

42. MARC, Bunting Mss, Robert Miller to Jabez Bunting, April 1, 1819. The precise figures were 501 families and 154 single men.

43. Ibid.

44. Whatever the alleged contribution of industrial conflict, economic destitution, political excitement, and social instability to the success of Methodist revivalism, they all had disastrous consequences for Methodist finances. See MARC, Bunting Mss, Theophilus Lessey Sr. to Jabez Bunting, January 17, 1811; Thomas Hutton to Jabez Bunting, February 4, 1811; Joseph Butterworth to Jabez Bunting, May 23, 1812, George Morley to Jabez Bunting, October 29, 1814; Richard Waddy to Jabez Bunting, January 31, 1816.

45. One of the most radical solutions, proposed partly tongue in cheek, was "to raise money not by begging but by *borrowing*." See MARC, Bunting Mss, Jonathan Roberts to Jabez Bunting, July 17, 1818. The proposal was based on the assumption that Methodist "begging" was beginning to obey a law of diminishing returns and that some form of graduated loan system would deliver enough in interest payments to keep the Connection ticking over.

46. Ward, *Religion and Society*, 101.

47. Roderick Floud and Donald McLoskey eds., *The Economic History of Britain since 1700, Vol. 1: 1700–1860* (Cambridge: Cambridge University Press, 1994), 368–72. Their estimates suggest that real wages were probably stagnant between 1750 and 1815 and then improved substantially between 1815 and 1850.

48. Ward, *Early Correspondence of Jabez Bunting*, 4–8.

49. Wigger, *Taking Heaven by Storm*, 49.

50. William H. Williams, *The Garden of American Methodism: The Delmarva Peninsula, 1769–1820* (Wilmington, Del.: Scholarly Resources, 1984), 127–28.

51. David Sherman, *History of the Revisions of the Discipline of the Methodist Episcopal Church* (New York: Nelson & Phillips, 1874), 363.

52. Michael G. Nickerson, "Historical Relationships of Itinerancy and Salary," *Methodist History* 21 (1982): 43–59.

53. *Journals of the General Conference of the Methodist Episcopal Church, Vol. 1, 1796–1836*, 35. The majority voting for an increase in 1800 was only five.

54. Wigger, *Taking Heaven by Storm*, 49.

55. New England United Methodist Historical Society, Manuscript History Collection (hereafter NEUMHS Mss), John Edwards Risley, Autobiography, Providence R.I., April 1883.

56. Ibid., 61–67.

57. NEUMHS Mss, Samuel Kelley (1802–83), semicentennial discourse delivered in the Methodist chapel at Quincy Point, Mass., March 17, 1872, 59 pps. Kelley gives a short account of the fate of all thirty-seven preachers received into conference with him in 1822. For an introduction to the early history of Methodism in New England, see Abel Stevens, *Memorials of the Early Progress of Methodism in the Eastern States* (Boston: C. H. Peirce, 1852); and George C. Baker, *An Introduction to the History of Early New England Methodism, 1789–1839* (Durham, N.C.: Duke University Press, 1941).

58. *Journals of the General Conference*, 20–22, 41–42.

59. Ibid. (1816), 148–49.

60. *An Account Shewing the amount of Collections and Disbursements in the New England Conference for the Year Ending June 20, 1821* (Boston: n.p., 1821).

61. A sum of $893 was received from central funds, made up of contributions from the book concern, the chartered fund, the widow's mite society, and conference collections. The modesty of this sum indicates the extent of the decentralization of Methodist fund-raising.

62. See, for example, *An Account Shewing the Amount of Collections and Disbursements in the New England Conference for the Year Ending June 7, 1825* (Boston: n.p., 1825), and *Minutes of the New England Annual Conference of the Methodist Episcopal Church for the Year 1835* (Boston: n.p., 1835).

63. NEUMHS Mss, William Gordon (1810–95), autobiography (copied and read before the society) and "Historical Incidents; Recollections of Methodism During a Half-century." Gordon's reminiscences are remarkably frank and unpretentious, especially concerning the subject of money.

64. NEUMHS Mss, Samuel Kelley (1802–83), "Semi-centennial Discourse." Kelley clearly found it hard to make ends meet in the first 15 years of his ministry. According to his own account, what sustained him, both in terms of money and morale, were odd jobs, hospitality in kind, and regular divisions of preachers' modest surpluses on the strict principle of equality.

65. *Journals of the General Conference*, 34.

66. This point may be more illustrative of the different chronologies of bureaucratic maturity in Britain and the United States than a real difference of substance between the two movements.

67. Hatch, *Democratization of American Christianity*, app.

68. *A Phonographic Report of the Debates and Addresses Together with the Essays and Resolutions of the New England Methodist Centenary Convention Held in Boston, June 5–7, 1866* (Boston: n.p., 1866), 172–74.

69. Ibid., 231–39. The closing address by Bishop Simpson focused on the prophetic significance of the year 1866 as the date of the American Methodist Centenary, the collapse of papal power in Italy, and the demise of "Mohammedan power."

70. Ibid., 85.

71. Ibid., 31. One delegate opined that the demise of the circuit system diminished the "social element" of Methodism, which was at the very heart of its evangelistic success.

72. Ibid., 171.

73. E. C. E. Dorion, *New England Methodism: The Story of the New England Convention of Methodist Men held in Tremont Temple, Boston, Mass., November 11–13, 1914* (New York: Methodist Book Concern, 1915), 95–96. See also G. A. Crawford, ed., *The Centennial of New England Methodism* (Boston: Crawford, 1891), 16–17, in which are presented eulogies of early Methodist itinerants who were not great in learning but who had "the tongue of fire."

74. Crawford, *Centennial of New England Methodism*, 16–17.

75. There is a rich collection of manuscript accounts and reminiscences of early Methodism in NEUMHS, Mss.

76. It must be borne in mind that the reminiscences on which this section is based are from old and relatively successful itinerant preachers eager to hold the attention of a later generation. Picaresque tales of heroic adventures mingle with more sober accounts of Methodist trials and tribulations in this literature. The accounts most pertinent to the subject of money are those by William Robert Clark, Daniel Dorchester (recounting the life of Moses Hill), William Gordon, Samuel Kelley, Daniel Clark Knowles (memoir of Nathaniel G. Ladd), Thomas Marcy, and John Edwards Risley.

77. For published versions of the kind of attitudes found among old New England itinerants in the 1880s, see Henry Boehm, *Reminiscences, Historical and Biographical of Sixty-four Years in the Ministry* (New York: Carlton & Porter, 1866); and James Jenkins, *Experience, Labours and Sufferings of the Rev. James Jenkins of the South Carolina Conference* (Columbia, S.C., n.p., 1842). I am grateful to Dr. John Wigger for these references and for his comments on an earlier version of this chapter.

78. *Journal of General Conference* (1820), 210–11. Resolution four stated "that the practice of building houses with pews, and the renting or selling said pews, is contrary to our economy; and that it be the duty of the several annual conferences to use their influence to prevent such houses from being built in future, and as far as possible to make those houses free which have already been built with pews."

79. For an insightful record of how pew rents were in fact operational in Methodist churches, see the pew rental records of the first Methodist church in Lynn, Mass., in BU STh Mss.

80. BU STh Mss, the church account book for the Methodist Ally Church in Boston, reveals that about half the money raised for the church building came from donations made to Jesse Lee in the Southern states. The rest came from subscriptions, collections, and borrowing. See also, Jesse Lee, *A Short History of the Methodists in the United States of America* (Baltimore: n.p., 1810), 164–65.

81. See Mark Smith, *Religion in Industrial Society: Oldham and Saddleworth 1740–1865* (Oxford: Clarendon Press, 1994).

82. For insightful discussions of these processes at work in Canada and Australia see William Westfall, *The Protestant Culture of Nineteenth-Century Ontario* (Kingston and Montreal: McGill-Queen's University Press, 1989); and Arnold D. Hunt, *This Side of Heaven: A History of Methodism in South Australia* (Adelaide: Lutheran Publishing House, 1985). See also Mark Noll, "Christianity in Canada: Good Books at Last," a review in *Fides et Historia* 23 (Summer 1991): 80–104.

83. Hempton, *Religion of the People*, 162–78.

84. See, for example, William R. Sutton, *Journeymen for Jesus: Evangelical Artisans Confront Capitalism in Jacksonian Baltimore* (University Park: Pennsylvania State University Press, 1998).

85. See Richard P. Heitzenrater, *The Elusive Mr. Wesley*, 2 vols. (Nashville, Tenn.: Abingdon, 1984); and Henry D. Rack, *John Wesley and the Rise of Methodism* (London: Epworth, 1989).

86. John Wesley, "The Use of Money" in *The Works of John Wesley*, ed. Albert C. Outler, Sermons II, 34–70 (Nashville, Tenn.: Abingdon, 1985), 2:263–80. See also his sermons "On Riches," "The Danger of Increasing Riches," "On Friendship with the World," and "On Worldly Folly."

87. Only more detailed research, society by society and occupation by occupation, would allow us to determine if Methodists became more or less generous with their resources (as a percentage of their family incomes) as their material stock in the world improved over time.

7

Benevolent Capital: Financing Evangelical Book Publishing in Early Nineteenth-century America

David Paul Nord

By the late 1820s, the American Bible Society (ABS) had become a major force in religious evangelism, in book publishing, and in the national organization of philanthropy in America. So productive was its modern printing plant in New York and so extensive was its national network of auxiliary societies that in 1829 the society launched its first "general supply," an audacious plan to place a Bible into the home of every family in the United States. For the managers of the society and its supporters, these were exhilarating, millennial times. But not everyone, not even all the members of the American Bible Society, were pleased. In 1830, a harshly critical pamphlet appeared, written by a disgruntled ABS member who believed that the society's charitable mission had been betrayed by a cabal of self-aggrandizing managers whose true goals were not evangelism but wealth, privilege, and power.[1]

The pamphlet, entitled *An Exposé of the Rise and Proceedings of the American Bible Society*, traced the society's fall from grace to a particular moment in 1819, when the managers had made a fateful decision: They had resolved to press their auxiliaries to sell Bibles, not simply to give them away. By this act, according to the writer of *Exposé*, the most prominent evangelical institution in America was transformed from a charity, whose mission was the free circulation of Scripture, into a commercial business, whose goal was profit. The writer scoffed at the society's continued use of a Bible passage long embraced by the Bible movement: the "waters of eternal life, without money and without price." After 1819, he declared, the reverse was true:

The Managers of the Parent Institution, in their Fourth Report [1820], mention the fact of their having sent a Circular to their Auxiliaries, urging them to use their influence and endeavours to *sell*, not to *give*, the "bread of life." This they have iterated and reiterated, year after year, with a pertinacity worthy of a better cause. The effect of this has been to make almost every pulpit in our churches a stall for the sale of their books, or clerical bookstores of temples of worship; and our pious young men have become travelling pedlers and hawkers, forcing their entrance into families which they had never before seen, and urging them to buy, at reduced prices, the books issued by this "National Institution." They district cities, towns, and villages, and scour them, either singly or in squads, seeking purchasers with money, not the indigent without it.[2]

Exposé focused on one event—the move to retail sales—and this was indeed a critical moment in the evolution of the ABS and of religious publishing more generally. But it was only one of a series of moments that transformed religious publishing societies from simple book-giving charities in the 1790s and 1800s into large-scale publishers, book manufacturers, and national booksellers by the 1830s and 1840s. This transformation was not simple or uniform across all societies, but some common patterns of change can be discerned. I see five critical moments:

1. The adoption in the 1780s of formal incorporation for publishing societies
2. The move in the 1810s to wholesaling books and tracts
3. The turn to manufacturing in the 1810s
4. The expansion of selling in the 1820s to include retail as well as wholesale
5. The creation of centrally administered national distribution systems in the 1840s

Each of these moments changed the character of the enterprise, in ways that this chapter will explore. To critics, these changes signaled betrayal of the spirit of charity and embrace of the spirit of commerce and capital. In one sense, this criticism was accurate. It is true that behind each change lay the desire for money, for the accumulation of capital, and for the market power that accumulated capital bestowed. But did this mean the end of charity, as the critics charged? Not in the minds of the managers of the publication societies. Through a half century of relentless change in the American economy and in their own work, religious publishers believed that they remained true to their ultimate goal: the universal circulation of the word. Universal, however, did not necessarily mean free, as they gradually came to understand. To give books freely to the indigent and indifferent, they needed revenues gained from sales to those who could and would pay. To manufacture books cheaply, they needed modern technology and centralized printing facilities. To deliver books widely, they needed networks of auxiliary societies and paid agents. And to do all of these things efficiently and on a grand scale, they needed capital. In many ways, then, the Jacksonian author of *Exposé* was right: "Wealth and power is their grand aim."[3] But wealth and power for what end? The salvation of souls, the publishers replied.

By 1830, the managers of the American Bible Society had become marketeers, capitalists, even would-be monopolists, just as the writer of *Exposé* had charged. But their goal was not profit, and their strategy not market-driven. They imagined their work to be in the market but not of it, for they believed that ultimately their business was not books at all; it was benevolence. It was the universal circulation of the word. Though the vessels that carried the word were the economic works of man, the word itself must be abundant and free, "without money and without price," as they liked to say. In their effort to wed the methods of commerce and capitalism to noncommercial goals, the ABS was typical of many not-for-profit religious publishing societies that sprang up in the early nineteenth century. They all struggled to work out a practical amalgamation of economics and religion, of scarcity and abundance, of business and benevolence. That struggle is the story of this chapter.

Incorporation

From the beginning, American Protestants, especially New Englanders, had been people of the word, believing with St. Paul that "faith cometh by hearing and hearing by the word of God." Only after Independence, however, did this belief flower into a systematic organizational commitment to evangelism by print. The first society after the American Revolution to move seriously into the distribution of religious publications was the Society for Propagating the Gospel Among the Indians and Others in North-America (SPGNA), founded in Boston in 1787. Its name notwithstanding, the SPGNA in its early years was interested less in Indians than in the poor white folk on the New England frontier. Though it supported missionaries and missionary schools, the society was conducted by men who believed in the power of print, and the distribution of books quickly became central to its mission. By 1795, the SPGNA was spending about $800 dollars per year; by 1798, it had distributed a total of 8,945 volumes. This was the modest beginning of the religious publication society in the new United States of America.[4]

Founded as an eleemosynary corporation, the SPGNA was chartered by the new state of Massachusetts. Almsgiving, including organized almsgiving, had always been part of the culture of Christianity, and the corporate form was not unknown in colonial America. In Massachusetts especially, the General Court from time to time had drawn on English models to charter corporations to manage towns, schools, parishes, and other public institutions. The SPGNA was modeled on the Society for the Propagation of the Gospel in Foreign Parts, chartered in England in 1701. After the Revolution, the formal, chartered corporation flowered lavishly in the newly independent United States, again especially in New England. In the nineteenth century, America would lead the world in adapting the corporation to private business enterprise. But in the early years of the Republic, state legislatures chartered corporations mainly for governmental, charitable, and other not-for-profit purposes.[5] A common purpose was religious evangelism. In the thirty years after

1787, New Englanders established at least 933 Bible, tract, missionary, and Sunday school societies, many of them holding corporate charters.[6] The first of these was the SPGNA.

The chief virtue of the corporate form was the power to accumulate, manage, and perpetuate capital. The trustees of the SPGNA sought to build a permanent endowment through bequests, special church collections, donations, and membership subscriptions. In the early 1790s, the society also received a multiyear grant from the state for missionary work in Maine. By 1810, the society's income derived in part from annual member subscriptions and private donations, but the bulk of it flowed from the endowment, which was invested in stocks, bonds, and mortgages.[7]

Although the SPGNA spent its income on missionaries, schools, and (increasingly) books, it did not print books itself. The managers procured them through donations, purchases from booksellers, and direct publishing contracts with printers. And they did not sell their books; they gave them away. This arrangement thrust on the society administrative burdens that commercial book publishers and sellers did not have to bear. With no price system to allocate books (they were given freely), someone representing the society had to evaluate the worthiness of each recipient. Someone had to decide whether to give or not to give. To do this, the SPGNA endeavored to engage "persons of integrity and virtue only" as salaried missionaries and local volunteer agents in distant places. The society required its missionary-agents to keep "a daily journal," including "an exact account of the manner in which you distribute the books entrusted to your care."[8] Such tools of accounting and management of employees in the field would become even more important for religious publishing societies in the nineteenth century.

The Society for Propagating the Gospel never grew much beyond its modest beginnings, but it had a permanent impact, nonetheless.[9] The model of organization pioneered by the SPGNA in the 1790s—the corporate form, organized on a base of members, administered by a board of managers, and funded by a combination of membership subscriptions, special donations, and legacies—flourished in the world of religious associations and publishing societies in nineteenth-century America.

The first American association devoted solely to religious books and tracts was the Massachusetts Society for Promoting Christian Knowledge (MSPCK), founded in 1803. Like the SPGNA, this society was patterned on a British model, the century-old Society for Promoting Christian Knowledge. The Massachusetts society also drew inspiration (as well as books and tracts) from the Religious Tract Society, founded in London in 1799. After 1811, the MSPCK supported missionaries, but before then the work was strictly books and tracts. The Society conducted its first "general distribution" in 1804, giving away 6,253 books and tracts. In a second major effort in 1806, it distributed 9,174. By 1815, the society had distributed 30,350 tracts and 8,224 bound volumes.[10]

In organizing the work, the MSPCK built on the example of the SPGNA. It contracted with a printer, William Hilliard of Cambridge (later Hilliard and Metcalf), for print work and for handling their stocks of books and tracts. They arranged with ministers and pious gentlemen in distant places to conduct distributions. To ensure that the publications were given only to deserving indigents, the MSPCK provided printed instructions to volunteers and required them to sign formal agreements and to submit detailed written reports. The society's annual reports included lists of all titles in print, tables of county-by-county figures on distributions, reports on income and expenditures, and printed forms for subscriptions and bequests. In its annual reports, the MSPCK also greatly expanded a feature that would become commonplace in organizational reports in the nineteenth century: the publication of excerpts from the reports of workers in the field on how the tracts and books were received.[11]

Like its predecessor, the was devoted from the beginning to fund-raising, especially the accumulation of a permanent capital fund. The society launched its work with an endowment of $1,165, contributed by the twelve original founders. Other donations and reinvested interest brought this fund to $2,983 by 1815. By then, however, most of the society's new money went directly into its "distributing fund" for current expenditures. Over its first twelve years of operations, the MSPCK spent $7,673, only $1,200 of which derived from endowment interest. The rest came from annual assessments of members, annual subscriptions by individuals, and special donations raised by small missionary societies, which were sprouting up all over New England. This gradual move to income derived from other societies and from expanded membership subscriptions, rather than from a permanent endowment, would become increasingly common among tract and book societies in the nineteenth century. Even more propitious, though insignificant at the time, was another entry tucked among the income figures for the period 1803–1815: Of the total income of $7,741, there appeared the modest entry of $191.65—"Receipts for books and pamphlets sold."[12]

Selling

The Massachusetts Society for promoting Christian Knowledge, like the SPGNA, was a charitable corporation whose alms were books and tracts. It was not organizationally different from the dozens of little missionary and Bible societies that appeared in America in the first decades of the nineteenth century.[13] But several of the founders of the MSPCK, notably the Reverend Jedidiah Morse, had a grander vision. They imagined a national tract society that would operate on a scale equal to the rapidly growing Religious Tract Society of London. Scale, however, was not the chief innovation of the New England Tract Society (NETS), which Morse and his colleagues founded in 1814. This was a new species of tract society in America; it was a kind of metasociety, whose mission from the beginning was to produce the tracts that others would give away. Their work was not alms but sales—sales at cost to members and affiliated societies.[14]

The New England Tract Society's managers made this plan clear in their first annual report: "This establishment, considered by itself, is not, at present,

to act as a Charitable Society in the *gratuitous* distribution of Tracts; but to
furnish to Tract Societies, on the easiest terms, the most abundant means of
accomplishing their designs." In other words, the founders of the NETS built
their organization on the business principles of centralization for economies
of scale in production coupled with localization in distribution. In a burst of
enthusiasm, the society's founders ordered the printing of 300,000 copies of
fifty different tracts even before its formal organization in May 1814. Within
a year of that date, the NETS had published 141,000 more and had established
formal relationships with several printers, mainly Flagg and Gould of
Andover, Massachusetts. Meanwhile, they had begun to link up with existing local tract societies and to form new ones around the country.[15]

To supply these far-flung societies, the New England Tract Society developed a "depository" system. The General Depository was located first in
Boston and after 1816 in Andover, at the offices of Flagg and Gould. By 1817,
thirty-three regional depositories had been established, most in New England
but several as far south as Charleston and as far west as Natchez. Operated
by agents out of their houses, offices, or churches, these "depositories" by
1817 held stocks of tracts worth $6,406, ranging from $1,072 (Boston) to $19.65
(Concord, N.H.). The tracts remained the property of the national society. That
is, the depositories held them on consignment, with full right of return, and
the agents received a 10% commission on sales to local societies.[16]

A national distribution system built on scattered regional depositories required considerable capital, but not capital vested in a permanent endowment,
in the tradition of the SPGNA. The managers of the New England Tract
Society, now operating as wholesale merchants, needed capital in the form
of tracts. And they never had enough. In report after report, the managers laid
out their needs, hopes, and desires: "A larger capital is needed to enable the
Committee to prosecute the business to the best advantage. . . . A large quantity of Tracts must constantly be kept on hand, in order to meet the wishes
and expectations of those who would purchase."[17]

By 1821, for example, the NETS had seventy-one depositories, and the
managers estimated that they needed working capital sufficient to supply
each with an average of $100 worth of tracts (100,000 pages, approximately).
To support these regional depositories, the General Depository required an
equivalent stock on hand. Thus, the society needed capital of at least $14,000
just to maintain its current operations efficiently. But it had less than half
that amount. The managers dreamed of a day when they could support 100
depositories, with stocks sufficient to circulate 6 million tracts a year. (The
actual number circulated in 1821 was 468,000.) Because these tracts would
be sold to other societies, the stocks could be replenished without reducing
the capital. This was the "principle of permanency," as the managers explained it to members and friends: "The funds which are given to this Society are never expended. They remain as in a bank, to be employed forever
in furnishing Tracts. And not only is the interest employed, as with permanent stock in other Societies, but the principal. And yet no part of it is ever
expended."[18]

This was working capital, and the managers of the New England Tract Society relentlessly sought to get it. Memberships provided some unencumbered income, though the society retained only one-fourth of each member's subscription; three-fourths was returned in the form of tracts. Auxiliary societies were also supposed to transmit one-fourth of their income to the national society, but few did. Miscellaneous donations and legacies raised some money, though never enough. In the rough financial year of 1819, donations amounted to a paltry $22.13. By 1820, the NETS finances were in "a state of much embarrassment," and "the publication of Tracts was much retarded, and almost suspended for a time." In desperation the managers hired an itinerant agent to solicit donations throughout New England. He was remarkably successful, raising nearly $4,000, though about half of that came in the form of loans.[19] By 1821, the society had incurred debts of $2,500, all sunk into tracts, with few prospects for repayment. One money-making scheme that did pay off for the society was *The Christian Almanack*, an annual publication that became a popular, steady seller and major source of income after 1821. Finally, the NETS launched the *American Tract Magazine* in 1824, a chief purpose of which was fund-raising.[20]

The managers of the New England Tract Society imagined that if they just had enough working capital their strategy of selling at cost would produce "permanency," a favorite word in annual reports. "The sum is never expended, or even diminished," they declared, "but, should Providence so order, it may continue in operation to the end of the world." Throughout its first decade of operations, however, the NETS never achieved that goal. The managers struggled with slow remittances from societies, overdue payments to printers, meager donations, and nagging debts. They never accumulated the capital they believed they needed to achieve the "delightful feature of permanency."[21] Yet rather than scale back their plans, the managers took a risky and fateful step in an entirely new direction. The annual report for 1824 announced: "The Committee have begun to use *stereotype plates*. This mode of printing will promote correctness, and at the same time add to the neatness of the impression, and a large advance of capital, will be required, yet your Committee are assured, that the pecuniary interests of the Society, will, in the progress of a few years, be essentially promoted by the measure."[22]

This was a move toward a new kind of capital and new kind of business for a tract society: manufacturing.

Manufacturing

The New England Tract Society was not the first American religious publisher to go into manufacturing. This turn was first taken, not in tract work, but in charity Bible work. The critical moment came in 1809, when the Bible Society of Philadelphia decided to invest in stereotype plates. This move was the first in a series of decisions that transformed the Philadelphia society and several later societies into large-scale book manufacturers. These decisions brought to the societies a new financial burden. Now they needed capital, not only for stock in trade, but for machines, factory buildings, and new technology as well.

The Bible Society of Philadelphia was the first Bible society organized in the United States, but it was just part of a broad wave of enthusiasm for charity Bible work at that time. The Philadelphia society was launched in December 1808, just months ahead of Bible societies in Connecticut, Massachusetts, New Jersey, and New York, all inspired by the success of the British and Foreign Bible Society, which was founded in 1804. By 1816, more than 100 Bible societies had sprouted up in the United States, most of them local book-giving charities. They dealt in Bibles, just as the Massachusetts Society for Promoting Christian Knowledge had dealt in tracts. They bought Bibles from commercial publishers and then offered them freely to the "destitute," a term they used to mean bereft of Scripture, and usually bereft of money as well.[23]

The Philadelphia society, however, quickly emerged as something more than the first among equals. In their initial surveys of Philadelphia and vicinity, the managers of the new society found the need for Bibles to be much greater than they had expected and their funds entirely inadequate to meet the demand. But rather than raise money to buy more Bibles, they decided to invest in a new, capital-intensive printing technology: stereotyping.

In Philadelphia in 1809 this was a daring and visionary plan. Stereotype printing was still a new process even in England, where the Cambridge University Press had just begun to adapt it to Bible printing in 1804–6. In the stereotype process, a plaster-of-paris mold was made of a page form of movable type, and in that mold a solid metal plate was cast. After a plate was made, the types could be redistributed and used again. In this way, a set of printing plates for an entire book could be made and then used, stored, and used again without the expense of keeping type "standing" or of resetting type each time a new edition was needed. Stereotypography was especially good for printing a book in many editions over time—a book such as the Bible.[24]

But stereotype plates were expensive. The set of plates that the Philadelphia society ordered from England cost about $3,500, an enormous sum for these fledgling publishers. The managers were wary but resolute:

> When they considered that the possession of a set of such plates would enable them to multiply copies of the Bible at the lowest expense, and thus render their funds more extensively useful; and still more when they reflected that it would put it in their power to give greater effect to the operations of other Bible Societies, which are springing up daily in every part of the country, the Managers did not hesitate to order the plates to be procured and forwarded from London as soon as possible. The expense is indeed great, when compared with the fund at their disposal; but they were willing to believe, that the obvious and high importance of the measure could not fail to draw from the public liberality a sum sufficient to counterbalance the heavy draught.[25]

This statement suggests the economic implications of stereotype printing. The society would henceforth be a publisher of Bibles, as well as a buyer and distributor of them. And to derive the full economies of scale from its capital investment in plates, it would have to become a *large* publisher, serving other societies' needs, as well as its own. The turn to capital-intensive manufacturing drove the Philadelphia society to centralize production while localizing distribution and fund-raising in a far-flung network of auxiliary societies. In the first three years of stereotype printing, the Bible Society of Philadelphia published 55,000 Bibles and New Testaments, most sold at cost to local societies, who then gave them away.[26]

The adoption of stereotyping by the Bible Society of Philadelphia was just the first phase of the move to manufacturing by religious book, tract, and Bible societies. Several other regional Bible societies, including New York and Baltimore, also acquired stereotype plates. By 1815, many Bible society promoters imagined a complete concentration of manufacturing and merchant capital in a single national operation, located in New York City. From these dreams was born, in 1816, the first truly national religious publishing house: the American Bible Society.[27]

The first order of business for the American Bible Society was the acquisition, not of Bibles, but of stereotype plates. In the months after the founding, the managers of the new society referred all requests for Bibles to the New York Bible Society, while they concentrated on more ambitious, long-range plans. The bulk of funds they had raised at the time of the founding and during the first year of operations would be sunk into manufacturing capital. In August 1816, the ABS ordered six sets of stereotype plates from D. & G. Bruce, one of a handful of stereotype foundries that had recently appeared in New York. Meanwhile, the New York Bible society transferred its plates to the ABS, and by the end of 1816 an edition of 10,000 Bibles was printed and ready to ship—the first imprint of the American Bible Society. Within three years the Society had printed 100,000 Bibles and New Testaments. Within eight years the ABS owned twelve sets of plates and had struck off 265,000 copies at a cost of $250,000.[28]

Though stereotyping was the most important new printing technology for Bible work, it was not the only one. In the 1820s, the ABS also promoted the development of machine-made paper in several American mills and was an early and important patron of the steam-powered printing press. In 1828–29, to support its first "general supply," the society installed sixteen Treadwell steam-powered presses in its newly expanded building. With these new presses, supplemented by twenty older hand presses, the society's printer had the capacity to turn out 300,000 volumes per year.[29] At that time, the ABS printer worked on contract, but in 1845 the Society made its printer a salaried employee. It did the same with its bindery and stereotype founding operations in 1848 and 1851. In 1853, the ABS opened opulent new offices and an integrated manufacturing facility in a six-story, $300,000 building that occupied a full city block in New York.[30] Charity Bible work had become a major American manufacturing enterprise.

Other religious publication societies followed the lead of the ABS into manufacturing. The most important were the American Tract Society, the American Sunday School Union, and the Methodist Book Concern. The

American Tract Society (ATS) was organized in New York in 1825, with the support of the American Tract Society of Boston (formerly the New England Tract Society). Its main purpose was to concentrate print work in one place, and New York by 1825 was the obvious choice. Like the ABS, the American Tract Society quickly became a major manufacturing concern and an early patron of stereotyping and steam-powered printing. The ATS took over the plates for 155 tracts from the New England and New York societies; by the end of the second year it had stereotyped 45 more tracts and had provided for a stereotype foundry in the basement of its new printing office. The society installed ten Treadwell presses shortly after its founding, the first in New York City.[31] The path into stereotyping and power printing was also taken by the American Sunday School Union, a major publisher of children's books by the late 1820s, and by the Methodist Book Concern, the largest of the denominational publishers. The American Sunday School Union, founded in 1824, employed stereotype printing from the beginning; the Methodist Book Concern, founded 1789, turned to stereotyping and power printing as part of a modernization effort in 1828–29.[32]

Raising money to fund these capital investments preoccupied all of these societies for decades. The largest stream of revenue flowed from sales, but because these sales were at cost or below, they kept the treasury at a steady level at best. The national societies needed unencumbered donations to expand their capital. To get it, they cultivated all the sources of income pioneered by earlier tract and book societies, going back to the Society for Propagating the Gospel: membership subscriptions, legacies, and special donations. Large individual donations were increasingly important. In the 1820s and 1830s, the ABS and the ATS were especially successful in cultivating the nouveau riche businessmen of New York for donations for stereotype plates and new buildings. The pious reformer and silk merchant Arthur Tappan, for example, donated $5,000 in 1825 for steam presses for the ATS, gave another $5,000 in 1830 toward the ABS's "general supply," and contributed many more anonymous gifts along the way.[33] But for most of their income—donations and sales revenue—the big publishing societies had to rely on their local auxiliaries scattered across the country. They were the key, as one of the ABS's founders, James Milnor, stated in 1829: "The machinery of a mill may be mechanically perfect in all its parts, but not a wheel will move without the impetus of water. And so those stereotype plates, giving so much facility to the art of printing, and those power-presses, multiplying with such unexampled rapidity impressions of the sacred pages, to produce their expected results, must be supplied, and for these means, the occupants of these plates and presses must be dependent on their Auxiliaries."[34]

The auxiliaries did not always come through, however. Like their parent societies, these local groups were usually strapped for funds because they bought tracts and books from the national societies but then usually gave them away for free. Dependent entirely on contributions for their own local work, they rarely could raise direct donations for the parent societies, and often they struggled even to pay their bills for the Bibles and tracts they supplied. As the scramble for money intensified, a new source of income gradually crept into common use in the 1820s, though it horrified those who believed that the ultimate goal of charity Bible and tract work was "gratuitous" distribution: Local societies, especially local Bible societies, began to sell books to individuals rather than simply to give them away.

Retail Selling

For a manufacturer, selling the product is the obvious route to revenue. For a commercial business, that is axiomatic. But for a charity, it is problematic. Wholesale operations—that is, selling at cost to local societies, who then gave the product away—were easily justified by the managers of the early national societies. Though the constitution of the Bible Society of Philadelphia seemed to sanction only free distribution, for example, the managers had no doubt that wholesaling was permitted as long as the work of the local societies remained pure charity. "Let it be remembered," they declared, "that the sole object to which this money is to be applied—the sole object to which by our charter we can apply it—is the purchase and printing of the Bible, to be ultimately bestowed as a free gift."[35] The New England Tract Society was founded on a similar principle. Both societies believed that a large-scale, centralized business enterprise could be built on a system of pure charity at the local level.

Given the roots of tract and Bible work in the local eleemosynary tradition, it should not be surprising that retail sales posed a more nettlesome problem. The concept of "selling alms" has an odd ring to it. But the pressure to sell was enormous. The move into manufacturing, beginning but not ending with investment in stereotype plates, forced society managers to think hard about efficient return on capital. The economic logic was compelling, and the Philadelphia Bible Society was only the first to struggle with it. In 1812, the Philadelphians eased into retail sales to individuals while still stressing charity at the local society level. Their reasoning reveals the seductive power of economies of scale in stereotype printing:

> The copies of the sacred scriptures, from your press, it is expected, from the excellence and beauty of the type, will be much superior to those which are generally in our market; and the managers have, at several meetings, deliberated on the question, Whether it be their duty to use the means which Providence has put in their hands for increasing your funds (all of which must be expended in a gratuitous distribution of the sacred volume) by selling, at a moderate gain, to other persons, as well as to Bible societies, who may prefer their copies, and send orders. After mature consideration of this question, they have resolved, that . . . it is both their duty and their interest, to supply any orders that may be sent to them for Bibles.[36]

Most of the Philadelphia society's Bibles continued to be distributed without charge, but the precedent had been set. The new American Bible Society made retail sales to the ultimate recipient central to its marketing strategy.

The ABS had begun its work in 1816 on the model of the Bible Society of Philadelphia and the New England Tract Society: selling at cost to auxiliary societies, who would in turn give Bibles away for free. In 1819, the ABS changed course and began to urge auxiliaries to sell Bibles, as well as to donate them. Not surprisingly, the managers were at pains to justify this new course. Their rationale is nicely summarized in their annual report for 1821:

> The plan recommended by the Managers, of selling Bibles and Testaments at cost or at reduced prices, where persons are able and willing to pay, has been highly approved by all the Auxiliaries from whom accounts have been received; and has been carried into effect, in many instances, with unexpected and very pleasing success. Those who needed Bibles have usually preferred to give something for them; and the process of distribution has not been impeded, if it has not been accelerated, by the measure referred to. The Auxiliaries have found their ability enlarged by it; and they have been enabled to supply more fully the necessity of those who were not possessors of the Sacred Volume, and yet could not, or would not, purchase it. For it should be distinctly understood, that the Managers were very far from designing, by the plan, to diminish the circulation of the Scriptures: they designed rather to add to it. They were satisfied that many persons would gladly become possessors of a Bible by paying the full, or a reduced price, whose feelings of independence revolted from receiving it as the gift of charity. On the plan which the Managers have recommended, the Scriptures are still given freely to the destitute who are without means, or without disposition to pay for them; while receiving the whole, or a part of the cost from such as are willing to pay, the funds are rendered more availing, and a degree of security is obtained, that the volume which has been purchased will be prized, preserved, and used.[37]

I have quoted this passage at length because it is thick with thoughts about economics, charity, and readers' psychology. The immediate goal of retail sales certainly was to get more revenue, but the ultimate goal was to distribute more Bibles. If some Bibles were sold, the managers reasoned, the local societies would have more resources for "gratuitous distribution." Put into economic terms, this was a policy of "price discrimination," or, as I like to call it, "differential pricing"—that is, charging each buyer the price he or she is willing or able to pay, all the way down to zero. Such a pricing strategy was clearly the most efficient one when the goal was universal circulation. Thus, the managers affirmed that selling was consistent with charity. They also argued that many people would value the Bible more if they had paid for it. In short, the managers of the American Bible Society were able to turn an economic imperative into an evangelical virtue.[38]

The annual reports of the auxiliaries suggest that the local societies, not surprisingly, liked the new policy of selling Bibles. For many small societies, especially in the West, outright donations were scarce and sales revenue was almost their only source of income. A society in Dearborn County, Indiana, for example, reported: "We have received many applications for Bibles and Testaments, from persons who were unable to purchase them; but, in owing to our very limited means, we have adjudged it unadvisable to give away a single copy, until we are able to pay for the stock we have received."[39] In the more prosperous East, societies usually tried to achieve the mix of selling and giving that the national ABS had in mind. Sometimes, though, societies took the new policy as license to grant no free books at all. This approach went too far, according to the ABS managers. "The principle of the Parent Society seems to have been misapprehended," they wrote about one such case. "It is designed that the Scriptures shall be furnished gratuitously to those who are truly poor, and will faithfully use them. We would cheerfully *give* in such cases, and trust Providence to furnish means for the future."[40]

Following the lead of the American Bible Society, other national religious publishers also urged their auxiliaries to sell, as well as give. The American Tract Society, for example, began to publish full-size books in 1827, and its reports were soon praising the virtues of retail sales. But, the managers hastened to add, "*it is no scheme of pecuniary profit. . . . It is, from beginning to end, purely an effort of benevolence*, to tender Divine truth in these interesting and durable forms to those who would not come after it—who have but a feeble desire to obtain it; but who, without it, may perish eternally." Like the ABS, the ATS told its auxiliaries that a system of differential pricing provided the best mixture of evangelism (universal circulation) and economics (permanent capital and revenue flow):

> It is most clear, that the tremendous influence of the public press in our country *may not be left* solely to the operation and influence of sales for the purposes of gain. The most valuable books must be *prepared in an attractive style, and furnished at cost, or less than cost*, and Christian efforts must be put forth all over the land *to place them in the hands of the people*—by sale, if it can be done—gratuitously, if it cannot. . . . To every donor to the Society there is also this encouragement, that as the volumes are chiefly *sold*, the amount of each donation returns with every sale; is sent out again, and again returns; and thus continues to revolve, and may, and probably will revolve long after the benevolent donor shall be sleeping in dust.[41]

Critics denounced, on several grounds, this turn to retail sales. The author of *An Exposé of the Rise and Proceedings of the American Bible Society* implied that any retail sales mocked the claim of publishers to furnish Scripture "without money and without price." He argued that sales allowed the managers to accumulate vast stores of capital in the form of stereotype plates, buildings, and real estate. "Rich in this world's goods," he said, "they are supercilious and arrogant." But more important, by spurning the indigent in favor of those who could buy, the society stole the market for Bibles from honest commercial book publishers and booksellers. In a later polemic, another critic levied a similar complaint against all "charity publication societies," especially the American Tract Society. This author charged that charitable donations subsidized the low prices of ATS books; private enterprise, therefore, could not compete. He argued that charity publishing was unnecessary and wrong when people were willing to buy. "What business have Christians to give their charity to do that which business enterprise and capital would do, if let alone, quite as well and cheaply?" Helpfully answering his own question, the author added, "None at all."[42]

The managers of the ABS and the ATS usually ignored criticism, but gradually they came to realize that the strategy of retail sales undermined the charitable nature of their mission. Under such a policy, books naturally flowed to where the money was. Rich regions produced more sales; more sales produced rich auxiliaries; rich auxiliaries produced more remittances for the national office. But the founding mission of the publishers was altogether different. It was to "tender the message of the Gospel to *all*—high and low, rich and poor."[43] Ironically, the turn to retail sales, which was designed to produce universal circulation, not profit, had entrapped the societies in market forces they had been founded to resist. To circulate books *everywhere*, the societies could not depend on the financial condition of local auxiliaries. They needed some form of centrally administered, national distribution system.

Administered Distribution

In the early years, the auxiliary society system, which most religious publishers used, seemed ideal. Auxiliaries gave the national organization geographical reach for distributing the product, for enlisting volunteer labor, and for raising money through both sales and donations. Indeed, an effective policy of differential pricing depended on local distributors because only local people could draw on local knowledge to decide whether a recipient was worthy of a grant or discounted price.[44] Furthermore, a system of auxiliary societies ostensibly involved less risk for the national office and less capital tied up in product stocks than a regional depository system such as the one set up by the New England Tract Society. Rather than holding books on consignment, auxiliaries purchased them outright (at a discount). The auxiliary network, then, functioned as a market—a wholesale market—even when the final recipient received the product for free. If such a system worked, it certainly would be the most efficient way to mesh centralization in production with localization and differential pricing in distribution.

But the auxiliary system did not work—at least it did not produce the results the publishers had expected. The American Bible Society, for example, had more than 600 auxiliaries by 1829–30, when it launched its first "general supply." The managers were confident that they could manufacture enough Bibles to supply the entire country, but they had grave doubts that the auxiliaries could deliver them to the entire country. The ABS managers were highly critical of their auxiliaries. Many local societies failed to pay their bills yet continued to order Bibles on credit, thus plunging the national office into a debt of some $36,000 on the very eve of the "general supply." Some societies, the managers complained, "seem to use their Institution too much as a matter of personal convenience, procuring a few books every year for family use, while little or nothing is done toward looking out and supplying the destitute around them, or in aiding the general cause by their surplus contributions."

Other societies were virtually dead, living on in name only. And perhaps worst of all for a "general supply," some regions of the country had no auxiliaries at all, alive or dead. Thus, just as they feared, the "general supply" failed.[45]

The problem was that the auxiliary system did not produce the geographical reach or the cross-class circulation that the ABS managers had hoped for. Some areas and some people were well supplied; others were neglected. Given the market structure of the auxiliary system, this stands to reason. The managers routinely had to remind the auxiliaries that the goal of charity Bible work was not just to sell books: "But it ought to be remembered, that while these Associations are intended to bring the Bible, of good quality and at low price, to those in good circumstances, this was not the prominent design of their formation. Persons in such circumstances could have been supplied, as formerly, through the book-merchant. The great design of these Associations is to carry the Bible to those who would otherwise never possess it."[46]

The American Tract Society ran into similar problems in its efforts to use the auxiliary system to achieve universal circulation. Like the ABS, the ATS in the early 1830s launched its own version of a "general supply," a campaign to place at least one small evangelical book into every American household. But the wholesale basis of the auxiliary system worked against this great work. The ATS sold most of its tracts and books to auxiliaries in the richer areas of the East, areas that were already better supplied with Bibles and religious publications, not to mention churches, ministers, and Sunday Schools.[47] The society's managers frequently warned of the dangers of market forces, within the organization, as well as without:

> It would be far easier and less expensive to hover around the communities now enjoying some of the means of grace, and where a ready circulation of books could be effected; but the Committee cannot disregard the more pressing wants of the border population, and they believe that the friends of the Society will cheerfully sustain them in the effort to convey the gospel to the "poor," whether they are found in the wilds of Nebraska or among the gold-heaps of California.[48]

Societies employed several strategies to deal with the market bias of the auxiliary system. One strategy was simply to give free books to poor auxiliaries. Throughout the 1820s, the American Bible Society regularly issued grants to new and "languishing" societies. In 1831, in the midst of their "general supply," the ABS gave more than 50,000 Bibles to auxiliaries, mainly in the trans-Appalachian West.[49] To pay for these grants, the society depended on donations from wealthy individuals and auxiliaries and some sales at a profit. A second strategy was to employ field agents to organize auxiliaries in new areas, to stir up enthusiasm in old auxiliaries, and to solicit donations. The ABS hired its first agent in 1821; by 1828 it had twelve agents in the field, with its first agent into the Mississippi Valley in 1825; by 1830 it had nine agents at work in the West and four more in the East.[50]

Agents were employed to found, to support, and to admonish auxiliaries, not to do their work for them. The ABS relied increasingly on agents paid and supervised by the national office. In 1850, after more than thirty years of frustrating experience with auxiliaries, the managers wrote that "we are satisfied that over the whole country new societies will not be formed, funds will not be collected, nor the work of exploration and supply prosecuted to any adequate extent . . . without the influence of active, enterprising agents." But the ABS did not abandon the auxiliary system. Despite the problems, the national leaders believed that the society could not survive without the volunteer labor and the flow of funds (sales and donations) that the auxiliaries provided.[51]

In contrast, the American Tract Society gradually developed a system of distribution and fund-raising that did not depend on auxiliaries and, therefore, did not depend on market incentives. In the 1840s, the ATS resolved that its major work must be removed from forces beyond its administrative control.[52] It set out to build a national distribution system based on salaried distributors, geographical administrative divisions, and a hierarchy of salaried managers. This was the beginning of the American Tract Society's famous system of colportage.

The word *colporteur* was a French term for an itinerant hawker of religious tracts or books. Though the roots of colportage reached back to the Reformation, the ATS was the first to develop the system on a large scale in America. In August 1841, the society commissioned its first two colporteurs and dispatched them to Indiana and Kentucky. From that small beginning, the project grew rapidly. At the end of five years, the ATS had 175 colporteurs in the field; after ten years, more than 500. In the first ten years, Tract Society colporteurs visited more than 2 million families (11 million individuals), nearly half the population of the country. They sold 2.4 million books, donated 650,000 books, and gave away "several million" tracts.[53]

The colportage campaign did not realize the society's grandest hopes— "to visit *every abode*," to provide "the gospel for everybody." Colporteurs and agents in the West and South frequently wrote that the needs far exceeded their resources.[54] But the work that was done was impressive. By 1856 the executive committee could boast that "when we record the fact, from carefully kept statistics, that more than five millions of families have been visited at their firesides on a gospel errand, and that alone—more than are embraced in the census returns of the United States—it implies toil for Christ, such as no other nation on the globe ever witnessed." Under the colportage system, the committee declared, tracts "have fallen like snow-flakes over the land."[55]

The principle of colportage was simple. In the first full report on the project, in the annual report for 1843, the executive committee explained the idea: "Colporteurs go indiscriminately to every family; and whenever a family is found in need of a volume to guide them to heaven, and is unable to purchase, one is furnished gratuitously."[56] "Indiscriminately to every family"—that was the key. Whether a family would be likely to buy a book was beside the point. To make such a system work, colporteurs had to be salaried employees (no commissions or discounts), their travel expenses had to be paid, they had to be supplied in the field with sufficient materials, they had to have detailed information about their territories, they had to be trained and motivated, they had to keep good records and file detailed reports, and they had to be closely supervised. In other words, colportage required systematic management, from top to bottom, on a national scale.[57]

No religious publication society moved as fully into nationally administered distribution and fund-raising as did the American Tract Society. Colporteurs employed by other societies typically were not on salary; they had to earn their subsistence through commissions on sales.[58] Though unable to raise sufficient funds to support a fully administered distribution system, these societies understood that their commission systems necessarily left colporteurs enmeshed in the market. Evangelistic fervor notwithstanding, the economic incentive of colporteurs on commission was to sell. The American Baptist Publication Society and the Presbyterian Board of Publication both tried to follow the lead of the ATS into salaried colportage. Though never very successful, they knew what was at stake. The managers of the former wrote, "The masses of unconverted people, and of errorists, go not to stores to buy religious books; *they must be carried to their doors*, pressed upon their attention, sold when they can be, and given away where there is inability to purchase." The managers of the Presbyterian Board put the matter even more sharply: "Religious knowledge is a benefit, of which men less feel the need the less they possess of it. Here the demand does not create a supply, for the demand may not exist, however extreme the necessity. The gospel provides for its own dissemination. It was never contemplated that men would 'seek,' and hence the command is 'to send,' 'to go,' and 'to preach.'"[59] Though a market–sales system was needed to generate revenue and capital, it could never complete the mission of evangelical publishing, which was to deliver the word to all.

Scholars seem generally agreed that American religion plunged into the market world of the early nineteenth century. The terminology of commerce and marketing is now commonplace in religious history. R. Laurence Moore writes of the "commodification" of religion and the competition of religious groups and movements in "the marketplace of culture." Roger Finke and Rodney Stark frame their sweeping overview of American religious history as a story of breakneck competition in "a free market religious economy." John H. Wigger's account of the rise of American Methodism is a story of the "replacement of state-sponsored churches with a religious free market." In his account of the bureaucratization of the American Bible Society, Peter Wosh writes that "religion had become a problem of effective marketing."[60]

This linking of early nineteenth-century religion to "the market revolution," as it has come to be called, has been a useful, clarifying trend in the historiography of American religion. Religious organizations most certainly were awash in a sea of commerce, as well as a sea of faith.[61] And, as manufacturers and distributors of manufactured goods, the publishing societies especially had no choice but to sail on that rough sea. But to say that religious publishers entered the commercial marketplace is to say too little and too much. On the one hand, the leaders of the Bible and tract movements were utterly exhilarated by the possibilities of the commercial culture. The expansion of business and the extension of market relations, along with improvements in transportation, technology, and access to capital, allowed the Bible and tract entrepreneurs to imagine and to build truly national institutions (some of the first modern business firms, really) with national reach. By the late 1820s, the millennial dream of reaching everyone with books and tracts seemed immanent, and the evangelical publishers became leading innovators and entrepreneurs of print. On the other hand, they viewed the market as their most wily and dangerous foe, for it was private enterprise that was poisoning the nation with the cheap trash of print culture, the literature of wickedness, sensation, dissipation, and error. Private enterprise, obedient to the law of supply and demand, was the great problem of America. As charity publishers, the Bible and tract societies pursued a goal that was the opposite of the private enterprise. They tried to turn the market on its head, to deliver a product to *everyone*, regardless of ability or even desire to buy.[62]

The publications of the American Tract Society, for example, brim with examples of this manichaean vision of modern commercial society. "The impossibilities of a century ago are the easy achievements of the age of steam," the society's newspaper gushed. "For practical purposes oceans are bridged and mountains leveled, and continents spanned. The most subtle and powerful elements of nature are tamed by science and harnessed to the car of human progress." One ATS writer imagined the voice of God enjoining Christians to take up the new technology and to make it their own: "I the Lord have given you power and wealth, mountains of iron and valleys of gold, a boundless territory and a free government. . . . I have added the ocean steamer, and the rail-way, and the steam printing-press, and the telegraph; employ all these for my glory and for the establishment of my kingdom!"[63]

But the ATS's managers were also appalled by what the market revolution had wrought. The commercial press was largely a "satanic press." "The plagues of Egypt were tolerable," the society declared, "compared with this coming up into our dwellings of the loathsome swarms of literary vermin to 'corrupt the land,' to deprave the hearts, and ruin the souls of our citizens."[64] Private enterprise was the cause of the evil print culture of America; it could not be the cure. This theme was emphasized routinely in ATS publications:

> No nation on the globe, perhaps, has so large a reading population; and in none is the press more active, or more influential. What the reading matter prepared for such a nation would be, if left solely to private enterprise, may be inferred from an examination of the catalogues of some of the respectable and even Christian publishing houses. Self-interest would shape the supply to the demand; and the mightiest agent God has given to the world for moulding public opinion and sanctifying the public taste, would be moulded by it, and be made to reflect its character, were there no conservative, redeeming influences.

The society's newspaper argued that "if the public taste be wrong, the press with its indescribable power perpetuates and extends the injury thus inflicted on vital interests. The question is, *What will sell*? and as in other shambles and markets, so here, supply responds to demand, although souls are included in the traffic."[65] The American Tract Society proposed the reverse: Make supply drive demand.

Some historians have argued recently that the market revolution should not be conflated with the rise of capitalism. "In point of fact," Michael Merrill writes, "there are many different market economies, not all of which ought to be called capitalist." Many artisans and small producers eagerly embraced "commercial society," says Merrill, while resisting the rise of "capitalism."[66] The evangelical publishing societies took the opposite tack. They eagerly embraced capitalism while resisting the rise of commercial society. They launched their products into the currents of commerce, but against the flow. Modern business methods and technologies were the allies of the publishers, but commercial culture itself was the enemy. The managers of the noncommercial Bible and tract societies made themselves practical businessmen, savvy marketeers, large-scale manufacturers, and grasping capitalists in order to save the country from the market revolution.

Notes

1. *An Exposé of the Rise and Proceedings of the American Bible Society, During the Thirteen Years of Its Existence, by a Member* (New York: n.p., 1830), 3, 15–17.

2. Ibid., 13–14. The author's quotation of Isaiah 55:1 is a paraphrase. In the King James translation, the passage actually reads as follows: "Ho, every one that thirsteth, come ye to the waters, and he that hath no money; come ye, buy, and eat; yea, come, buy wine and milk without money and without price."

3. Ibid., 17. For a similar critique, focusing mainly on the American Tract Society, see [Herman Hooker], *An Appeal to the Christian Public, on the Evil and Impolicy of the Church Engaging in Merchandise; Setting Forth the Wrong Done to Booksellers, and the Extravagance, Inutility, and Evil-Working, of Charity Publication Societies* (Philadelphia: King & Baird, 1849).

4. *Brief Account of the Society for Propagating the Gospel among the Indians and Others in North-America* ([Boston]: n.p., 1798), 2–6; *A Brief Account of the Present State, Income, Expenditures, &c. of the Society for Propagating the Gospel among the Indians, and Others, in North-America* ([Boston]: n.p., 1795), 1. See also James F. Hunnewell, ed., *The Society for Propagating the Gospel among the Indians and Others in North America, 1787–1887* (Cambridge, Mass.: Society for Propagating the Gospel, 1887).

5. Pauline Maier, "The Revolutionary Origins of the American Corporation," *William and Mary Quarterly*, 50 (January 1993): 51–84. See also the classic article on the subject, Oscar Handlin and Mary F. Handlin, "Origins of the American Business Corporation," *Journal of Economic History* 5 (1945): 1–23.

6. Conrad Edick Wright, *The Transformation of Charity in Postrevolutionary New England* (Boston: Northeastern University Press, 1992), 56. See also Peter Dobkin Hall, *Inventing the Nonprofit Sector and Other Essays on Philanthropy, Voluntarism, and Nonprofit Organizations* (Baltimore: Johns Hopkins University Press, 1992), chap. 1.

7. *Brief Account of the Present State*, 1–2; *Brief Account of the Society*, 2–3; Jedidiah Morse, *Signs of the Times: A Sermon Preached Before the Society for Propa-*

gating the Gospel among the Indians and Others in North America, at Their Anniversary, Nov. 1, 1810 (Charlestown, Mass.: Samuel T. Armstrong, 1810), 67–68.

8. *Brief Account of the Society*, 4–6.

9. Hunnewell, *Society for Propagating the Gospel*, 7–8.

10. *An Account of the Massachusetts Society for Promoting Christian Knowledge* (Cambridge, Mass.: William Hilliard, 1806), 17–18, 31–32; *An Account of the Massachusetts Society for Promoting Christian Knowledge* (Andover, Mass.: Flagg & Gould, 1815), 13–15. See also *An Address to Christians Recommending the Distribution of Cheap Religious Tracts, with an Extract from a Sermon, by Bishop Porteus, Before the Yearly Meeting of the Charity Schools, London* (Charlestown, Mass.: Samuel Etheridge, 1802). On British tract work, see William Jones, *The Jubilee Memorial of the Religious Tract Society* (London: Religious Tract Society, 1850).

11. *Account of the Massachusetts Society* (1806), passim; (1815), passim.

12. *Constitution of the Massachusetts Society for Promoting Christian Knowledge* (Charlestown, Mass.: Samuel Etheridge, 1803); *Account of the Massachusetts Society* (1815), 75–76.

13. Wright, *Transformation of Charity*, chap. 4.

14. "The Constitution" and "To the Friends of Religion in New England," in *Proceedings of the First Ten Years of the American Tract Society, Instituted at Boston, 1814* (Andover, Mass.: Flagg & Gould, 1824). This volume includes the first ten annual reports of the New England Tract Society. It is reprinted in facsimile in *The American Tract Society Documents, 1824–1925* (New York: Arno Press, 1972). On Morse, see Joseph W. Phillips, *Jedidiah Morse and New England Congregationalism* (New Brunswick, N.J.: Rutgers University Press, 1983); and Richard J. Moss, *The Life of Jedidiah Morse: A Station of Peculiar Exposure* (Knoxville: University of Tennessee Press, 1995).

15. "To the Friends of Religion" and "First Report, 1815," in *Proceedings of the First Ten Years*, 9, 27, 29. On the formation of auxiliary societies, see "Second Report, 1816," in *Proceedings of the First Ten Years*, 38–39.

16. "Third Report, 1817," ibid., 42–43; Seventh Report, 1821," ibid., 78.

17. "Fifth Report, 1819," ibid., 55.

18. "Seventh Report, 1821," 80–85

19. "Constitution," ibid., 6; "Second Report, 1816," 39–39; "Fifth Report, 1819," 62; "Sixth Report, 1820," ibid., 64; and "Seventh Report, 1821," 74–75.

20. "Seventh Report, 1821," 76; "Eighth Report, 1822," 92; and "Tenth Report, 1824," 129–31; *The Christian Almanack, for the Year of Our Lord and Saviour Jesus Christ, 1824* (New York: American Tract Society and Religious Tract Society of New York, [1823]), 3. The New England Tract Society changed its name in 1823 to the American Tract Society. It retained that name even after the founding of the larger American Tract Society in New York in 1825.

21. "Ninth Report, 1823," in *Proceedings of the First Ten Years*, 112–13; "Seventh Report, 1821," 85.

22. "Tenth Report, 1824," 128.

23. [William Jay], *A Memoir on the Subject of a General Bible Society for the United States of America* (N.J.: n.p., 1816). For overviews of the pre-1816 Bible societies, see Eric M. North, "The Bible Society Movement Reaches America," *ABS Historical Essay*, no. 7, pt. 1 (New York: American Bible Society, 1963); Eric M. North, "The Bible Societies Founded in 1809 in the United States," *ABS Historical Essay*, no. 7, pt. 2 (New York: American Bible Society, 1963); and Rebecca Bromley, "The Spread of the Bible Societies, 1810–1816," *ABS Historical Essay*,

no. 8, pts. 1 and 2 (New York: American Bible Society, 1963). On the British and Foreign Bible Society, see Leslie Howsam, *Cheap Bibles: Nineteenth-Century Publishing and the British and Foreign Bible Society* (Cambridge: Cambridge University Press, 1991).

24. Howsam, *Cheap Bibles*, 77–79. On the stereotype process, see George A. Kubler, *A New History of Stereotyping* (New York: Little & Ives, 1941); and George A. Kubler, *Historical Treatises, Abstracts, and Papers on Stereotyping* (New York: Brooklyn Eagle Press, 1936). Two very early pamphlets on the process have been reprinted. See Charles Brightly, *The Method of Founding Stereotype*, and Thomas Hodgson, *An Essay on the Origin and Progress of Stereotype Printing* (New York: Garland, 1982). Brightly's pamphlet was originally published in 1809, Hodgson's in 1822. This reprint volume has a useful introduction by Michael L. Turner.

25. Bible Society of Philadelphia, Second Annual Report (Philadelphia: Fry & Krammer, 1810), 10–11; Third Annual Report (Philadelphia: Fry & Krammer, 1811), 7. The plates arrived in October 1812 and were turned over to the Philadelphia printer William Fry, who immediately struck off an edition of 1,250 copies, the first stereotyped Bible in America. See also Margaret T. Hills, *The English Bible in America: A Bibliography of Editions of the Bible and the New Testament Published in America 1777–1957* (New York: American Bible Society and New York Public Library, 1961), 37.

26. Bible Society of Philadelphia, Third Annual Report (1811), 4; *Eighth Annual Report* (1811), 3–4. I explore the business history of the Philadelphia Bible Society in more detail in David Paul Nord, "Free Grace, Free Books, Free Riders: The Economics of Religious Publishing in Early Nineteenth-Century America," *Proceedings of the American Antiquarian Society* 106 (1996): 241–72.

27. *Constitution of the American Bible Society . . . Together with Their Address to the People of the United States* (New York: G. F. Hopkins, 1816), 16. See also Elias Boudinot, *An Answer to the Objections of the Managers of the Philadelphia Bible-Society, Against a Meeting of Delegates from the Bible Societies in the Union* (Burlington, N.J.: David Allinson, [1815]); [Samuel Mills], "Plan of a General Bible Society," *The Panoplist*, October 1813; [Jay], *a Memoir*. See also Eric M. North, "The Pressure toward a National Bible Society, 1808–1816," *ABS Historical Essay*, no. 9 (New York: American Bible Society, 1963). For a recent history of the ABS, see Peter J. Wosh, *Spreading the Word: The Bible Business in Nineteenth-Century America* (Ithaca, N.Y.: Cornell University Press, 1994).

28. American Bible Society, First Annual Report (Newark, N.J.: J. Seymour, 1817), 10; *Third Annual Report* (1819), 11; Hills, *English Bible*, 50–51. I explore manufacturing technology at the ABS in David Paul Nord, "The Evangelical Origins of Mass Media in America, 1815–1835," *Journalism Monographs*, no. 88 (1984).

29. American Bible Society, *Twelfth Annual Report* (1828), 28–29; *Address of the Board of Managers of the American Bible Society to the Friends of the Bible of Every Religious Denomination, on the Subject of the Resolution for Supplying All the Destitute Families in the United States with the Bible in the Course of Two Years* (New York: J. Seymour, 1829), 13; *An Abstract of the American Bible Society, Containing an Account of Its Principles and Operations* (New York: Daniel Fanshaw, 1830), 12. On the Treadwell press, see James Moran, *Printing Presses: History and Development from the Fifteenth Century to Modern Times* (Berkeley and Los Angeles: University of California Press, 1973), 113–16.

30. Margaret T. Hills, "Production and Supply of Scriptures, 1831–1860," *ABS Historical Essay*, no. 18, pt. 3 (New York: American Bible Society, 1964), 20–26, 32; Wosh, *Spreading the Word*, 9–10, 252.

168   FINANCE AND THE EXPANSION OF AMERICAN PROTESTANTISM

31. American Tract Society, *First Annual Report* (1826), 11–12, 18; *Second Annual Report* (1827), 10; *American Tract Magazine* 1 (April 1826): 275. On manufacturing technology at the ATS, see Nord, "Evangelical Origins."

32. *Address of the Managers of the American Sunday School Union, to the Citizens of Philadelphia* (Philadelphia: I. Ashmead, 1826), 8–9; American Sunday School Union, *Third Annual Report* (1827), iii; *Fourth Annual Report* (1828), iv–v; James Penn Pilkington, *The Methodist Publishing House: A History, Vol. 1, Beginnings to 1870* (Nashville, Tenn.: Abingdon, 1968), 172–74, 182, 197–201, 215, 218. See also Nathan Bangs, *A History of the Methodist Episcopal Church*, vol. 4 (New York: Carlton & Porter, 1857), chap. 16. Not all publication societies moved into manufacturing. The American Baptist Publication Society is an example of one that remained a wholesale merchant, as did the old NETS. The ABPS never grew to the scale of the manufacturing societies or to the scale its managers dreamed of. See J. Newton Brown, *History of the American Baptist Publication Society, from Its Origin in 1824, to Its Thirty-Second Anniversary in 1856* (Philadelphia: American Baptist Publication Society, [1856]); and Daniel Gurden Stevens, *The First Hundred Years of the American Baptist Publication Society* (Philadelphia: American Baptist Publication Society, [1924]).

33. Bertram Wyatt-Brown, *Lewis Tappan and the Evangelical War Against Slavery* (Cleveland, Ohio: Press of Case Western Reserve University, 1969), 49–50.

34. James Milnor, speech text, in *Monthly Extracts of the American Bible Society*, no. 17 (1829): 239. See also *A Brief Analysis of the System of the American Bible Society, Containing a Full Account of Its Principles and Operations* (New York: Daniel Fanshaw, 1830), 34–35.

35. *An Address of the Bible Society of Philadelphia to the Friends of Revealed Truth in the State of Pennsylvania* (Philadelphia: Fry & Kammerer, 1810), 6, 8–9. See also Nord, "Free Grace," 251–52.

36. Bible Society of Philadelphia, *Fifth Annual Report* (1813), 11–12.

37. "Circular Letter of the Committee on Auxiliary Societies," in American Bible Society, *Third Annual Report* (1819), 74; *Fifth Annual Report* (1821), 30.

38. I discuss pricing strategies of charity Bible work more fully in Nord, "Free Grace."

39. American Bible Society, *Eighth Annual Report* (1824), 89; *Seventh Annual Report* (1823), 24; *Ninth Annual Report* (1825), 63.

40. American Bible Society, *Eleventh Annual Report* (1827), 37.

41. American Tract Society, *Eleventh Annual Report* (1836), 41–44. See also *Proposed Circulation of the Standard Evangelical Volumes of the American Tract Society to the Southern Atlantic States* (New York: American Tract Society, 1834).

42. Exposé, 13, 17; [Hooker], *Appeal to Christian Public*, 5–6.

43. American Tract Society, *Fifth Annual Report* (1830), 28–29.

44. See Nord, "Free Grace," 260–61, 270–72.

45. *Address of the Board of Managers*, 12; American Bible Society, *Thirteenth Annual Report* (1829), 6–7, 22–23, 42–44; *Fifteenth Annual Report* (1831), 6–7, 17–18; *Seventeenth Annual Report* (1833), 16–17; *Twenty-fourth Annual Report* (1840), 25–26. For narrative overviews of the general supply, see Creighton Lacy, *The Word Carrying Giant: The Growth of the American Bible Society* (South Pasadena, Calif.: William Carey Library, 1977), chap. 4; and Henry Otis Dwight, *The Centennial History of the American Bible Society* (New York: Macmillan, 1916), chapter 12.

46. American Bible Society, *Thirteenth Annual Report* (1829), 43–44.

47. American Tract Society, *Second Annual Report* (1827), 19–23; *First Annual Report* (1826), 20; *Third Annual Report* (1828), 14. See also David Paul Nord, "Systematic Benevolence: Religious Publishing and the Marketplace in Early Nineteenth-Century America," in *Communications and Change in American Religious History*, ed. Leonard I. Sweet (Grand Rapids, Mich.: Eerdmans, 1993), 249–51.

48. American Tract Society, *Twenty-Fifth Annual Report* (1850), 65.

49. American Bible Society, *Fifteenth Annual Report* (1831), 8–10.

50. American Bible Society, *Seventh Annual Report* (1823), 20–21, 74; American Bible Society, *Twelfth Annual Report* (1828), 30–31; *Abstract*, 18–19, 38; *Brief Analysis*, 27–30.

51. American Bible Society, *Thirty-Fourth Annual Report* (1850); *Forty-Fourth Annual Report* (1860), 21–22. See also Eric M. North, "The Great Challenge: Distribution in the United States," *ABS Historical Essay*, no. 14, pts. 2 and 3 (New York: American Bible Society, 1963–64); Mary F. Cordato, "The Relationship of the American Bible Society to Its Auxiliaries: A Historical Timeline Study," *ABS Historical Working Paper Series*, no. 1991–1 (New York: American Bible Society, 1991); and Nord, "Free Grace," 269–70.

52. In its report to a special auditing committee appointed in 1857, the American Tract Society's executive committee admitted that by 1841 the system of auxiliaries, agents, and monthly distributions had failed and that "millions of the most destitute, neglected, and needy of our population were not thus reached." See *Report of the Special Committee Appointed at the Annual Meeting of the American Tract Society, May 7, 1857, to Inquire into and Review the Proceedings of the Society's Executive Committee* ([New York: American Tract Society, 1857]), 15. See also Nord, "Systematic Benevolence," 248–54.

53. American Tract Society, *Twenty-First Annual Report* (1846), 21; *Twenty-Sixth Annual Report* (1851), 46–47, 64–65. Each annual report after 1841 reviewed the colportage effort for the year. For general accounts of colportage, see the special report "Ten Years of Colportage in America," in American Tract Society, *Twenty-Sixth Annual Report* (1851), 45–72; *The American Colporteur System* (New York: American Tract Society, [1843]), reprinted in American Tract Society Documents; [R. S. Cook], *Home Evangelization: View of the Wants and Prospects of Our Country, Based on the Facts and Relations of Colportage* (New York: American Tract Society, [1849]); and [Jonathan Cross], *Five Years in the Alleghenies* (New York: American Tract Society, 1863).

54. [Cook], *Home Evangelization*, 19; *American Messenger*, September 1851, 34. For examples of laments from the South and West, see American Tract Society, *Twenty-Third Annual Report* (1848), 70, 73.

55. American Tract Society, *Thirty-First Annual Report* (1856), 41–42; *Twenty-Seventh Annual Report* (1852), 53.

56. American Tract Society, *Eighteenth Annual Report* (1843), 28.

57. I develop the idea of systematic management in Nord, "Systematic Benevolence." See also Peter Dobkin Hall, "Religion and the Organizational Revolution in the United States," in *Sacred Companies: Organizational Aspects of Religion and Religious Aspects of Organizations*, ed. N. J. Demerath III, et al. (New York: Oxford University Press, 1998).

58. Brown, *History of the American Baptist Publication Society*, 156–57, 170–71; American Baptist Publication Society, *Sixth Annual Report* (1845), 38–39; *Seventh Annual Report* (1846), 17–18; Abel Stevens, comp., *Documents of the Tract Society of the Methodist Episcopal Church* (New York: Carlton & Phillips, 1853), 22; Meth-

odist Episcopal Church, Journal of the General Conference (1852), 120, 123; (1856), 229.

59. American Baptist Publication Society, *Principles and Purposes* (Philadelphia: n.p., n.d.), 18; *Principles and Plans of the Board of Publication of the Presbyterian Church in the United States of America* (Philadelphia: Presbyterian Board of Publication, [1854]), 19–20.

60. R. Laurence Moore, *Selling God: American Religion in the Marketplace of Culture* (New York: Oxford University Press, 1994), 5–6; Roger Finke and Rodney Stark, *The Churching of America, 1776–1990: Winners and Losers in our Religious Economy* (New Brunswick, N.J.: Rutgers University Press, 1992), 5–6, 59; John H. Wigger, *Taking Heaven by Storm: Methodism and the Rise of Popular Christianity in America* (New York: Oxford University Press, 1998), 5; Wosh, *Spreading the Word*, 250. The economic approach to the study of religion is reviewed in Laurence R. Iannaccone, "Voodoo Economics? Reviewing the Rational Choice Approach to Religion," *Journal for the Scientific Study of Religion* 34 (1995): 76–89; and Steve Bruce, "Religion and Rational Choice: A Critique of Economic Explanations of Religious Behavior," *Sociology of Religion* 54 (1993): 193–205.

61. Jon Butler, *Awash in a Sea of Faith: Christianizing the American People* (Cambridge, Mass.: Harvard University Press, 1990), chap. 8; Nathan O. Hatch, The Democratization of American Christianity (New Haven, Conn.: Yale University Press, 1989), chap. 5. On the "market revolution," see Charles Sellers, *The Market Revolution: Jacksonian American, 1815–1846* (New York: Oxford University Press, 1991); Melvyn Stokes and Stephen Conway, eds., *The Market Revolution in America: Social, Political, and Religious Expressions, 1800–1880* (Charlottesville: University Press of Virginia, 1996); "A Symposium on The Market Revolution," *Journal of the Early Republic* 12 (1992); Sean Wilentz, "Society, Politics, and the Market Revolution, 1815–1848," in *The New American History*, rev. ed., ed. Eric Foner (Philadelphia: Temple University Press, 1997); and Daniel Feller, The Jacksonian Promise: America, 1815–1840 (Baltimore: Johns Hopkins University Press, 1995).

62. I develop this theme more fully in Nord, "Systematic Benevolence" and "Free Grace." For thoughtful discussions of the complex responses of religious Americans to the market revolution, see Daniel Walker Howe, "Charles Sellers The Market Revolution, and the Shaping of Identity in Whig-Jacksonian America," and Richard Carwardine, "Charles Sellers's 'Antinomians' and 'Arminians': Methodists and the Market Revolution," in this volume.

63. "Prospectus," in *American Messenger*, January 1843, 1; October 1851, 38; June 1851, 22; [Cook], *Home Evangelization*, 140.

64. [Cook], *Home Evangelization*, 41; *American Messenger*, February 1844, 6.

65. Ibid., 107 and 5, respectively.

66. Michael Merrill, "Putting 'Capitalism' in Its Place: A Review of Recent Literature," *William and Mary Quarterly*, 3rd ser., 52 (1995): 325; Merrill, "The Anticapitalist Origins of the United States," *Review: A Journal of the Fernand Braudel Center* 13 (1990): 468–69. See also "Special Issue on Capitalism in the Early Republic," *Journal of the Early Republic* 16 (1996).

8

Philadelphia Presbyterians, Capitalism, and the Morality of Economic Success

Richard W. Pointer

In a lecture to young people on the Eighth Commandment ("Thou shall not steal") in 1830, a prominent Philadelphia minister pronounced that its essential implication was the Christian's duty to increase his own and his neighbor's "worldly prosperity." Fulfilling that duty required choosing the right calling and pursuing it diligently. Industry or hard work was the key to success in any business. Young people simply had to look around and "see who are the men of wealth." Virtually all of them "began the world with little—often with nothing but their hands and their industry." Fortunately, the minister concluded, "the same way to wealth" was still "equally open to all."[1]

During the 1840s a Protestant tract, circulating among American workingmen, claimed to provide *True Philosophy for the Mechanic*. Written by the editor of a denominational board of publication, the four-page pamphlet related the story of a Mr. Wiggins, a master cabinetmaker, and his blacksmith neighbor, Mr. Sledge. After hearing Wiggins's complaints about economic misfortune and domestic strife, Sledge confessed how he, too, had suffered hard times and how they had sparked him to find a solution. Following his wife's example, he had turned to a "book of philosophy" and began reading it. Ever since, his work and family life had been successful. Wiggins quickly realized that Sledge's book was none other than the Bible and promptly committed himself to trying its philosophy. As argued, there now was not "a more quiet, orderly, and prosperous man in all the neighborhood than Mr. Wiggins."[2]

In 1851 the clergyman-president of Washington College hailed the fruits of American technology in an address titled *The Progress of the Age*: "What mighty changes have been wrought in the economy of human industry, by the application of the power of steam! . . . Machines of countless forms, and for a thousand various purposes, impelled by steam, perform like things of life, with magic rapidity and exactness, the processes of production, which once demanded long continued and patient toil." Not only did the "whirl and clatter of machinery" demonstrate man's progress in controlling nature, it alleviated "the curse of labor"; technological improvements, as part of the progressive spirit of the age, had "marvellously lifted from man the weight of that curse which was pronounced upon him when driven from Eden's bliss."[3] Such evidence indicated clearly that "our race, as a whole, has been progressing rapidly toward a millennial state."[4]

The Christian's duty to prosper, self-help as the way to wealth, Christian piety as an asset to temporal success, the ordering and disciplining effects of Christian morality, the material and moral benefits of the machine, technological progress as a sign of the approaching millennium—all these themes were common fare for American Protestant ministers in the second quarter of the nineteenth century. Here there is no surprise. Historians have long recognized the growing convergence of Protestant and American middle-class values in the mid-nineteenth century, and they have demonstrated that evangelical Protestants' ascendancy was due in part "to their willingness to allow their message to be accommodated to the spirit of the culture."[5]

Nowhere was that accommodation more complete, according to many studies, than in evangelicalism's embrace of emergent capitalism. One prevailing interpretation suggests that the new revivalist theology—promoted by such ministers as Charles G. Finney, Lyman Beecher, and Albert Barnes and embodied in progressive denominations like New School Presbyterianism—emphasized individual human freedom and responsibility in attaining salvation, a view consonant with a market economy predicated on individual acquisitiveness.[6] The moralism of this Arminianized evangelicalism included a work ethic of self-reliance and self-control.[7] Such an ethic proved ideally suited to the needs of middle-class entrepreneurs bent on maintaining moral authority over workers despite fundamental changes in employee-employer social relations.[8] By midcentury, nothing less than a coherent theory of "Christian capitalism" had developed to give sweeping religious and moral sanction to the existing economic order.[9]

In light of this scholarship, particularly on the links drawn between the new revivalism and entrepreneurial capitalism, what may be surprising about the preceding examples is that they all come from accounts made by men thoroughly opposed to the "new measures" and (in their view) Pelagian theology associated with Finney's evangelicalism. The three authors—Ashbel Green, William Engles, and George Junkin—were Old School Presbyterian ministers, staunchly committed to the theological tenets of traditional Calvinism. Collectively, they are perhaps best known as the accusers in the heresy trials of their fellow Philadelphia pastor Albert Barnes, during the 1830s.[10] Aligned with what scholars have invariably labeled as the "ultraconservative" wing of Presbyterianism, these men were largely responsible for precipitating the denominational split into formal Old and New School branches in 1837.[11]

That split, and particularly its causes, have received considerable attention by historians. The most thoroughgoing account to date has argued that theological differences were at the heart of the division.[12] Dogmatic Old Schoolers insisted that the New School party held heretical views on the nature of unregenerate man, original sin, and the imputation of Adam's guilt. Debates on these issues helped reveal other points of disagreement, including attitudes toward subscription to the Westminister Confession, Presbyterian polity, voluntary societies, methods of revivalism, and slavery. In general, scholars have depicted the New School as demonstrating a confidence in the dignity, freedom, and ability of man, which blended well with the main currents of thought in Jacksonian America.[13] Its revised Calvinism might have even been a conscious doctrinal adjustment to the nation's new market economy.[14] In contrast, Old School members have been seen as spokesmen for the old religious and social order, promoting a theology and worldview increasingly out of step with the beat of antebellum culture.[15]

Yet in the writings described above, each Old School minister espoused economic views remarkably similar to that which historians have associated with New School "progressives" such as Barnes and Henry Ward Beecher. Were these examples mere aberrations from the Old School norm? Was the Old School's hard-line Calvinism paralleled by an equally old-fashioned social perspective? Or were these statements representative of the reaction of Calvinist evangelicals to antebellum capitalism? How did they and their New School opponents look upon the new economic order and what moral advice was given on how to live within it?

By examining the economic views of both Old School and New School Presbyterians within a particular setting—in this case, Philadelphia—in the thirty years (1825–1855) in which their theological and ecclesiastical cleavage was greatest, it is possible to explore these questions. Such an inquiry also will highlight the economic ethic worked out by members of both schools amid the onset of industrial capitalism in the city of Brotherly Love.

Philadelphia's demographic and economic character changed dramatically in the first half of the nineteenth century. An urban population of less than 70,000 in 1800 grew to over 500,000 by the consolidation of 1854, thanks to heavy in-migration from other parts of the United States and, after 1840, to large-scale foreign immigration from Ireland and Germany.[16] Philadelphia's expanding populace generally found ample job opportunities because of the growth and transformation of the city's manufacturing sector.[17] The port's once vital foreign commerce stagnated in the early part of the century, but the establishment of such industries as textiles, machinery, and precious metals compensated for the decline in the export trade. Each industry fed off the strong demand for its goods from domestic markets in eastern Pennsylvania and adjoining states. Rapid industrial growth resulted in the urbanization of most of the county by the 1840s.[18] Thus, well before the Civil War, Philadelphia was familiar with the twin byproducts of enterprising capitalism in the 1800s: industrialization and urbanization.

Presbyterianism thrived in that antebellum setting. In a city experiencing impressive population growth, Presbyterian gains were even more impressive. The tenfold rise in Presbyterian membership (from 500 to 5,000) between 1800 and 1830 dwarfed the 133% increase in the county's population.[19] Communicant growth remained strong in the next quarter century, so that total Presbyterian church membership in Philadelphia topped 12,000 in 1855.[20] In addition, since communicants constituted perhaps of only a fifth of those attending Presbyterian churches in the mid-nineteenth century, there may have been as many as 60,000 Presbyterian adherents in the city by the outbreak of the Civil War.[21] The denomination's institutional expansion in Philadelphia was equally noteworthy, as its congregations ballooned from a total of four in 1800 to over fifty by the mid-1850s.[22] That growth, along with Presbyterianism's long tradition in the area, helped make Philadelphia the most Presbyterian of any of America's major cities. Little wonder, then, that the General Assembly of the Presbyterian church met there almost every year during the antebellum period or that most prominent northern Presbyterian ministers held pastorates in the city at some point in their careers.[23] If, as one historian has suggested, the first half of the nineteenth century was "the greatest age of Presbyterianism in America," the same can be asserted about Presbyterianism in Philadelphia.[24]

To what extent Presbyterian success was directly tied to Philadelphia's demographic and economic changes has never been conclusively determined.[25] What is clear is that members of both schools in Presbyterianism were aware of these changes in the years from 1825 to 1855. Presbyterian religious periodicals routinely noted the growth of American manufacturing from the 1820s on, citing increases in the number of factories and their output.[26] The corresponding rise in the number of industrial workers also occasioned comment. As early as 1828 the Presbyterian-dominated Philadelphia City Sunday School Union recognized the expansive working class: "Mechanics . . . form a large part of our community, and are increasing in a more rapid progression than any other class. They will soon, to all appearance, become the vital part of this community—the spring and life of activity."[27]

During the 1830s and 1840s, Presbyterian ministers paid attention to other vital shifts in Philadelphia's business life. In the 1830s, the New School's Albert Barnes and the Old School's Cornelius Cuyler observed that investment opportunities were on the rise, tempting entrepreneurs and workers alike with dreams of instant wealth.[28] In the 1840s, James W. Alexander (Old School) expressed concern about the ill effects of the decline of the old master-apprentice relationship and its replacement by the wage relation of factory owner and workingman.[29] And in 1851, Henry Boardman (Old School) suggested that the pace and scale of the city's commercial activity had increased so much in the past generation that if a businessman who had finished his career in 1821 was brought back to work, "he might almost imagine himself in another planet."[30]

Like most evangelical Protestants, Philadelphia's Presbyterians responded to the city's (and nation's) economic development with a mixture of enthusiasm and fear. No one was more enthusiastic than Old School minister George Junkin. As editor of *The Religious Farmer* in the later 1820s, Junkin told his rural readers that they should welcome greater domestic industry because the manufacturer was the farmer's "most sure market."[31] To illustrate industry's other benefits, he reprinted a North Carolina State legislative report that hailed northern manufacturing for "diffusing wealth and prosperity, and improving the moral condition of society."[32] The following two decades of industrial growth heightened Junkin's optimist. At midcentury, he celebrated industry for making life easier and more comfortable through new forms of clothing, housing, and domestic appliances—all democratic improvements that served "to elevate and bless the masses rather than the few."[33]

Junkin's ebullience was balanced by the more tempered views of other Presbyterians, who found plenty to worry about in the economic trends of their day. Among the most pressing concerns of Presbyterian leaders was how to keep members of the rapidly increasing working class from "vicious" indulgences. Philadelphia's industrial and commercial expansion was bringing thousands of young, single males to the city to work in factories and merchant houses. Separated from the morally uplifting influences of home and family in the countryside and confronted by the city's numerous "seductive allurements," these young mechanics and clerks were supposedly extremely vulnerable to vice and immorality.[34] If unchecked, their moral transgressions would have devastating effects on everything from Philadelphia's social order to its spiritual ardor and economic prosperity.

Prosperity itself was not an unmixed blessing, according to some Presbyterians. Speaking before the Mechanics' and Workingmen's Temperance Society in 1835, Albert Barnes insisted that the prevalence of intemperance in America was due to the nation's prosperity. People simply had too much money to spend on drink and other vices.[35] Cornelius Cuyler expanded the point after the panic of 1837. He suggested that when American prosperity had seemed limitless, "there was an expansion of grasp, of desire, and of hope, which saw neither end nor limit to the acquisition of this world's goods. Few doubted their ability to obtain their desires, and few were careful to confine their desires within reasonable or moderate bounds."[36] This excessive devotion to material gain did not abate with the depression but instead continued in the 1840s and 1850s, constituting in Barnes's words America's "besetting sin."[37] He and his Old School theological opponents joined voices in those years in repeatedly decrying Americans' unrestrained passion for wealth.[38]

As if the moral vulnerability of the working class and American materialism were not enough to fret about, Philadelphia Presbyterians also found time to bewail the growth of speculation: "An intense eagerness for large and quick profits" had infected the business community, and a get-rich-quick mentality permeated workers of all social ranks. The speculative spirit was troublesome

not only because it shifted the object of men's labor from healthful employment and a decent living to amassing wealth but also because it made them
impatient with the "true" path to economic success—honest, persistent toil.[39]
In the process, speculation threatened to turn men into "practical Atheists"
by persuading them that their business fortunes were more dependent
on their skill at exploiting current economic circumstances than on God's
providence.[40]

Shared by members of both parties in the city, these anxieties combined
with a general optimism about capitalist progress to form the immediate intellectual and emotional backdrop for Presbyterian moral advice on work and
wealth. Between 1825 and 1855, ministers and laymen used sermons, tracts,
newspapers, lectures, books, and magazines to offer ethical guidance on economic matters to any and all Philadelphians who would listen or read. They
were confident that saints and sinners alike could benefit from heeding com-
monsensical moral wisdom mined from the pages of Scripture and applied to
everyday life in nineteenth-century America.[41]

At the heart of any economic counsel that Philadelphia Presbyterians provided
was the promotion of a set of individual virtues reminiscent of both the Puritan work ethic and Benjamin Franklin's plan of moral perfection. Industry,
thrift, frugality, sobriety, honesty, charitableness—these were the qualities
that brought distinction to a man in the workplace and readied him for success. Their opposites—idleness, intemperance, prodigality, sloth, extravagance—led to economic ruin and poverty.

Labor historians Paul Faler and Bruce Laurie have seen these work values
as part of a "new industrial morality" for America's laboring people. Arising
alongside the growth of manufacturing in the Northeast in the first half of
the nineteenth century, this morality was avidly supported by middle-class
Arminian evangelicals as part of their moral reformism.[42] In Philadelphia,
Laurie says, New School Presbyterians and Methodists led the way in indoc-
trinating artisans with the importance of work and self-discipline.[43] Men like
Albert Barnes "lauded the sober, hard-working middle class" and lashed out
at whatever "abetted idleness and profligacy."[44]

But so, too, did Old School Presbyterians. And whereas Barnes revived
images of Cincinnatus, the virtuous yeoman farmer, to illustrate the nobility
of work to his urban audiences, conservative Calvinist Thomas Beveridge
appealed to the far more relevant example of Jesus Christ, the ancient artisan, whose toil as a carpenter until the age of 30 "ennobled labour" forever.[45]
Beveridge and other Old School sympathizers left little doubt about what the
Bible enjoined regarding work: "The Scriptures give no tolerance to idleness,
no countenance to carelessness respecting our worldly concerns. Industry was
the duty and happiness of man in a state of innocence."[46] They left equally
little doubt about the rewards of obedience: "The sleep of the labouring man
is sweet—the bread of industry is pleasant and healthful, while the idle are
dull, discontented, devoured by care, and sinful lusts, wearied by time, and
oppressed by the load of existence."[47]

To industry were to be added frugality and economy. Saving and thrift, said Ashbel Green, were the surest means of increasing one's property. Wise economy of personal resources rather than rapid gains held the greatest promise for long-term success.[48] This did not imply a parsimonious lifestyle, for miserliness was as much to be avoided as prodigality.[49] Instead, the ideal that Calvinist Presbyterians set out was one of respectable, middle-class enjoyment of God's providential blessings:

> To provide for a household, is not to heap up riches without using them. There is nothing more foolish than to deny ourselves every thing comfortable for the present, that we may guard against want in the future. A kind Providence is a better security than all the property you may collect, or the precautions you may adopt. . . . Should we not . . . freely use what Providence bestows, trusting that while we are *Diligent in business and fervent in spirit*, he will never fail nor forsake us?[50]

Where Old Schoolers implored a more rigid self-denial was in the use of strong drink. Although assuming a variety of positions on the question of total abstinence, ministers Cornelius Cuyler, John McDowell, Thomas Hunt, Ashbel Green, Thomas McAuley, George Junkin, James Alexander, William Engles, Henry Boardman, and William Neill were all active temperance advocates in Philadelphia in the 1830s.[51] Their reform efforts were joined by such Old School laymen as Robert Ralston, Alexander Henry, and Matthew Newkirk—prominent members of Second Presbyterian Church and Central Presbyterian Church and wealthy representatives of the city's commercial and manufacturing sectors.[52] Collectively, they saw temperance as a "Christian duty" incumbent upon men of every occupation and social rank. Masters were responsible for setting an example of "rigid temperance" and encouraged to dismiss employees who spent their leisure hours drinking since "to say that a man is often seen hanging about the tavern porch, under whatever pretence of business, is to say that his work is neglected, his habits declining, and his company detestable."[53] For their part, young laborers were to realize early in life the benefits of sobriety and thereby avoid the almost inevitable consequences of drink: poverty, disease, crime, intemperate children, heartbroken wives, early death, and everlasting torment in hell.[54]

Honesty and benevolence rounded out the set of economic virtues most often extolled by Old School men. Great commercial magnates, as well as lowly domestic servants, were to practice "inflexible integrity" in their business dealings. Anything less violated God's law and led eventually to financial and personal ruin.[55] All people were also to practice charity to the poor and needy. Whether possessing little or much, the "good worker" undertook selfless acts of kindness and exercised an individual Christian stewardship aimed at relieving human suffering.[56]

Whether sitting in the pews of Albert Barnes's First Presbyterian Church or in those of one of the city's Old School congregations, Philadelphians heard ministers and lay leaders celebrate similar work values. To be sure, there were some differences in emphasis or tone between New Schoolers and their opponents. Both sides, for example, implored their listeners to be industrious, but conservatives usually framed this admonition in the context of duty to employer and family, whereas their New School counterparts spoke more often of industriousness as a key to preserving republican virtue and as an asset for building a favorable reputation.[57] Likewise, each school preached charity to the needy, but Old Schoolers talked of the poor as an inevitable part of the hierarchical (and legitimate) social order, whereas Barnes and his colleagues saw the "*lower stratum* of society" as an obstacle to revivals. From the New Schoolers' perspective, a narrowing of the gap between rich and poor might lead to a greater harvest of souls.[58] These differences are important and should not be overlooked. If Robert Doherty is correct in asserting that the two schools had distinct social bases, such differences may have stemmed from efforts to appeal to different audiences.[59] Yet, when looking at the big picture, these disparities are overshadowed by the sweeping convergence of Old School and New School thought, not only on work values, but also on economic ethics in general.

How was this convergence possible in light of the two parties' strong ideological differences on other matters? One aid to understanding may be the realization that these economic virtues and attitudes toward work were hardly new to Presbyterians in the second quarter of the nineteenth century. From at least 1800 on (and perhaps much sooner than that), Presbyterian and other Reformed ministers had emphasized the importance of industry and frugality in averting national decay and ruin; hard work and sensible economy had to overcome idleness, luxury, and extravagance for the Republic to survive.[60] When Old School and New School ministers echoed these themes in the 1830s and 1840s, therefore, they were merely testifying to the two schools' common Presbyterian heritage and inheritance.

Moreover, members of both schools operated out of a single set of intellectual assumptions concerning how to think about economic life. Those assumptions were rooted in the republicanism and commonsense philosophy that permeated America in the early nineteenth century. From republicanism, Presbyterians derived the notion that the structures of American society were virtuous, regardless of how often individual or group action failed to uphold those structures. Commonsense philosophy prompted them to believe that economics was a sphere of life in which it was more necessary to work out a set of commonsense principles than a full-fledged theology. Consequently, what proceeded from Presbyterians of both schools in the antebellum era were piecemeal Christian responses to developments in an economic order assumed to be essentially sound. Neither Old Schoolers nor New offered or even saw the need for any rigorous theological analysis of the foundations of American economic life.

If there was anything new in the Presbyterians' message after 1825, it was the bolder association of traditional economic virtues with individual economic success rather than simply with the national welfare. Whereas in the eighteenth century private interest and public good had often been depicted as being in conflict, by the mid-1820s what was good for the individual was also good for the nation and vice versa. Most likely, both New School and Old School ministers learned to apply the ties between virtue and success and vice and poverty to the individual through the moral philosophy courses they took in college or seminary. Whether at Princeton, Union, Jefferson, or some other institution of higher learning, these men had been taught to believe that the universe operated according to a divinely established moral law. God's governance through that law extended to all spheres of human activity, including the economic, and was evidenced best by the preestablished harmony between virtue and happiness.[61] As Old School pastor and professor Archibald Alexander put it in his own textbook on moral philosophy, by the "laws of nature, virtuous conduct is generally productive of pleasure and peace of mind; and immoral conduct is generally a source of misery."[62] Alexander and other academic moral philosophers trained this generation of Presbyterian pastors to think that God's moral ordering of the universe included making *both* individual success and social prosperity dependent on virtue.[63]

That ministers of each party learned that lesson well is evident in how Philadelphia Presbyterians interpreted the depression of the late 1830s. For all their other disagreements, Albert Barnes and Cornelius Cuyler saw eye to eye on why America was beset with an economic crisis. Widespread individual moral failure had provoked divine judgment on the whole nation in the form of financial panic and economic hardship. Barnes emphasized the idleness and intemperance of workers as contributing factors. Cuyler highlighted Americans' covetousness and prodigality: "Let us . . . think of this evil which we are suffering as a divine infliction upon us for the ardour with which we have loved, and the eagerness with which we pursued the world."[64] For both men, however, the bottom line was the same: Vice had reaped its just reward—individual and national calamity.

In arguing that current American sins led to divine punishment, Presbyterian ministers asserted that God acted on the basis of his moral law to direct people to him. Their statements suggested that the United States was not at the mercy of an arbitrary, inscrutable divine providence. Instead, God's intervention in American history was rational and even predictable to those who were attuned to his natural laws. Fred Hood and James Turner have examined how that largely mechanistic conception of God's actions robbed the divine of much of its former mystery and awe.[65] Equally important is what it implied about human responsibility and control over the temporal fortunes of individuals and nations. If personal success and national prosperity depended on human virtue, human beings were on their own to determine their earthly fate within the bounds of the moral law.

Adopting that wide view of human autonomy in an era of burgeoning democracy, individual self-reliance, and capitalist expansion would hardly have been shocking, even for American Calvinists. Yet the truth is that Philadelphia Presbyterians recoiled from this conclusion in favor of a continued insistence that God alone ultimately controlled the earthly, as well as heavenly, lot of human beings. Even while subscribing to the newer view of divine providence, Presbyterians of both schools maintained their belief that

God was sovereign over worldly wealth and free to bestow or withhold it as he saw fit.

The surprising point here is not that only staunch Old Schoolers but that Albert Barnes also could be heard saying, "In temporal things, we say that riches, and health, are given according to his good pleasure."[66] As recent historians have correctly suggested, Barnes placed the responsibility for finding and practicing a calling squarely on human shoulders,[67] but so, too, did Ashbel Green and James Alexander.[68] None of them, however, concluded that people governed their own earthly fortunes. Barnes readily admitted that "men make great changes themselves, and do much to effect their own destiny," but he was clear about who ultimately directs the course of human lives: "We deem our dwellings fixed, and think we could draw on a map the great outlines of our course. Yet who knows what a day may bring forth? And could our fancied chart be laid beside the *real* one how faint would be the resemblance!"[69]

For Barnes and other Philadelphia Presbyterians, believing that God reigned over humankind's temporal affairs provided a means for explaining why human reality in economic matters did not always conform with human expectation. Given God's sovereignty and mysterious providence, there was an explanation for why the wicked sometimes prospered and the righteous sometimes suffered, economically and otherwise. As New School pastor Daniel Carroll described it, "The most unlikely, and, to all appearance, the most undeserving" were often crowned with "the richest blessings of a bountiful Providence," while others more deserving were denied these temporal blessings and forced to endure poverty and affliction.[70] These aberrant cases did not disprove natural law; they merely indicated that the links between virtue and prosperity and vice and poverty were *generally but not absolutely* true. Furthermore, they served as reminders of God's overruling authority, for even the prosperity of the unrighteous was "the direct result of God's sovereign providence."[71]

Thus, God's unyielding control over the things of this world stood alongside middle-class work values and the moral law in the economic ethic of antebellum Philadelphia Presbyterians. Although anxious to hold men accountable for working hard and saving earnestly, both New School and Old School devotees were reluctant to sacrifice the traditional Calvinist emphasis on divine sovereignty. As a result, their economic perspective was fraught with the same kind of tension between human responsibility and divine prerogative as permeated their various understandings of human salvation.

Nowhere was that tension more evident than in the argument that men were responsible to practice the virtues that brought success but any success achieved was to be understood as wholly the gift of God. If a man acquired wealth through diligent, frugal, laborious effort, "It was [still] God that did all this."[72] Apparently, not all lay Presbyterians found that mystery easy to accept. At least that is the impression Ashbel Green left in one charity sermon: "But in regard to their *worldly substance*, perhaps gradually acquired, and in the acquisition of which their contrivance and management, their laborious efforts and persevering industry, have been constantly exerted, they are not so sensible of the truth. They do not at least, so deeply and constantly realize that whatever they possess in this way, cometh as truly of God as if he had given it to them by the most remarkable and extraordinary dispensation of providence."[73]

Much else in what Old School and New School Presbyterians said on the morality of economic success was similarly tension-riddled. Probably all Presbyterian ministers at one time or another preached on the biblical admonition to be content in all things, including one's present economic lot. To envy and covet the wealth of others was idolatrous and an implicit denial of God's authority over worldly affairs. On the other hand, laypersons were told that contentment did not mean inactivity or passive acceptance of the status quo. Rather, they must try to improve their social and economic circumstances even while accepting their current condition graciously.[74] The task set before them, then, was to exhibit economic ambition and economic satisfaction simultaneously.

Underlying that ambiguous if not contradictory message were conflicting Presbyterian convictions about the moral legitimacy of economic gain and the immoral zeal with which antebellum Americans were seeking it. Neither Old School nor New School pastors questioned the moral integrity of wealth acquired through honest means. Henry Boardman stated plainly: "I look with no disfavor upon legitimate accumulation. I greatly honour the man who secures, by honest means, a competent or opulent estate, and employs it in doing good."[75] Ministers, as well as lay people, were entitled to have material goods beyond the simplest "necessaries of life," according to New School sympathizer Ezra Stiles Ely. Owner of a sizable estate himself, Ely refuted the notion that Christians, and especially clergymen, were obligated by scriptural authority or Christian values to avoid all luxuries. On the contrary, he said, buying luxury goods helped ensure the continued employment of thousands of artisans who produced them and thereby fulfilled the Christian duty to aid one's neighbor.[76] Albert Barnes similarly blessed the amassed possessions of the well-off, insisting that the economically successful were not required by Christianity to deny themselves the "ordinary comforts" attached to their social rank.[77]

These endorsements of material acquisition were at least partially countermanded by laments over America's insatiable thirst for wealth. As noted earlier, concern with the nation's growing materialism was widespread among Philadelphia Presbyterians throughout the second quarter of the nineteenth century. If, as one historian has remarked, Albert Barnes's sermons and lectures may be read as "celebrations of acquisitive man," many of them may also be read as jeremiads calling the Republic's citizens back from their all-consuming pursuit of money.[78] In one memorable passage reflecting his eastern bias, Barnes acerbically described those who had succumbed to mammon:

> This passion [for wealth] goads on our countrymen, and they forget all other
> things. They forsake the homes of their fathers; they wander away from the
> place of schools and churches to the wilderness of the west; they go from the

sound of the Sabbath-bell, and they forget the Sabbath, and the Bible, and the place of prayer; they leave the places where their fathers sleep in their graves, and they forget the religion which sustained and comforted them. They go for gold, and they wander over the prairie, they fell the forest, they ascend the stream in pursuit of it, and they trample down the law of the Sabbath, and soon, too, forget the laws of honesty and fair dealing, in the insatiable love of gain.[79]

When Presbyterians aimed to strike a balance in their message between the value and vanity of economic advance, they usually resorted to the injunction that Christians were to have a "holy superiority" to the world's possessions even while being free to enjoy any happiness they afforded.[80] The believer was to have "a spirit and temper *above* the things" that influenced others. If Christians were blessed with wealth, they were to make sure that their "affections" were not "supremely fixed on it."[81] A healthy detachment from material goods, in other words, was the proper Christian attitude toward the things of this world.

Presbyterian leaders of both schools recognized that refusing to be preoccupied with temporal gains took a large dose of divine grace in a society absorbed with economic prosperity. Only true religion could keep a busy public man from getting caught up in the frenzied pace and enticing allurements of the modern business community. A genuine conversion and the practice of Christian piety were necessary to steel a person against the seductive temptation of a life aimed solely at worldly indulgence.[82]

Alongside their case for Christian piety as a defense against selfish materialism, Presbyterians ironically juxtaposed arguments promoting piety as an essential asset in achieving economic success. Typical was Ashbel Green's claim that "true religion has no tendency to diminish, but on the contrary, a direct tendency to increase, the stock of present fruition."[83] In a city afflicted with capitalist fever, these evangelical Protestants were not about to suggest that Christianity was an obstacle to earning a good living. On the contrary, any hint that godliness among young men would "interfere with their secular business, and defeat their worldly anticipations" was quickly denied.[84]

Among the most avid promoters of this idea was Henry Boardman. Educated at Yale College and Princeton Seminary, Boardman assumed the pastorate of Philadelphia's Tenth Presbyterian Church in 1833 at age 25. He spent the next forty-three years ministering to that large Old School congregation located in the midst of the city's rapidly expanding business sector (Twelfth and Walnut). Boardman publicly advocated the theme "Piety Essential to Man's Temporal Prosperity" for the first time in a sermon before the Philadelphia Young Men's Society in 1834.[85] Taking as his text 1 Timothy 4:8, "Godliness is profitable unto all things," he set out to show that "the best interests of every young man for the present life, will be greatly promoted by personal piety." In his view, Christian faith aided the temporal pursuits of young men by clarifying what was worthy in life and producing "the most valuable qualities for the management of business."[86] In the 1840s, Boardman outlined the benefits of personal religion for the professional careers of doctors and lawyers, and then in the early 1850s did the same for the city's merchants.[87] Overall, he evangelized thousands of Philadelphians with the word that Christian piety would enhance rather than undermine their chances for a happy, successful life.

Here again, Presbyterians sounded a mixed message. A life of Christian discipline kept people from conforming to America's current passion for wealth, but at the same time it heightened their likelihood of reaping large earthly rewards. Such a message was not a crass gospel of success. Yet it did reflect how extensively emergent capitalism was shaping the proclamation of Christianity in antebellum Philadelphia.

A final ingredient in what Presbyterians said on the morality of economic success was their denial that wealth was a sure sign of individual salvation or God's blessing. Recent arguments that Albert Barnes preached that "success in one's calling was a sign of regeneration" are not borne out by the evidence. Neither Barnes nor any other Presbyterian minister equated material prosperity with Christian fruitfulness.[88] Instead, they claimed that the "highest degree of worldly prosperity" did not prove the "favor of God."[89] Daniel Carroll typified Presbyterian opinion on how temporal riches provided no assurance of spiritual health: "When you have actually reached the highest pinnacle to which your worldly aspirations can carry you, you have no security that you can stand there a moment! Your very *success* may prove your [spiritual] *ruin!*"[90]

Still, like the other parts of their economic ethic, the Presbyterians' witness on this point was not without its inconsistencies. The tension here lay not in the rhetoric itself but in an apparent conflict between what was said and what was practiced. For all the disclaimers of automatic parallels between man's economic and spiritual estates, Presbyterian ministers and congregations acted as though prominence in the business world was the primary prerequisite for leadership in the church. That, at least, was the claim of Philadelphia industrialist and Old School layman Stephen Colwell. He sharply criticized the churches for treating their wealthy members as a spiritual elite, regardless of whether they were "real disciples of Christ." In Colwell's mind, "the spirit of business" had invaded Protestantism to the point that financial contribution rather than spiritual devotion dictated who were the men of influence.[91] Churchly behavior, in this case, belied the testimony of ministers against associating material blessings with divine pleasure.

From this survey of Presbyterian ethical thought in antebellum Philadelphia, four major conclusions are possible. The first and most obvious is that despite their important theological differences and formal ecclesiastical separation, Old School and New School Presbyterians in the city promoted a virtually identical set of moral precepts for Americans to follow within the economic realm. Certain subtle variations in emphasis did exist. Nevertheless, the two schools' thinking was sufficiently close that it is entirely legitimate to speak of a single Presbyterian economic ethic in the years from 1825 to 1855. That is not to suggest that all Presbyterians supported or opposed the same economic theories and policies. Preferences on issues such as protective tariffs or internal improvements cut across both Presbyterian and Old School–New School boundaries. But members of both schools offered remarkably similar wisdom on the morality of economic success.

Second, in Philadelphia at least, the New School was not so radically accommodating and the Old School was not so thoroughly tradition-bound in their economic views as historians have implied. Instead, both groups sought to endorse, support, and protect the market economy without sacrificing the distinctions or integrity of their religious faiths. Sometimes they succeeded; sometimes they failed. But in either case, they could be Arminian proponents of the new revivalism or stalwart advocates of five-point Calvinism. One did not have to be a disciple of Finney to champion burgeoning capitalism or the new industrial morality. Nor did one have to believe in the imputation of Adam's guilt to question material acquisitiveness. Rather, the ideologies of both Presbyterian schools encouraged men and women to attach themselves to emergent capitalism in a wholehearted but tempered manner.

Why theological disagreement did not in this case translate into more divergent economic perspectives is puzzling. Given the respective progressive and traditional characters of their theologies, New School Presbyterianism and Old School Presbyterianism would appear to have been ripe for holding vastly different views on the new economic order. That kind of divergence, of course, is what previous historians have assumed and postulated. Uncovering a high degree of convergence in what Presbyterians actually said, then, comes as a striking surprise. How is their agreement to be explained? Although any full explication will have to await additional investigation into the intellectual and social character of Philadelphia Presbyterianism, a partial answer lies in a recognition of the common intellectual heritage of the two schools. On the one hand, they both drank at the waters of republicanism and commonsense philosophy. On the other hand, the collective wisdom on economic morality offered by Presbyterian greats—the John Calvins, John Knoxes, John Witherspoons, and Samuel Stanhope Smiths—filtered down to both parties. Since the main issues (with the possible exception of slavery) that divided the two sides did not call any of their heritage into question, both Old Schoolers and New Schoolers could perpetuate their collective heritage even after splitting denominationally.

The fact that they did perpetuate that tradition points to a third major conclusion: Rather than manifesting an innovative or novel character, mid-nineteenth-century Presbyterian moral thought was decidedly "conservative" in that its main emphases were carryovers from the early Republic. As Donald Meyer has suggested, Presbyterians and other evangelicals in antebellum America were not looking to create a new morality but only "to find new ways to justify and sustain basic ethical values in an emerging capitalist economy."[92] In their view, industry, frugality, thrift, temperance, honesty, and charitableness had always been and would always be the appropriate values for the workplace. And they were sure that these values were equally relevant for the seaman, the merchant clerk, the mechanic, the lawyer, the minister, and every other calling. Likewise, God's sovereignty over temporal affairs was a constant, even if humans now saw the divine governing more indirectly through the operation of his moral law. That law, fully expounded by 1800, made clear the innate connections between virtue and prosperity and vice and misery, and it provided the means for three different generations of Presbyterians to explain the panics of 1819, 1837, and 1873 in precisely the same way.[93]

Fourth, Presbyterian borrowing from the past during the second quarter of the nineteenth century did not mean that their thought was immune from the changing social and economic realities of antebellum Philadelphia. On the contrary, the tensions that riddled the Presbyterian economic ethic derived largely from the pressures exerted by the new economic order. Capitalist emphasis on self-determination and individual freedom pushed Presbyterians toward affording men and women more (if not ultimate) responsibility for their worldly fortunes. Similarly, the age's absorption with upward mobility and material prosperity encouraged Presbyterians to posture themselves as the friends of economic gain. It also led them to promote Christianity as a business asset for aspiring merchants and industrialists. Meanwhile, the market economy's rewards to the wealthy of power and prestige carried over into the distribution of influence within Presbyterian congregations. At the same time, the economic fluctuation between prosperity and depression reminded Presbyterians of the temporal quality of worldly riches. And the speculative spirit and exorbitant lust for wealth that consumed so many Americans sparked a renewed emphasis on the virtue of godly contentment. All in all, then, Presbyterians were both powerfully drawn and occasionally repulsed by emergent capitalism. In a very real sense, they were torn between embracing it fully and keeping it at arm's length. For the time being they tried to tread a middle road. But the development in the postbellum era of the Gospel of Wealth, on the one hand, and the Social Gospel, on the other hand, suggests that Presbyterians could live with that tension only so long.[94]

Notes

1. *Christian Advocate* 8 (November 1830): 557–59. This lecture was also published in Ashbel Green, *Lectures on the Shorter Catechism of the Presbyterian Church in the United States of America: Addressed to Youth* (Philadelphia: A. Finely, 1829).

2. [William M. Engles], *True Philosophy for the Mechanic* ([Philadelphia]: n.p., n.d.), 1–4. This tract was printed and distributed by the Old School Presbyterian Board of Publication.

3. George Junkin, *The Progress of the Age* (Philadelphia: n.p., 1851), 8–9.

4. Ibid., 21.

5. George M. Marsden, *The Evangelical Mind and the New School Presbyterian Experience* (New Haven, Conn.: Yale University Press, 1970), 230–31.

6. James H. Moorhead, "Charles Finney and the Modernization of America," *Journal of Presbyterian History* 62 (1984): 101–5; Clifford E. Clark, Jr., "The Changing Nature of Protestantism in Mid-Nineteenth Century America," *Journal of American History* 57 (1971): 832–38, 843–46; Bruce Laurie, *Working People of Philadelphia* (Philadelphia: Temple University Press, 1980), 34–39; and Robert Doherty, "Social Bases for the Presbyterian Schism of 1837–1838: The Philadelphia Case," *Journal of Social History* 2 (Fall 1968): 76–78. This is also the general stance of Sean Wilentz, *Chants Democratic: New York City and the Rise of the American Working Class, 1788–1850* (New York: Oxford University Press, 1984); and as discussed in chapters 3 and 4 in this volume by Richard Carwardine and Daniel Walker Howe of Charles Sellers, *The Market Revolution: Jacksonian America, 1815–46* (New York: Oxford University Press, 1991).

7. Paul Faler, "Cultural Aspects of the Industrial Revolution: Lynn, Massachusetts, Shoemakers and Industrial Morality, 1826–1860," *Labor History* 15 (1974): 367–80; Bruce Laurie, "'Nothing on Compulsion': Life Styles of Philadelphia Artisans, 1820–1850," *Labor History* 15 (1974): 337–66; and Clark, "Changing Nature of Protestantism," 840–44.

8. Paul E. Johnson, *A Shopkeeper's Millennium: Society and Revivals in Rochester, New York, 1815–1837* (New York: Cambridge University Press, 1978), 38–61, 116–41; and Anthony F. C. Wallace, *Rockdale: The Growth of an American Village in the Early Industrial Revolution* (New York: Norton, 1978), 297–350.

9. Wallace, *Rockdale*, 394–97.

10. Earl A. Pope, "Albert Barnes, *The Way of Salvation*, and Theological Controversy," *Journal of Presbyterian History* 57 (1979): 27–32; and Marsden, *Evangelical Mind*, 53–58.

11. Marsden, *Evangelical Mind*, 59–87.

12. Ibid., 59–67. Marsden's book remains the definitive study of this subject. He summarizes the historiography on the Old School-New School division in an appendix, 250–51.

13. Ibid., 58, 67–103, 230–44.

14. Laurie, *Working People of Philadelphia*, 35.

15. For this interpretation of Old School Presbyterians in Philadelphia, see Doherty, "Social Bases for the Presbyterian Schism," 79; and Laurie, *Working People of Philadelphia*, 34–35.

16. Sam Bass Warner, Jr., *The Private City: Philadelphia in Three Periods of Its Growth* (Philadelphia: University of Pennsylvania Press, 1968), 51; Howard Gillette, Jr., "The Emergence of the Modern Metropolis: Philadelphia in the Age of Its Consolidation," in William W. Cutler III and Howard Gillette, Jr., eds., *The Divided Metropolis: Social and Spatial Dimensions of Philadelphia, 1800–1975* (Westport, Conn.: Greenwod, 1980), 3–25. The Consolidation Act of 1854 expanded the city limits of Philadelphia to include the entire county of Philadelphia.

17. Diane Lindstrom, *Economic Development in the Philadelphia Region, 1810–1850* (New York: Columbia University Press, 1978), 23–27.

18. Ibid., 23–24, 29–40, 91.

19. William M. Rice, "Introduction," in William P. White and William H. Scott, *The Presbyterian Church in Philadelphia* (Philadelphia: Allen, Lane, & Scott, 1895), xix. These numbers apply to the congregations affiliated with the Presbyterian Church in the United States of America (hereafter PCUSA). If augmented by the membership growth in congregations belonging to other Presbyterian denominations (Re-

formed Presbyterian, Associate Presbyterian, Associate Reformed), they would be
even more impressive.

20. This figure is a composite total for Old School and New School congregations in the city. *Minutes of the General Assembly of the Presbyterian Church in the
United States of America . . . 1855* [Old School] (Philadelphia: Presbyterian Board
of Publication, 1855), 365–68; *Minutes of the General Assembly of the Presbyterian
Church in the United States of America . . . 1855* [New School] (New York: n.p., 1855),
102–4.

21. Rice, "Introduction," xix.

22. Ibid.; Marion L. Bell, *Crusade in the City: Revivalism in Nineteenth-Century
Philadelphia* (Lewisburg, Pa: Bucknell University Press, 1977), 254. Bell says that
there were sixty-two Presbyterian congregations in the city in 1857, but this figure
includes churches not affiliated with the PCUSA. My figure is the total of Old School
and New School congregations.

23. Lefferts A. Loetscher, *Facing the Enlightenment and Pietism: Archibald
Alexander and the Founding of Princeton Seminary* (Westport, Conn.: Greenwood,
1983), 91, claims that in the early 1800s, Philadelphia "was virtually the capital city
of American Presbyterianism." At the same time, Presbyterian prominence in the
city had grown. According to Bell, *Crusade in the City*, 44, by the late 1820s "the
city was . . . undeniably Presbyterian. Presbyterians had in fact become the important religious presence in Philadelphia."

24. Marsden, *Evangelical Mind*, xi

25. Works that explore aspects of this question include Bell, *Crusade in the City*,
73–75; and Laurie, *Working People of Philadelphia*, 34–52.

26. Periodic references to the growth of manufacturing can be found in issues of
*The Philadelphian, The Presbyterian, Christian Advocate, Philadelphia Christian
Observer, and The Religious Farmer.* Usually, these notices were short reprints from
other periodicals and appeared in the "Literary and Philosophical Intelligence" sections of the journals.

27. *The Philadelphian* 4 (January 1828): 14. For the Presbyterian role in the city's
Sunday School Union, see Anne M. Boylan, "Presbyterians and Sunday Schools in
Philadelphia, 1800–1824," *Journal of Presbyterian History* 58 (1980): 304–10.

28. Albert Barnes, *The Choice of a Profession* (Amherst, Mass.: J. S. & C. Adams,
1838); Cornelius Cuyler, *The Signs of the Times: A Series of Discourses Delivered in
the Second Presbyterian Church, Philadelphia* (Philadelphia: Martien, 1839), 87–93,
195–96.

29. James W. Alexander, *The Working Man* (Philadelphia: The Working Man,
1843), 114–19.

30. Henry A. Boardman, *The Bible in the Family: Or, Hints on Domestic Happiness*, 7th ed. (Philadelphia: Lippincott, 1853), 183.

31. *The Religious Farmer* 2 (January 1829): 30.

32. Ibid., 1 (January 1828): 45–46.

33. Junkin, *Progress of the Age*, 8–10.

34. *The Philadelphian* 9 (January 1833): 6; William M. Engles, *A Plea for Reli-
gion* (Philadelphia: J. Young, 1833), 21–23; Engles, *The Wages of Unrighteousness.
A Sermon Delivered in the Seventh Presbyterian Church, Philadelphia, to the Young
Men of the Philadelphia Institute* (Philadelphia: Russel & Martien, 1834), 8–17; Henry
A. Boardman, *Piety Essential to Man's Temporal Prosperity: A Sermon Delivered
. . . before the Philadelphia Young Men's Society* (Philadelphia: William F. Geddes,

1834), 5–7; Alexander, *Working Man*, 117–18; Henry A. Boardman, *Suggestions to Young Men Engaged in Mercantile Business* (Philadelphia: Lippincott, Grambo, 1851), 7–14.

35. Albert Barnes, *The Connexion of Temperance with Republican Freedom* ([Philadelphia]: Boyle & Bennedict, 1835), 6.

36. Cuyler, *Signs of the Times*, 93.

37. Albert Barnes, "Vindication of Revivals, and Their Influence on This Country," *American National Preacher* 15 (January 1841): 23.

38. For Old School examples, see James W. Alexander [Charles Quill], *The American Mechanic* (Philadelphia: Henry Perkins, 1838), 30; [William M. Engles], *Certain Rich Men* ([Philadelphia]: n.p., n.d.), 3–11; Henry A. Boardman, *The Bible in the Counting-House: A Course of Lectures to Merchants* (Philadelphia: Lippincott, Grambo, 1853), 103–4.

39. Boardman, *Bible in the Counting-House*, 106–20, 164; Barnes, *Choice of a Profession*, 11–12; Cuyler, *Signs of the Times*, 195–96; John Mears, *The Bible in the Workshop; or Christianity the Friend of Labor* (New York: Scribner, 1857), 15–18; Alexander, *Working Man*, 188–96.

40. Boardman, *Bible in the Counting-House*, 137–38; Mears, *Bible in the Workshop*, 16.

41. Both schools of Presbyterians were strongly influenced by Scottish common-sense realism. See Mark A. Noll, "The Irony of the Enlightenment for Presbyterians in the Early Republic," *Journal of the Early Republic* 5 (1985): 149–75; and Noll, "Common Sense Traditions and American Evangelical Thought," *American Quarterly* 37 (1985): 216–38.

42. Faler, "Cultural Aspects of the Industrial Revolution," 367–71; Laurie, *Working People of Philadelphia*, 35–39.

43. Laurie, *Working People of Philadelphia*, 36–52.

44. The first quote is from Doherty, "Social Bases for the Presbyterian Schism," 77; the second is from Laurie, *Working People of Philadelphia*, 39.

45. Barnes, *Choice of a Profession*, 11–12; Albert Barnes, "The Hinderances to Revivals There," *American National Preacher* 15 (March 1841): 51–58; Thomas Beveridge, *A Sermon on the Duties of Heads of Families*, 3rd ed. (Philadelphia: n.p., 1830), 5. Beveridge was pastor of the First Associate Presbyterian Church in Philadelphia from 1827 to 1835. The Associate Presbyterians were a small Calvinist denomination sympathetic to the Old School side in the PCUSA split. For a New School book that argues for the legitimacy of work on the basis of Christ's example and the rest of the Bible, see Mears, *Bible in the Workshop*, 19–50.

46. Beveridge, *Sermon on Duties*, 4. Cf. Boardman, *Bible in the Family*, 256–79.

47. Beveridge, *Sermon on Duties*, 5.

48. *Christian Advocate* 8 (November 1830): 559.

49. *Religious Farmer* 2 (December 1828): 14; Alexander, *Working Man*, 57–58, 184–85.

50. Beveridge, *Sermon on Duties*, 5–6. Cf. Christian Advocate 7 (November 1829): 488.

51. Othniel A. Pendleton, Jr., "Temperance and the Evangelical Churches," *Journal of the Presbyterian Historical Society* 25 (1947): 24–40; *Christian Advocate* 7 (October 1829): 428; D. X. Junkin, *The Reverend George Junkin* (Philadelphia: George Junkin, 1871), 98, 134–35; Boardman, *Piety Essential to Prosperity*, 21–22; William P. White, "A Historic Nineteenth-Century Character," *Journal of the Presbyterian Historical Society* 10 (1920): 165–69; Engles, *Wages of Unrighteousness*, 8–15; Wil-

liam Neill, "Temperance: Its Necessity and Obligation," in *Autobiography of William Neill, D. D., with a Selection from his Sermons*, ed. Joseph H. Jones (Philadelphia: Presbyterian Board of Publication, 1861), 167–79.

52. Pendleton, "Temperance and the Evangelical Churches," 27, 34. All three were members of Second Presbyterian Church, but Henry and Newkirk switched to Central Presbyterian after it was established in 1832.

53. Alexander, *Working Man*, 178–80.

54. Alexander, *American Mechanic*, 12–15.

55. Boardman, *Bible in the Family*, 157–58; Boardman, *Bible in the Counting-House*, 25, 84–92.

56. *Christian Advocate* 8 (December 1830): 613–14; Alexander, *Working Man*, 258; [Stephen Colwell], *New Themes for the Protestant Clergy* (Philadelphia: Lippincott, 1851), 163–90.

57. *Christian Advocate* 9 (January 1831): 5–6; Alexander, *Working Man*, 233–37; Barnes, *Connexion of Temperance with Republican Freedom*, 11, 17; Albert Barnes, *The Desire of Reputation: An Address before the Mercantile Library Company of Philadelphia, December 3, 1851* (Philadelphia: n.p., 1851), 9–11.

58. Alexander, *American Mechanic*, 146–47; Alexander, *Working Man*, 123–25, 197–201, 220–22; James W. Alexander, et al., *The Man of Business, Considered in His Various Relations* (New York: Randolph, 1857), 3–5, 23, 30–31, 36–37; Barnes, "Hinderances to Revivals," 51; Laurie, *Working People of Philadelphia*, 146. See [Colwell], *New Themes for the Protestant Clergy*, 173–76, for a critique of Protestant views on poverty.

59. Doherty, "Social Bases for the Presbyterian Schism," 69–79, argues that New School Presbyterians were "a relatively homogeneous group of well-to-do middle-class entrepreneurs who were not old family Philadelphians." In contrast, the Old School had a higher percentage of artisans and laborers. For a critique of this interpretation, see Ira R. Harkavy, "Reference Group Theory and Group Conflict and Cohesion in Advanced Capitalist Societies: Presbyterians, Workers, and Jews in Philadelphia, 1790–1968" (Ph.D. diss., University of Pennsylvania, 1979), 271–81.

60. Fred Hood, *Reformed America: The Middle and Southern States, 1783–1837* (University: University of Alabama Press, 1980), 40–42.

61. Donald H. Meyer, *The Instructed Conscience: The Shaping of the American National Ethic* (Philadelphia: University of Pennsylania Press, 1972), 89–99. Among those ministers for whom biographical information is readily available, virtually all attended college and most graduated. A minority attended seminary, and many of the older pastors had studied theology privately with another minister.

62. Archibald Alexander, *Outlines of Moral Science* (New York: Scribner, 1852), 254.

63. Meyer, *Instructed Conscience*, 99–107. For a contemporary example in sermonic literature, see Joel Parker, *Invitations to True Happiness, and Motives for Becoming a Christian* (New York: Harper, 1844), 66–97.

64. Laurie, "Life Styles of Philadelphia Artisans," 355–56, 365–66, discusses Barnes and other New School Presbyterians who interpreted the depression in these terms. Cuyler, *Signs of the Times*, 97–102, 274–75, 289, 304–5 quotation from 102.

65. Hood, *Reformed America*, 33, 44–47; James Turner, *Without God, Without Creed: The Origins of Unbelief in America* (Baltimore: Johns Hopkins University Press, 1985), 34, 82–113.

66. Albert Barnes, *A Discourse on the Sovereignty of God, Delivered at Mos-Town, N.J., June 21, 1829* (Cortland Village, N.J.: Mann, 1829), 9. This address was

190 FINANCE AND THE EXPANSION OF AMERICAN PROTESTANTISM

given shortly before Barnes came to Philadelphia. I have found no evidence that Barnes shifted from this position later in his career.

67. Doherty, "Social Bases for the Presbyterian Schism," 77; Laurie, *Working People of Philadelphia*, 36.

68. *Christian Advocate* 8 (November 1830): 557; James Alexander, *American Mechanic*, 106–8.

69. Barnes, *Discourse on the Sovereignty of God*, 7.

70. Daniel L. Carroll, *Sermons and Addresses on Various Subjects* (Philadelphia: Lindsay and Blakiston, 1846), 282.

71. Ibid., 281.

72. *Christian Advocate* 12 (August 1834): 340.

73. Ibid., 338–39.

74. For examples of the admonition to be content economically, see *Christian Advocate* 9 (June 1831): 281–85; Carroll, *Sermons and Addresses*, 274–75; William Neill, "Self-Denial a Christian Duty," in *Autobiography of William Neill*, 89–102. For examples of the admonition to be seeking economic improvement, see *Christian Advocate* 8 (November 1830): 557–60, and 9 (January 1831): 1–6; Barnes, *Desire of Reputation*, 3–27.

75. Boardman, *Bible in the Counting-House*, 288.

76. *The Philadelphian* 7 (December 1831): 195, 202–3, 206.

77. Albert Barnes, *The Rule of Christianity, in Regard to Conformity to the World* (Philadelphia: n.p., 1833), 38–39.

78. Laurie, *Working People of Philadelphia*, 36.

79. Barnes, "Vindication of Revivals," 23.

80. *Christian Advocate* 7 (October 1829): 487–88.

81. Barnes, *Rule of Christianity*, 46–48.

82. James Alexander, *American Mechanic*, 27, 30, 275–86; Henry Boardman, *The Vanity of a Life of Fashionable Pleasure* (Philadelphia: n.p., 1839), 31–41; [Thomas Brainerd], *Influence of Theatres. A Lecture* (Philadelphia: Presbyterian Board of Publication, 1840), 5–10; William Ramsey, *Church Debts; Their Origin, Evils, and Cure* (Philadelphia: R. E. Peterson, 1851), 79–86.

83. *Christian Advocate* 7 (October 1829): 488. Cf. *The Philadelphian* 9 (January 1833): 6

84. Boardman, *Piety Essential to Prosperity*, 8.

85. Ibid. This society was founded in 1833 to provide "rational and useful recreation for leisure hours, and, generally, to promote the moral and intellectual improvement of young men in this city." It also aimed to "cultivate correct religious feelings" among young men. Ibid., 5–7.

86. Ibid., 7–18.

87. Henry Boardman, *The Claims of Religion upon Medical Men* (Philadelphia: Book & Job, 1844); Boardman, *The Importance of Religion to the Legal Profession* (Philadelphia: Martien, 1849); Boardman, *Bible in the Counting-House*.

88. Laurie, *Working People of Philadelphia*, 36. Cf. Doherty, "Social Bases for the Presbyterian Schism," 77.

89. Cuyler, *Signs of the Times*, 107. Cf. *Christian Advocate* 9 (June 1831), 284.

90. Carroll, *Sermons and Addresses*, 276.

91. [Colwell], *New Themes for the Protestant Clergy*, 128–31. On Colwell, see Milton C. Sernett, "Stephen Colwell and the 'New Themes' Controversy," ms., Presbyterian Historical Society, Philadelphia; and Bruce Morgan, "Stephen Colwell: Social Prophet Before the Social Gospel," in *Sons of the Prophets: Leaders in; Protestant-*

ism from Princeton Seminary, ed. Hugh T. Kerr (Princeton, N.J.: Princeton University Press, 1963), 123–47.

92. Meyer, *Instructed Conscience*, 109.

93. Hood, *Reformed America*, 40–42; Charles D. Cashdollar, "Ruin and Revival: The Attitude of the Presbyterian Churches Toward the Panic of 1873," *Journal of Presbyterian History* 50 (1972): 229–44.

94. For Presbyterian economic views in the late nineteenth century, see Gary Scott Smith, *The Seeds of Secularization: Calvinism, Culture, and Pluralism in America 1870–1915* (Grand Rapids, Mich.: Christian University Press, 1985), 126–40.

III

THE ECONOMICS
OF SECTIONAL STRIFE
AND REVIVAL

9

Trauma in Methodism: Property, Church Schism, and Sectional Polarization in Antebellum America

Richard Carwardine

Agonizing conflicts tore apart most mainstream Protestant churches in antebellum America. Historians of the onset of the Civil War, the remorseless
process of alienation between North and South, have rightly treated these
ecclesiastical schisms as early, limited expressions of a wider ideological
polarization.[1] The breakdown of a church consensus over Christian slaveholding—with immediate abolitionists fashioning a scriptural assault on the
peculiar institution and with radical Southern religious leaders pushing
towards a proslavery millennialism—left religious institutions open to fracture. No case has been more often cited to show the issue's convulsive power
than the experience of the Methodist Episcopal Church (MEC). At the General Conference of 1844 in New York, a majority of delegates called on Bishop
James O. Andrew, recently made a slaveholder by inheritance and remarriage,
to stop exercising his episcopal office for as long as he held slaves. That action propelled a sequence of events that brought about the division of what
was the largest Protestant denomination in the country. Contemporary political and church leaders alike blamed divergent attitudes toward slavery for
the split and reflected on its potentially somber implications for the Union.[2]

In much modern historiography, this stress on ideological confrontation
has tended to obscure or neglect the material concerns that each of the great
church schisms aroused, notably over the disposal of property and the control of funds. This emphasis has a long pedigree. Charles Elliott, a protago-
nist in the dispute and later the compiler of a gargantuan contemporary

chronicle of the breach in Methodism, an essential source for all subsequent studies, concluded in 1855 that though financial matters were involved, "yet the mere money is not the thing in question on either side."[3] But, if he was right to stress that church assets were not the underlying cause of the schism, no reader of the torrent of Methodist publications in these years can fail to be struck by the ferocity and bitterness with which Methodists fought to control denominational assets during the decade after the General Conference of 1844.

At stake were contested church buildings in the borderlands. So, too, were the Chartered Fund (a modest supply of money for the support of "worn-out" ministers and their wives, widows, and orphans) and the denomination's chief glory, the Book Concerns in New York and Cincinnati. Methodists' publishing operations had developed from modest beginnings in 1789 to become a massive business by the 1830s. When fire almost totally destroyed the New York headquarters in 1836 (a substantial property on Mulberry Street, which housed editorial offices and the printing plant), the stock and buildings were valued at $250,000; the weakness of the city's insurance companies, following the great fire two months earlier, had left the enterprise largely uninsured, and only about $25,000 was recovered from that source. Mostly because of public subscriptions, the restored Book Concern quickly resumed its place as the jewel in the crown of a publishing and distribution empire unrivaled in the United States. By 1844, due especially to its range of popular newspapers, magazines, and journals, the business was worth $750,000, and the steady growth of the Western Book Concern put the value of the combined businesses at close to $1 million dollars. These were not just stunningly profitable operations and prodigious assets but also the potent emblems of Methodists' enterprise and their arrival as a force in the land.[4]

The story of this million dollar struggle, though not neglected by denominational historians, has rarely received the wider attention its significance deserves.[5] It does not help that the tale is messy and complicated. The bitterness of the polemics, the tedium of charge and countercharge, and the slipperiness of the details have no doubt helped to deter historical analysis. But the story is intrinsically fascinating and offers a narrative full of drama, emotion, and poignancy. Beyond that, the property dispute impinges on larger questions. What were the balance and interplay of financial and ideological interests in the process of polarization and schism within Methodism? What was the significance of the conflict for the larger process of sectional conflict in antebellum America? How far did Northern and Southern Methodists bring different sectional perspectives to bear on matters of church finance and wider economic issues? And what can we learn about their posture toward an increasingly market-orientated world?

Methodist Schism

The failure of Northern and Southern members of the MEC General Conference of 1844 to agree on the issue of ministerial and episcopal slaveholding, as raised in the cases of Francis Harding and James O. Andrew, marked the beginning of a protracted process of separation and disposition of assets that would not be finally resolved until 1854.[6] For clarity's sake, that process can be broken down into a number of phases, in none of which was either side in danger of losing sight of the financial issues involved.

The General Conference of 1844 and the Plan of Separation

By the later stages of the General Conference of 1844 it was clear that delegates were unable to find a workable compromise to resolve the issue of episcopal slaveholding. They instructed a Committee of Nine "to devise, if possible, a constitutional plan for a mutual and friendly division of the church." It submitted a scheme whose implementation would be contingent on the subsequently judgment of the Southern conferences that it was necessary "to unite in a distinct ecclesiastical connection." This was the Plan of Separation. Proffered in a spirit of self-sacrifice, it appealed to "Christian kindness and the strictest equity," and its terms mostly addressed the division of the finances and property of the MEC. Its twelve resolutions implicitly accepted the justice of the Southern claim to a proportionate share of all of the church's joint property. Delegates gave each resolution either unanimous or overwhelming support.

They recognized that the bulk of MEC property in the South belonged solely to the membership, under the control of lay trustees who were beyond the jurisdiction of either the General Conference or any of the Annual Conferences that would ultimately form the Southern Church. The General Conference accordingly resolved that all MEC property "in meeting-houses, parsonages, colleges, school, conference funds, cemeteries, and of every kind within the limits of the southern organization, shall be for ever free from any claim set up on the part of the Methodist Episcopal Church." The Book Concern and the Chartered Fund and their prized dividends, however, belonged to the MEC ministers, who could make dispositions through constitutional majorities in the General and Annual Conferences. The Plan, as accepted, thus provided for the transfer to the Southern church of a share of the capital and produce of the Book Concern in proportion to the number of traveling preachers. It involved compensating the southern conferences with their dividends from the Chartered Fund and delivering up to the Southern church "all notes and book accounts . . . all the real estate, and . . . all the property, including presses, stock, and all right and interest connected with the printing establishments at Charleston, Richmond, and Nashville." It also acknowledged the Southern church's "common right" with the MEC "to use all the copyrights in possession of the Book Concerns at New-York and Cincinnati, at the time of the settlement." Delegates appointed commissioners (Nathan Bangs, George Peck, and James B. Finley) to carry out these arrangements in conjunction with Southern commissioners yet to be chosen. The Plan also set out the means for establishing a line of division between Northern and Southern Methodists. [7]

Many Northerners voted for this report of the Committee of Nine on the understanding that it might never be needed and that its real function was to offer the South an olive branch, in the form of a reassurance that Northerners had no intention of depriving Southern Methodists of their ecclesiastical rights. George Peck considered it a provisional scheme, to be implemented only if the annual conferences in the South later judged separation inevitable. William Capers of South Carolina told Peck that he hoped there would be no division, intimating that he would go home by way of Washington to consult with John C. Calhoun about the continuation of Southern Methodists in the General Conference.[8] In fact, no sooner had delegates adjourned than Capers and other Southerners met, without even leaving New York, to compose an address to the annual conferences and the members they represented. They declared that a separation ("*not* schism . . . *not* secession") was inevitable, and they invited delegates to gather in Louisville, Kentucky, in May 1845.

Northern reactions would be critically important. To effect the Plan of Separation, as it pertained to the proceeds of the Book Concern and the Chartered Fund, a change in the Methodist *Discipline* was required. Such a constitutional revision (amending the sixth restrictive article, one of the rules limiting the powers of the General Conference) needed the support of three-quarters of all the annual conferences. The Northern conferences voted sequentially from June 1844 to the spring of 1845 against a background of increasing hostility to the Plan and furious debates in the Methodist press.[9] Opponents tended not to emphasize the financial interests at stake, though as we shall see these were never far from the surface. Instead they focused on issues of trust, constitutionality, evangelistic mission and slaveholding in the church. Had the Southern delegates acted in good faith when they agreed to divide the church before they had even returned home to sound out their annual conferences? Was it right that there had been no vote of all the preachers, societies and circuits within those conferences? Did the General Conference have the power to transfer societies and property to the jurisdictional authority of another body? Had not the original donors of the funds that built Southern meeting houses, parsonages, colleges, and cemeteries intended to benefit the MEC, not seceders from it? Was it right to turn a deaf ear to the Southern minority who wanted to stay loyal to the mother church? Was it proper to connive at a division of the MEC that would limit its geographical area of operations and defy Christ's command to preach the gospel to all the world? Was it wise to create an unfriendly rival to the MEC, one likely to engage in what Abel Stevens described as "a battery of unceasing hostility and abuse against ourselves." Most pointedly of all, should the MEC offer any aid to an institution that would operate as a bastion of slavery? Enough northern Methodists gave a negative answer to these questions to leave the Plan of Separation short of the 75% majority it required.[10] Nonetheless, the near universal view in the South was that the concurrence of the Northern conferences was not needed to give validity to the Plan or legitimize division; those conferences in voting on the sixth article, were not voting on the *entire* Plan but only on the fair distribution of common property. The vote was a consequence, not a condition, of separation.[11]

The Border Conflict from 1845

The Plan offered a basis for drawing a 1,200–mile line through those border conferences of the MEC that covered nonslaveholding areas but also penetrated into the slave states of Maryland, Virginia, Kentucky, Arkansas, and Missouri. This was terrain where antislavery Methodists confronted devoted members of the separating Southern church. Under the terms of the Plan, "societies, stations, and conferences" along the line between slave and free states (but not "interior charges") could vote on their allegiance; thereafter the authorities of the other church would respect that choice and refrain from forming their own societies. This arrangement particularly upset antislavery hardliners who could see no moral case for limiting their activities in areas where, they were convinced, thousands of conscientious Methodists would want to continue their membership in the unstained MEC and not be dragooned into a "slavery church." The Plan also gave rise to considerable uncertainty. If, following a vote, the line separating churches were to be redrawn, was the society newly abutting the border also allowed to vote to determine its allegiance? If Henry B. Bascom were correct and the border took on the character of neutral common ground until loyalties were established, what would stop a perpetual, unsettling campaign of recruitment? A Virginian would complain of the aggressive proselytism of the Methodist Episcopal Church, South: "They have declared the border a movable line, so that when they have procured the secession of a society, station, or circuit, from the Methodist E. Church, the next one north becomes a border, and so on, *ad infinitum*."[12]

Thus the Plan gave rise to mounting frustration and anger on both sides, each regarding the other as "nullifiers" and predators. Congregations split. Cases of irregular voting and manipulation proliferated. Formal procedures were disregarded. Invective that had once been targeted only at the other section's radicals was now extended to embrace all the departed brothers and sisters. Each deplored the "tirade of abuse" that was emanating from the other. Southerners particularly blamed the breakdown of understanding and trust on free-state editors, especially Thomas Bond of the *Christian Advocate and Journal* and Charles Elliott of the *Western Christian Advocate*. Their papers, regarded as threats to social peace, were regularly seized and even burned by magistrates at the post offices, their actions sustained by a combination of statute, grand jury endorsement, and demands of vigilance committees.[13]

Pent-up feelings exploded into physical intimidation and violence. MEC preachers in Missouri were "threatened hard," seized, and told to go north. Armed young sympathizers of the Southern church camped for several weeks in the church at Clarksburg, Maryland, to deny access to MEC loyalists. (Churchwomen brought them their meals, and their beds were piled up at the back of the church during services.) Some of the most unpleasant violence occurred in Virginia, especially in the Kanawha Valley and on the eastern shore. In Northampton and Accomac Counties, social prestige and judicial power conspired with "mobbism" against the Northern church. In July 1846, Valentine Gray, a preacher of the Philadelphia Conference, was ejected from his pulpit and dragged from the church; a mob drove him away the next day when he turned up at the county court for redress. The same year a mob at Guildford broke up the service of another MEC preacher, James Hargis, by shooting, throwing missiles, and jeering; both the *Richmond Christian Advocate* and the local magistracy connived at the intimidation.[14]

Especially significant was the proliferation of heated disputes over the disposition and ownership of scores of church buildings, parsonages, and other assets. In many cases only protracted recourse to law brought a settlement. The most poisonous contests included those that developed over properties in Alexandria and Warrenton, Virginia, and Hannibal and St. Louis, Missouri. But most significant of all was the quarrel at Maysville, Kentucky, a small community on the Ohio River. The Methodist society of nearly 300 members, part of the Ohio Annual Conference, had been quietly harmonious in the months leading up to the Louisville Convention in May 1845, virtually no one supporting the idea of withdrawal from the MEC. But once the delegates at Louisville had constituted their annual conferences into the Methodist Episcopal Church, South (MECS), a number of Southern sympathizers in the Maysville society, including the presiding elder and the minister, pressed hard for separation. The members at church meeting narrowly voted in favor of tying the society to the South, but this view possibly would have been in the minority had a number of MEC loyalists not been absent because of a misunderstanding. A Northern sympathizer added the names of these absentees to the report of the meeting before dispatching it to the Ohio Conference, thereby constructing a majority for the connectional status quo. Each side thus had grounds on which to claim the church building. Each fought for control of the pulpit. Each turned to the law. Each sought the financial support of the parent church for an expensive but landmark judgment. Eventually the state court ruled in favor of the South: The separation and Plan of 1844 were valid, even though the annual conferences had failed subsequently to authorize a division of MEC property; the Southern church was not a secession from a continuing MEC but rather one of two new churches created out of a now defunct institution; and the Maysville church had legitimately elected to join the MECS, which was awarded exclusive use of the meeting house.[15] The leaders of the MEC denounced an opinion they considered to be the result of improper Southern pressure on the court, notably from Henry Bascom, president of Transylvania University at Lexington. But the MECS, fortified by a "fighting fund" built up by levies on the districts of the annual conferences, celebrated "a triumphant vindication" of its rights.[16]

The 1846 General Conference of the MECS

The first General Conference of the MECS, meeting at Petersburg, Virginia, in May 1846 and comprising largely the same ministers who had convened at Louisville in the previous year, established a Finance Committee designed to pursue the Southern church's interest in the Book Concerns in New York and Cincinnati and in the Chartered Fund. The full Conference listened grimly to a letter from the managers of the New York Book Concern which reported that they were under advice—given the votes in the annual conferences— not to pay any annual dividends to the Southern church. Instead they would hold on to a sum proportionate to the previous receipts of the Southern annual conferences, pending the next session of the MEC General Conference, in 1848.

The Finance Committee drafted an emotional, extended reply, designed to impress forcibly on the MEC its financial obligations to the Southern church under the Plan of Separation: "Whence was the far larger portion of the money collected, by which your [MEC] splendid edifices were erected? Whence have you realized the far larger portion of those profits by which you have divided from eight hundred to one thousand dollars per annum to the [Annual] Conferences, and swelled your capital to the gross sum of seven hundred thousand dollars?" There was a simple answer: "Southern and South Western men . . . and Southern and South Western trade mainly contributed to raise it [the Book Concern] to its present position of pecuniary profit and moral influence."[17] The Petersburg Conference also appointed three commissioners (Bascom, A. L. P. Green, and S. A. Latta) to act with the commissioners of the MEC (Bangs, Peck, and Finley) to secure a full financial adjustment. The only response came from Finley, however, who refused their overtures for a meeting, insisted on his constitutional impotence in view of the position taken by the annual conferences, and advised patience until the matter came before the forthcoming MEC General Conference in 1848.[18] This did nothing to allay Southern fears of Northern intentions and prompted talk of recourse to the law as the ultimate weapon.[19]

Denied access to its share of the Book Concern capital, the MECS at Petersburg considered how best to provide the publications that Southern Methodists needed. Given that supplying the necessary buildings, press, type, and other equipment required a sum of $100,000 or more, it was clear that a formal publishing house was, for the present at least, beyond the church's means. Instead the General Conference appointed an agent, John Early, to provide books for the MECS, to be secured mainly by purchase but in a few instances (the church's minutes, a revised *Discipline*, and a new hymn book) by publication. There were to be book depositories at Charleston, Richmond, and Louisville. The conference also approved the publication of weekly papers (*Nashville Christian Advocate*, *Richmond Christian Advocate*, and *Southern Christian Advocate*) and a quarterly review. The total cost of this more limited exercise was set at $62,000, a sum that was still likely to embarrass the

MECS. Though the Petersburg General Conference passed a resolution asking Southern editors to "allude as seldom as possible" to the controversy with
the North, this was a vain hope, not least as the authorities' appeals to the
churches to give generously were accompanied by the reminder that such
insistent requests arose "from the failure of the Northern Annual Conferences
to agree to a fair and amicable division of the Book Concern."[20]

The MEC General Conference of 1848 and
the Repudiation of the Plan of Separation

The tone at Pittsburgh in 1848 was set when the MEC General Conference
told Lovick Pierce, the delegate sent by the MECS, that it declined "to enter
into fraternal relations" with the Southern church. Then, after protracted debate, the Conference declared the Plan of Separation null and void and told
Bascom and the other Southern commissioners present that it lacked the
authority to negotiate. There were two sets of considerations at work. The
formal ground for repudiation, as presented in the report of the Committee
on the State of the Church (chaired by George Peck), was the failure of the
annual conferences to change the sixth restrictive rule.[21] The private correspondence of the bishops shows genuine bafflement about the proper course
of action in the face of developments unexpected when the General Conference had agreed to the Plan in 1844.[22] But the conferring ministers generally
held that the stance of the annual conferences had robbed Southern separation of its legitimacy, making the MECS a schismatic movement with no legal
rights to the property of the continuing MEC. Indeed, the term *repudiation*
seemed unwarranted: How could one repudiate a Plan that had never achieved
contractual status?

    The second, and more emotionally potent, consideration was "the hostile
movements" of the Southern church along the border and "the high official
sanctions," which sustained "a ruthless war, of mobs and infuriated partisans."
Peck was clear that the charge of "nullification" leveled at the General Conference carried no moral force from the lips of those whose own border activities "*practically nullified* the 'Plan,' while it was in force," and who simultaneously "professed to regard it with the sacredness of a 'treaty.'"[23] Elliott
correspondingly argued that the confiscation of MEC meeting houses and
parsonages by southern ministers represented the real act of repudiation.[24]

    There was, in fact, an internal contradiction in these two lines of argument:
If, through the actions of the annual conferences, the Plan had never possessed
contractual force, how could the MECS be breaking its terms? One may note,
too, as did John Norwood seventy-five years ago, that whatever the legal and
moral merits of the Northern Methodists' case (and they were considerable),
there was something of a problem in one-half of "the supposedly dissolved
partnership . . . assuming to act as judge in its own case."[25] It was to deflect
the charge of unfair dealing, and even of avarice, that the General Conference advised the agents of the Book Concern to submit the question to
arbitration.

Repudiation of the Plan doused whatever embers of charitable feelings remained from 1844. Southern editors and other church leaders branded the work at Pittsburgh "a most nefarious fraud," an act of abolitionist malignity, and a "breach of faith, allied to sacrilege."[26] None was more scathing than Bascom, the moving force among the Southern commissioners. His *Brief Appeal to Public Opinion* provided a sustained diatribe against the "disguised robbery" of one-third of $1 million by northerners who manufactured legal obstacles both in dividing the capital of the Book Concern and the Chartered Fund and in distributing the annual dividends from the sale of jointly owned books.[27] Yet few professed surprise. For some time they had been monitoring the annual conferences' election of delegates for Pittsburgh (which turned largely on attitudes to the property question), noting the absence of those, like Stephen Olin and Nathan Bangs, who were thought to be sympathetic to Southern claims. Thus, few of the faces of 1844 were present four years later.[28] Of the 142 delegates who participated in the vote to repudiate the Plan (by 132 to 10), only 41 had been present at the General Conference in New York to see it passed. Thirty of these had then voted for it, but 26 of these supporters now went for repudiation.[29]

The *Southern Christian Advocate*, twelve months before the delegates gathered in Pittsburgh, was already predicting that the Conference would find constitutional obstacles in the way of division of property. "In common with a large majority of people in the Southern Methodist Church, we believe that all . . . [these] professions about wishing the South to get its property, are moonshine," declared its editor, William M. Wightman. "Look one way, but keep rowing the other. . . . Wish [Southerners] . . . well, O yes! but fix the brand of secession upon their Organization. Insist upon the unconstitutionality of the Instrument under which they formed their connection, and keep the 'Medusa's head' of 'constitutional difficulties' always staring the good people of the North in the face, and then never fear the issue."[30] Bascom read the events of 1848, including the abrupt treatment of the elderly Lovick Pierce, entirely in this light, as the culmination of a four-year policy of denying Southerners "the rights of an equal contracting party" and depriving them of their share of property. The Conference had been "a studied and strenuous effort to avoid settlement with the South on any terms. . . . The most charitable construction that charity itself can put upon their conduct is, that they intended at least to hold our funds for four years longer, if not forever."[31] The MEC's proposed arbitration ("so completely made up of conditions, shifts, and evasions") seemed little more than a ruse to avoid censure and prevent the division of the funds. Few Southerners believed that Northerners had either the power or the appetite to carry out any rulings that might emerge.[32]

Recourse to the Federal Courts: The New York and Cincinnati Property Cases

With financial uncertainty limiting their institutional development, Southern Methodists lost patience with what they saw as Northern procrastination. When the New York agents told the Southern commissioners in the dying days of 1848 that they lacked the legal power to offer voluntary arbitration, Southerners felt a grim sense of vindication and pressed ahead with legal action against the MEC. They prepared three suits. One involved the Chartered Fund, on which a settlement emerged before trial. The other two, relating to the New York and Western Book Concerns, were filed in the summer of 1849.

The New York suit was heard before a U.S. Circuit Court in May 1851. Southern counsel insisted that the MECS had a right to the property that their ministers had earned through their years of active work; for the North, Rufus Choate argued that the Southern ministers' right to MEC funds persisted only for as long as they remained in that church.[33] Though the MEC was sympathetic to the advice of both counsels to pursue an out-of-court settlement, the correspondence between the two sides only established the unbridgeable chasm between them. The legal process continued. The South was jubilant when Judge Samuel Nelson ruled in late 1851 that the MECS was not a secession but a legal withdrawal; as such, it was entitled to a share of the church's assets. Many Northerners followed George Peck in believing Nelson's ruling to be legally flawed, for it appeared to leave all charitable trusts open to plunder. It certainly looked as if cotton influence were at work. But those who called for an appeal to the United States Supreme Court were overridden by men like Peck himself who judged that whatever the legal merits of the case, the North's sense of justice would be better served by an amicable, mutually agreed-upon outcome than by further litigation.[34] Under the chairmanship of Supreme Court Justice John McLean, an arbitrated settlement emerged at the end of 1853. The MECS got $275,000, consisting of $191,000 in cash; the printing houses at Richmond, Charleston, and Nashville; and the elimination of debts owed to the Book Concern.[35]

The Cincinnati case, instituted in the U.S. Circuit Court for the District of Ohio, elicited a wholly different response from Judge Humphrey Leavitt, who found for the MEC on every one of its points: "The Book Concern Funds being a charity for a special case," he ruled, "anyone withdrawing from that class ceases to be a beneficiary. . . . Any individual or section may withdraw from the church but can take with them no right to share in the property they enjoyed as members." Lodging an appeal to the Supreme Court under Chief Justice Roger Taney, the MECS mustered its arguments anew: Northern claims that it was a secession were contradicted by the MEC's own admission that church buildings and other property valued at millions of dollars were "justly and fairly held by the Southern Church"; the denomination's Publishing Fund had been built up over years by contributions from all over the South, as well as the North, and the present Book Concern had drawn "the vital juices of its prosperity" from these sources; in helping to make good the $250,000 loss inflicted by the fire of 1836, the South had played its part through liberal subscriptions and public donations: "Southern dollars built [the Book Concern] . . . up in no unstinted proportions. Southern markets consumed its productions. Southern preachers carried its books into every portion of the Southern country."[36] The MECS awaited the Court's ruling with some confidence. They had good cause. Its unanimous decision, written by Justice Nelson (as a Methodist, Justice McLean declined to sit), followed the logic of his decision in the lower court in the New York suit. The subsequent settlement delivered to the MECS $60,000 in cash and $20,000 in book stock. Whereas Southerners celebrated their vindication at the hands of the Court, many Northern Methodists found the decision, in Granville Moody's words, "astonishing, unparalleled and unjust."[37]

Interpretations

Some of the most forceful advocates of church division in 1844 argued that separation would remove the abrasions between Northern and Southern Methodists, but in practice the schism served only to aggravate old sores and generate new sources of hostility. Methodism, a major instrument of American national integration in the early Republic, ironically became a principal contributor to sectional alienation. The circumstances surrounding separation encouraged mutual sectional stereotyping that seriously corroded Methodists' sense of belonging to a political and ecclesiastical Union based on common values. The Methodists' experience, as Methodists, gave them a lens through which to interpret political developments beyond the narrowly ecclesiastical arena. The chronic and nasty warfare between the MEC and the MECS over their respective areas of jurisdiction along the border gave grassroots members a vivid understanding of the terms "slave power expansion" and "abolitionist intrusions." Leonidas Hamline was not alone in regarding developments as unique and ominous ("I have known nothing like it in America").[38] The language of the disputes was ugly and unedifying. Bascom's misnamed *Brief Appeal* was no more emollient than it was terse. Peck judged its sneering, bitter, and sarcastic tone a sad prostitution of the author's enormous talents and an agent of polarization: "The venom of his pages spread far and wide through the South . . . his splendid powers were abased to the fostering of sectional prejudice and hate. Thus he applied his giant strength to push the nation toward the bloody chasm of fratricidal war."[39] The outcome of the church suits over the disposition of property worked both to strengthen Southerners' confidence in the political conservatism of the highest federal court and to persuade antislavery Northern Methodists that it was controlled by King Cotton and proslavery influence.[40] Judge Nelson's 1851 decision in the U.S. Circuit Court prompted bitter Michigan Methodists to reflect that it was "about as easy 'for a camel to go through the eye of a needle,' as for justice to secure her rights when the rapacious maw of the slave interests yawns to receive them." In response, Southern Methodists berated those who for the first time in the nation's history had had the temerity to charge the high courts with corruption.[41] The Dred Scott opinions of the Taney Court, with its proslavery rulings on black citizenship and congressional authority in the territories, and the North's subsequent impugning of the highest judicial authority in the land, scarcely came as a surprise to either party.

The Methodists' ten-year debate over church assets and finances has to be set in this context. Each side found in the other's actions and language clear evidence of moral dereliction and flawed values, as well as confirmation of its own purity. For their part, Southerners united in castigating their Northern coreligionists for a materialistic self-interest that apparently prompted all their actions. One editor wrote of New England as "the land where the main chance is most religiously looked to in every thing," and of Ohioans bewitched by the "almighty dollar." Another reflected on the moral blindness induced by "the love of filthy lucre": "Our Yankee friends are keen sighted when money is the object of their vision."[42] It was believed that a simple calculation of increased dividends lay behind the language of principle and legalities used to block disbursements to the Southern conferences. An anonymous Georgian, stung by developments at Pittsburgh, called repudiation a simple act of greed: Northerners had once been so pleased at the prospect of shaking off the slave South that they had been ready, or so they said, to abandon the whole Book Concern, but "after taking the sober second thoughts for which the Yankees are remarkable, they seem to have concluded that as we could be got rid of on cheaper terms, it was as well to turn the thing to the most profitable account by holding on to the money." They invoked "numberless scruples of conscience" and constitutional violations, but it was what Wightman called "the bread and butter motive" that underlay their sophistry.[43]

Southerners linked the greed of the MEC with what they saw as the North's gallop toward intellectual madness, especially manifest in its strident abolitionism. The "vandal invasions of northern cupidity and fanaticism" were the product of a society that lacked reverence for the Word of God, led by those "preaching what Christ never preached." Lacking the anchorage of biblical principles and believing they had found "a 'north-west passage' to truth and Heaven," Northern Methodists had fallen into a mania. True religion had given way to "agitation, caprice, passion, and resentment." Revolutionary abolitionists had challenged the bishops, penetrated the middle levels of Methodist leadership as editors and agents, and had done so while living parasitically off the wealth of the church. Southern property was their natural target.[44]

Southern evangelicals were never wholly comfortable with the code of honor that so deeply influenced the section's social thought and behavior in the antebellum years: An ethic built around pride was not easily squared with the humility celebrated by the Christian gospel. Even so, the honor code left its mark on Southern Methodist interpretations of Northern behavior, as lacking all sense of shame. William Wightman considered the nullification of the Plan "a forfeiture of honour, and fair dealing, and Christian candour"; William Capers was in no doubt that Thomas Bond was using the single most influential mouthpiece of American Methodists, the *Christian Advocate and Journal*, for one end only, "to dishonor us, right or wrong"; Bascom, in an analysis shot through with similar terminology, believed that the hate shown by its Northern leaders covered the MEC "with shame and dishonor"; and Leroy M. Lee's *Richmond Christian Advocate*, measuring by two codes of

behavior, considered the North's stance "an outrage upon honor and conscience."[45] Nowhere, they argued, did this delinquency appear more clearly
than in the MEC's double standards over the profits of slavery. They pointed
to individual culprits: The abolitionist, Daniel Curry, for instance, who had
worked among Georgia Methodists, preaching to slaveowners and pocket-
ing their money to make his living, now damned them from afar, in defiance
of "common honor." But the guilt really attached to the whole institution for
holding onto the proceeds of blood money while its ministers preached antislavery. Northern Methodists refused to divide a Book Concern "whose very
foundations were laid in money, the result of slave-labour, whose walls are
stained with blood, and whose funds were made up, in no small part, by the
'curse.' . . . It is very much like finding stolen goods in the pockets of one
crying lustily, '*Stop thief!*'"[46]

The Southern Methodists' self-image was of a community moved by Christian principle, right behavior, and humanitarian concern, not materialism.[47]
They were engaged in an elemental and noble endeavor in which grubby issues of property had no place. "Men who struggle for existence cannot be
diverted from the last hold on life to preserve their money," reflected William
Capers in the aftermath of the confrontation over Bishop Andrew.[48] Leroy
Lee was equally sure that "dollars and cents can never be suffered to mingle
in a question of principle. The Southern Conferences will unquestionably
separate, money or no money—they go for principle not *interest*."[49] The local
revivals that multiplied across the Southern states in the fall of 1847 only
heightened the Methodists' sense of their own right behavior under divine
protection; their faith in the South's moral superiority was merely confirmed
by watching the MEC shrink in membership, despite (or possibly because of)
its control over finances that belonged to both churches.[50]

By the 1840s several of the ministers and lay leaders of Southern Methodism
were men and women of moderate, even considerable wealth. Yet there was
an edge to the South's criticisms of the MEC that had something to do with
their perception of the Northern church as a wealthy legatee of the schism,
one that still turned a deaf ear to the claims of the poor and suffering. An
Alabamian complained that whereas the MEC had bequeathed about a third
of its members to the MECS, and thus a third of its poor, it had held onto the
funds that might have done something to alleviate the condition of the neediest.
Veteran ministers, widows, and orphans, declared the Committee on Finance
of the South Carolina Annual Conference, were all victims of the straitened
circumstances in which the ministers of the Southern church had to operate.[51]
Without falling into the trap of economic reductionism, we can reasonably
conclude that the most intransigent and polemical Southern critic of the MEC,
the gifted Bascom, would surely have followed a less acerbic, more diplo-
matic course had his own personal circumstances been less desperate. Born
into poverty, he struggled with debt for the whole of his life. Financial worries
underlay his deteriorating health during the critical years of the Methodist
conflict. Biliousness, neuralgia, and (in 1845–46, at least) daily, six-hour
paroxysms of unendurable pain were all ascribed by his doctors to the men-

tal anxiety occasioned by the unstable finances of Transylvania University, in Lexington, Kentucky, where he was president; by the problems of the *Southern Methodist Quarterly Review*, which he edited, but for which he lacked the funds to pay suitable contributors; and by his onerous duties as Southern commissioner.[52] Fear of personal financial ruin was not the best of circumstances in which to judge soberly the course pursued by the MEC in the property dispute.

Southern Methodists' interpretation of events through the decade following schism was entirely congruent with the Old South's wider defense of its own socioeconomic order and its developing critique of Northern commercialism and industrial capitalism. There were, ministers conceded, evils connected with slavery, but these were the consequence of individuals who were abusing their positions as managers or owners: "As men are found depraved and unprincipled, as is the case everywhere, the rights of master and slave are shamefully abused and trampled upon." In the North, however, waged labor—supposedly free but unprotected from capitalist greed and economic downturns—was the victim of an enslavement of another sort. "There *society* is the great *slave-holder*, and millions are crushed and victimized beneath the weight and cruelty of its Juggernaut oppressions, without appeal or remedy of any kind," Henry Bascom asserted. "The enlightened student of human nature and the Bible will perceive but little difference in the moral aspects of the one system and the other."[53] When, during the years of the Methodist quarrel, Southern religious journalists produced editorials on "The Worldly Spirit" or "The Uncertainty of Riches," in which they discussed the dangers to simple republicanism that were lurking in the desire for wealth, and when they wrote of an intensifying passion for property, the target of their concern lay partly in their midst but even more to the world north of Mason-Dixon.[54] When Southern political leaders warned of threats to Southern order from Yankee acquisitiveness, their Methodist hearers knew exactly what they meant.

Northern Methodists proved less united than the South in their responses to the property dispute. Many, deeply hostile to slavery, rejected any idea that withholding funds was unethical. On the contrary, the evil of slavery and Southern Methodists' self-interested defense of their propertied interests in the peculiar institution, demanded that the MEC retain all assets. James Finley, a Northern commissioner and the leader of the movement against Bishop Andrew, resolutely opposed dividing the property, believing that Southerners had split the church to allow Methodists "to breed slaves for the market, to separate husband and wife, parents and children, in the face of the laws of God and nature; and because some of the Southern preachers have 'a hundred poor negroes to sweat for them.'" These funds had been entrusted to the church for the promulgation of the true gospel; it would be "a gross perversion" of the intentions of the donors, both dead and living, to betray a solemn trust by splitting the money.[55] Since the time would come when the funds could be used to replant authentic Methodism in the South, the ethical policy was to refuse to release them.

Some antislavery Methodists in the North conceded that the law was on the South's side but refused on principle to cooperate over the assets: According to Alexander Green, the MECS commissioner, one of them told him that "the property we [Southern Methodists] were contending for was ours of right, and that he hoped we would get it in the end; but as he believed that we were going to appropriate it to an unholy purpose—the support of a slavery Church—he would have nothing to do with handing it over to us."[56] Other Northeners believed that both law and morality were combined on the Northern side. True Methodists in the South, they said, had not been deprived of their financial inheritance: If superannuated ministers there wanted help from the Book Concern and Chartered Fund, they were free to join a Northern conference. Moreover, the South was owed nothing by the Book Concern: As book vendors, ministers had secured a return of up to a third on their purchases; as editors, they had enjoyed the use of presses built up by MEC funding and then employed to secure an unscriptural division in the church: Since 1844 the sales and income of the New York establishment had fallen with the loss of Southern subscriptions.[57] A few skeptical Northerners noted that Southern Methodists had in the 1830s themselves refused a payout to departing Canadian Methodists in circumstances similar to those in which they now found themselves: What price has consistency of principle?

At the same time, however, other Northern spokesmen energetically argued that the agreement of 1844 was a moral compact whose betrayal would expose the MEC to legitimate excoriation. So contended Stephen Olin, who feared Northern disgrace, and William Raper, who insisted that—whatever the constitutional obstacles—natural justice sustained the Southern case.[58] The editor of the *Christian Advocate and Journal*, Thomas Bond, may have been demonized by Southern hardliners, but he proved to be one of the most scrupulous advocates of fairness in splitting the property. Bond noted that the intended Southern beneficiaries of Methodist assets had been deprived of their proper rewards through no fault of their own; that the Book Concerns had been built up by the joint labors of both sections of the church: "If my neighbor held with me an equitable interest in a tract of land, and he was to commit upon me an assault and battery, it would not abate by an iota his right in the land, or justify me in keeping him out of the possession, if I had the power. Nor with a Christian, should it make any difference, though there were a legal defect in my neighbor's title, provided justice was on his side. A Christian must be more than *law honest*. He must be morally and religiously honest." Whatever the persecution of border Methodists by Southern aggressors, the MEC had an obligation to Methodist widows and orphans in the South. Bond was not alone in fearing the accusation of mercenary motives and hypocrisy that would inevitably face those who voted for repudiation. His successor in the editorial chair likewise insisted that although the law favored the North, the issue could not "be dodged with an exclamation about a 'thriftless controversy about money' leaving the other fellow to shift for himself and collect if he could."[59] Even the Scottite abolitionists, the True Wesleyans, earlier seceders from the MEC, considered that good faith demanded that the property be shared (though we may speculate about their motives: A division of assets would establish a useful example, to be applied retrospectively to their own case).[60]

The embarrassment these Methodists felt at the sight of their fellow religionists arguing over mammon turned to deeper discomfort, and even disgust, as the two churches turned to the courts. The moral picture grew even more confused. The *Christian Advocate and Journal* recognized that "litigation between two religious bodies, in relation to money, has a rather bad appearance, and always gives occasion for scandal."[61] Reflecting on the whole "humiliating transaction," an upstate New Yorker expressed even greater regret: "Better, in our opinion, if the million of dollars had been swallowed up in the deep."[62] Here was a variation on the theme developed by one of the most earnest voices of the MEC during Methodism's decade of trauma, Thomas Bond, who had announced that he would "rather a thousand times that the property should be consumed by fire and his children remain penniless than that they should have bequeathed to them property so tainted."[63]

Conclusion

Issues of church finance had not caused the midcentury schism in Methodism, but after the fracture money matters infected the wound. Jurisdictional disputes would probably have poisoned relationships between the two parties after 1844 anyway, regardless of the property question. But it was that issue as much as any other that prevented a clean break and ensured that the persisting mistrust, suspicion, and abrasions between the two halves of Methodism contributed more than their fair share to the broader sectional alienation.

The debate over the Book Concern and other MEC assets also tells us something about the extent and limits of Methodists' embrace of the rapidly developing capitalist culture in antebellum America. From an essentially rural, "primitivist" perspective—that is, from the vantage point of such religious groups as the Primitive Baptists and the Disciples—the Methodists' aggressive institution building, elaboration of denominational machinery, meetinghouse ostentation, and accumulation of wealth demonstrated not just their numerical power but also their contamination by "money religion" and the materialist influences of the market. Their whole apparatus of publishing was a measure of the ease with which ministers and the socially prominent laity had adapted to the changing economic and technological order. They prized their Book Concerns not just because of their functional importance to evangelism and church consolidation, nor only because of the significance of the dividends they yielded to old ministers and their needy families, but because of their role as visible symbols of Methodists' religious energy, economic power, and increasing social status. These materialist attitudes clearly shaped the ways in which the conflicting parties responded to unfolding events and were evident enough to some clear-eyed protagonists. Thomas Bond told his fellow ministers that they might believe they were acting from pure motives in rejecting Southern claims but that each party had good reason "to fear the bias of self-interest."[64] George Peck was equally aware that Methodists had not immunized themselves against the materialist ambitions that, consciously or unconsciously, contaminated "the reasoning processes of poor frail humanity."[65]

Yet, as these words suggest, there was within Methodism a continuing and pronounced strain of concern over the dissipating, enervating effects of wealth on church activity and moral integrity. Limpet-like attachment to proceeds of the Book Concern and Chartered Fund worried those who believed that Methodist churches "would be more liberal, if there were no reliance, or counting upon, these dividends." Capital accumulations, by removing the need for church members to support their traveling preachers, were "a snare and a curse to us, not a blessing." Church history demonstrated "that spirituality and vital piety decline, whenever, and wherever, there ceases to be a mutual sense of dependence between the Church and her ministry." The Book Concern was important as a means of diffusing religious truth but was of dubious value if prized principally as a source of wealth. Most notably, ministers like Thomas Bond feared that their colleagues' determination to hold onto what had come to them by wrongful means reeked of the morality of the business corporation: "We are aware how difficult it is to impress aggregative or corporate bodies, with the same sense of moral responsibility which an individual person would feel under like circumstances. It has come into the currency of a proverb that 'corporations have no souls,' and . . . it is well it is so, for it is certain they have no conscience, and it would be a lamentable thing to have a soul, and no conscience."[66] As participants in the burgeoning commercial marketplace, then, antebellum Methodists could not insulate themselves from the moral complexities of the advancing capitalist order. They engaged in its operations even as they sought to address the ethical issues it prompted.[67]

Whether Southern Methodists were in practice more cautious than their Northern counterparts in their embrace of American capitalist development is not clear; there were plenty of Southern economic modernizers in the mold of William Gannaway Brownlow and David R. McAnally, whose Methodism was as pronounced as their Whiggish pursuit of material and commercial improvement.[68] But the language of those who spoke for the MECS suggests that they certainly *perceived* themselves to be less driven by the materialism and dollar worship that they castigated in the Northern church. This, along with their conclusion that the MEC was in the grip of fanatical, unscriptural abolitionism, gave Southerners a potent belief in their ideological superiority. And when ideological certainty fuses with a deep conviction of material injustice, as occurred in the church property case, the ensuing sense of outrage becomes almost palpable. The resolution of the legal suits naturally took much of the heat out of the conflict, and the MECS was able to deploy the first payments from the MEC, along with the products of its own money-raising efforts, proudly to establish its own publishing house at Nashville and to increase the number of subregional *Christian Advocates*. One of these newspapers, at Richmond, welcomed the potential for "a season of tranquillity" in the churches that would promote "all goodness, righteousness and truth."[69] But the damage wrought by the conflict over property would not easily be undone. In practice, the very publications to which Southerners now turned as a means of inaugurating a more peaceable era would only help consolidate their sense of purity and moral superiority. To that extent the publications played, wittingly or not, into the hands of the Southern nationalists and political separatists. The Methodist property dispute, even in its resolution, contributed in its own way to the advent of the Civil War.

Notes

1. The schism is treated in its wider denominational and political context in Charles Baumer Swaney, *Episcopal Methodism and Slavery: With Sidelights on Ecclesiastical Politics* (Boston: R. G. Badger, 1926); Donald G. Mathews, *Slavery and Methodism: A Chapter in American Morality, 1780–1845* (Princeton, N.J.: Princeton University Press, 1965); C. C. Goen, *Broken Churches, Broken Nation: Denominational Schisms and the Coming of the Civil War* (Macon, Ga.: Mercer University Press, 1985); Richard J. Carwardine, *Evangelicals and Politics in Antebellum America* (New Haven, Conn.: Yale University Press, 1993), 133–74; Mitchell Snay, *Gospel of Disunion: Religion and Separatism in the Antebellum South* (New York: Cambridge University Press, 1993), 113–50.

2. Bishop Leonidas Hamline wrote to a ministerial friend on New Year's Day, 1845: "I ventured to you the opinion last January, that within ten years this confederation would be dissolved. You thought not. It seems to me you will now consent that possibly the *prophecy* (forgive the word) will be fulfilled. May be the whole South will not go: but if South Carolina is not out of the Republic in nine years I shall be surprised." F. G. Hibbard, *Biography of Rev. Leonidas L. Hamline, D.D., Late One of the Bishops of the Methodist Episcopal Church* (Cincinnati, Ohio: Jennings & Pye, 1880), 165.

3. Charles Elliott, *History of the Great Secession* (Cincinnati, Ohio: n.p., 1855), 606.

4. Nathan Bangs, *A History of the Methodist Episcopal Church*, 4 vols. (New York: T. Mason & G. Lane, 1838–41), 4:6–15, 424–55; W. F. Whitlock, *The Story of the Book Concerns* (Cincinnati, Ohio: Jennings & Pye, 1906), 31–58. In 1850, before the final division of the property, the Chartered Fund's capital stood at $36,000; the two Book Concerns' assets were estimated at nearly $830,000. Douglass Gorrie, *Episcopal Methodism, as It Was, and Is: An Account of the Origin, Progress, Doctrines, Church Polity, Usages, Institutions, and Statistics, of the Methodist Episcopal Church in the United States* (Auburn, N.Y.: Derby & Miller, 1852), 334.

5. John Nelson Norwood's even-handed account of the separation and its aftermath, *The Schism in the Methodist Episcopal Church: A Study of Slavery and Ecclesiastical Politics* (Alfred, N.Y.: Alfred Press, 1923), has not been superseded, though Arthur E. Jones, Jr., contributes a fine chapter, "The Years of Disagreement 1844–1861," in *The History of American Methodism*, 3 vols., ed. Emory Stevens Bucke, et al. (New York: Abingdon, 1964), 2:144–205. Much of the story can be followed in the pages of the *Journal of the General Conference of the Methodist Episcopal Church* and the *Journal of the General Conference of the Methodist Episcopal Church,*

South, published after the quadrennial General Conferences of each church. The most rewarding sources are the weekly Methodist newspapers, North and South, on which Elliott drew generously for his valuable *History of the Great Secession*. The Southern case is laid out by Henry B. Bascom, A. L. Greene, and C. B. Parsons in their *Brief Appeal to Public Opinion, in Series of Exceptions to the Course and Action of the Methodist Episcopal Church, from 1844 to 1848, Affecting the Rights and Interests of the Methodist Episcopal Church, South* (Louisville, Ky.: J. Early, 1848).

6. Harding had been expelled by the Baltimore Conference for refusing to free the slaves whose legal ownership had fallen to him through marriage. The General Conference upheld that decision.

7. *Journals of the General Conference of the Methodist Episcopal Church* (New York: Carleton & Phillips, 1856), 2 (1844): 217–27. The dividends from the Book Concern and Chartered Fund were divided proportionately among the annual conferences to help provide for the support of "superannuated preachers, and the widows and orphans of preachers." (Annual conferences also drew on them to make their contributions toward the bishops' stipends and to make up any shortfall in the allowances of the regular ministry.) In all, the dividends provided about half the sum available to the conferences for these purposes, the remainder being raised by yearly collections on the circuits. In the mid-1840s some $500 to $800 annually reached each conference from the Book Concern, some $50 from the Chartered Fund. Depending on the number of claimants, individual beneficiaries might receive as little as a dollar or two or well over $100. The average return to each claimant in 1850 was a fraction over $44. These were not princely amounts, but they could make a material difference to families on the borderline. *Minutes of the Annual Conferences of the Methodist Episcopal Church* (New York: T. Mason & G. Lane, annually); Gorrie, *Episcopal Methodism*, 317–19.

8. George Peck, *The Life and Times of George Peck, D.D.* (New York: Nelson & Phillips, 1874), 247.

9. The hardening of attitudes was most clearly seen in Charles Elliott's dramatic shift from emollience to repudiation, one that prompted bewilderment and bitterness among Southern observers. Bascom, *Brief Appeal to Public Opinion*, 45–46.

10. *Zion's Herald*, July 3, and August 28, 1844; Peck, *Life and Times*, 319–20; Elliott, *History of the Great Secession*, 595–96. Southern conferences piled up 971 votes for the recommendation of the General Conference, with a mere 3 against. A majority of voters in the Northern conferences also supported the change (1,164 to 1,067), but the margin was too small to ensure a 75% majority across both sections. Swaney, *Episcopal Methodism and Slavery*, 174.

11. Edward H. Myers, *The Disruption of the Methodist Episcopal Church, 1844–1856: Comprising a Thirty Years' History of the Relations of the Two Methodisms* (Nashville, Tenn.: Redford & Burke, 1875), 113–27.

12. Jones, "Years of Disagreement, 1844–61," 159–76; William I. Fee, *Bringing the Sheaves: Gleanings from the Harvest Fields in Ohio, Kentucky and West Virginia* (Cincinnati, Ohio: Cranston & Curts, 1896), 242–43; Bascom, *Brief Appeal to Public Opinion*, 93–106, 127–34; *Christian Advocate and Journal* (New York; hereafter cited as *CAJ*), January 12 and May 24, 1848.

13. Hibbard, *Biography of Hamline*, 192–16; *Knoxville Whig*, July 22, 1846; *CAJ*, November 3, 1847, May 24, 1848; *Western Christian Advocate* (Cincinnati; hereafter cited as *WCA*), September 27, and October 25, 1848; October 9, 1850; *Southern Christian Advocate* (Charleston, S.C.; hereafter cited as *SCA*), June 9, 1848.

14. Lorenzo Waugh, *A Candid Statement of the Course Pursued by the Preachers*

of the Methodist Church South, in Trying to Establish Their New Organization in Missouri (Cincinnati, Ohio: n.p., 1848), 60–61; Waugh, *Autobiography of Lorenzo Waugh* (Oakland, Calif: Pacific Press, 1883), 164–66; *Knoxville Whig*, August 25, 1849; Hibbard, *Biography of Hamline*, 211–15; J. Thompson to T. E. Bond, October 30, 1846, and Twiford to T. E. Bond, November 24, 1846, T. E. Bond Papers, Dickinson College, Carlisle, Penn.: *CAJ*, October 5 and 12, and November 11, 18, and 25, 1846; January 20 and 27 and February 3, 10, and 24, 1847.

15. Norwood, *Schism in the Methodist Episcopal Church*, 141–46.

16. *SCA*, August 13, 1847, January 14 and 21 and March 10, 1848.

17. *Journal of the General Conference of the Methodist Episcopal Church, South*, 1846, 33–34, quoted in Nolan B. Harmon, "The Organization of the Methodist Church, South" in Bucke, *History of American Methodism*, 2:135.

18. *WCA*, November 13, 1846.

19. *SCA*, January 15, 1847.

20. *SCA*, August 7 and 14, 1846; January 22, March 12, and April 16 and 30, 1847.

21. Peck, *Life and Times*, 320–21.

22. Hibbard, *Biography of Hamline*, 219–22.

23. *CAJ*, December 2, 1846; July12 and 19 and October 4, 1848.

24. Elliott also maintained that other acts of repudiation by the MECS involved its seizure of the offices of three papers, the *Christian Advocates* at Charleston, Nashville and Richmond, all of them MEC property in law and equity; and the appropriation for its own use of the missionary funds allocated to the Indian schools in the United States by the MEC. Elliott, *History of the Great Secession*, 620–21.

25. Norwood, *Schism in the Methodist Episcopal Church*, 125.

26. *CAJ*, August 16, 1848; *SCA*, June 23 and July 28, 1848.

27. Bascom, *Brief Appeal to Public Opinion*, 40–41, 91–92.

28. *SCA*, 28 May 1847; Bascom, *Brief Appeal to Public Opinion*, 54; Swaney, *Episcopal Methodism and Slavery*, 177.

29. Myers, *Disruption of the Methodist Episcopal Church*, 144.

30. *SCA*, June 11, 1847. Wightman and others saw evidence in the MEC press of a "better tone" and more accommodating approach in late 1847; they took Abel Stevens's proposal for making books available to the South at cost as evidence that even in hostile New England there was some desire to secure a settlement. *SCA*, December 31, 1847; January 21, 1848. But the underlying anxiety and uncertainty remained.

31. Bascom, *Brief Appeal to Public Opinion*, 108, 160, 176, 179.

32. *CAJ*, July 26, August 16, 1848; Bascom, *Brief Appeal to Public Opinion*, 109–18, 123. The *Nashville Christian Advocate*, May 26, 1848, sarcastically summarized the General Conference's position: "We have your money and wish to keep it; if you will, like good Christians, allow us to keep it all, and make no ado about it—that is, pay us, as the price of our friendship, some $250,000[,] we will readily and cordially fraternize with you; but should you . . . attempt to coerce it out of us, it will be impossible to recognize you as a legitimate branch of the Methodist family." Quoted in Swaney, *Episcopal Methodism and Slavery*, 178.

33. R. Sutton, *The Methodist Church Property Case: Report of the Suit of Henry B. Bascom, and Others, vs. George Lane, and Others, Heard before the Hon. Judges Nelson and Betts, in the Circuit Court, United States, for the Southern District of New York, May 17–29, 1851* (New York: Lane & Scott, 1851).

34. Peck, *Life and Times*, 346–47.

35. The decision not to pursue further the legal route was also influenced by considerations of "the strictest economy." *Journal of the General Conference of the Methodist Episcopal Church, held in Indianapolis, Ind., 1856* (New York: n.p., 1856), 276.

36. *SCA*, October 27, 1848.

37. Norwood, *Schism in the Methodist Episcopal Church*, 164–68; Myers, *Disruption of the Methodist Episcopal Church*, 151; Jones, "The Years of Disagreement," 177–81; Granville Moody, *A Life's Retrospect: Autobiography of Rev. Granville Moody* (Cincinnati, Ohio: Cranston & Stowe, 1890), 227. Despite Moody's words, the decision scarcely came as a shock. The editor of *Zion's Herald* expected the worst, "knowing how completely the atmosphere at Washington is filled with the miasma of slavery." Swaney, *Episcopal Methodism and Slavery*, 183.

38. Hibbard, *Biography of Hamline*, 239–40 (journal entry: April 5, 1847).

39. Peck, *Life and Times*, 253–54; *CAJ*, October 18, 1848.

40. Peter Cartwright and W. P. Strickland, *Autobiography of Peter Cartwright, the Backwoods Preacher*, (New York: Carlton & Porter, 1856), 454.

41. Swaney, *Episcopal Methodism and Slavery*, 180. Bishop Hamline judged Nelson's decision an appalling commentary on the changed character of the federal judiciary since the days of the Marshall Court; "incorruptible and unflinching integrity" had given way to partiality and special pleading. Hibbard, *Biography of Hamline*, 342.

42. Bascom, *Brief Appeal to Public Opinion*, 41, 63; *SCA*, July 12, 1844, September 17, 1847; Elliott, *History of the Great Secession*, 604.

43. *SCA*, June 23, 1848; March 21 (quoting *Zion's Herald*) and April 4, 1845.

44. Bascom, *Brief Appeal to Public Opinion*, 59–67, 169.

45. *SCA*, August 23, 1844; March 7, 1845; Bascom, *Brief Appeal to Public Opinion*, 5, 9, 10, 37–39, 51, 86, 127, 142, 148, 150, 168, 176; Elliott, *History of the Great Secession*, 504.

46. *SCA*, September 17, 1847; June 9 and 23 and October 27, 1848.

47. *WCA*, January 16, 1846.

48. *SCA*, July 26, 1844.

49. *Richmond Christian Advocate*, March 27, 1845, quoted in Norwood, *Schism in the Methodist Episcopal Church*, 119. Cf. Swaney, *Episcopal Methodism and Slavery*, 175.

50. *SCA*, October 8 and November 12, 1847; Bascom, *Brief Appeal to Public Opinion*, 63.

51. *SCA*, November 19, 1847; January 28, 1848.

52. Moses M. Henkle, *The Life of Henry Bidleman Bascom* (Nashville, Tenn.: Publishing House of the Methodist Episcopal Church, South, 1856), 286–97.

53. Bascom, *Brief Appeal to Public Opinion*, 163. See also chapter 10 by Kenneth Startup in this volume.

54. *SCA*, April 25, 1845; July 2 and December 24, 1847.

55. *SCA*, September 17, 1847; Elliott, *History of the Great Secession*, 598; *WCA*, January 16, 1847; January 14, 1848.

56. William M. Green, *Life and Papers of A. L. P. Green, D.D.*, ed. T. O. Summers (Nashville, Tenn.: Southern Methodist Publishing House, 1877), 519.

57. *WCA*, January 16, 1847; *CAJ*, April 26, 1848; Elliott, *History of the Great Secession*, 529. Between 1844 and 1848, the *Christian Advocate and Journal* lost nearly 25% of its subscribers (down from 25,000 to 19,000) and the *Methodist Quarterly Review* nearly a third (from 3,100 to 2,100). New York Book Concern sales dropped by $45,067. *Zion's Herald*, May 17, 1848; *SCA*, June 2, 1848.

58. *SCA*, May 28, 1847; Elliott, *History of the Great Secession*, 598.

59. *CAJ*, March 1, and July 26, 1848; May 31, 1849.

60. *WCA*, March 6, 1846; January 16, 1847. Even staunch antislavery ministers within the MEC were alert to the moral obligations of the MEC to the Southern church. This was implicit in Abel Stevens's proposal that the MEC Book Concern provide books at cost for MECS. *CAJ*, March 8, 1848; *Zion's Herald*, May 24, 1848.

61. *CAJ*, July 17, 1851.

62. *Buffalo Christian Advocate*, quoted in Norwood, *Schism in the Methodist Episcopal Church*, 168.

63. *CAJ*, March 1, 1848; *Zion's Herald*, June 21, 1848.

64. *Zion's Herald*, June 21, 1848.

65. Peck, *Life and Times*, 326. For the early equivocation of Methodism toward the market and the complexity of the movement's responses, see Richard Carwardine, chapter 4 in this volume; John H. Wigger, *Taking Heaven by Storm: Methodism and the Rise of Popular Christianity in America* (New York: Oxford University Press, 1998), 8, 11–12, 50, 173–95; Cynthia Lynn Lyerly, *Methodism and the Southern Mind 1770–1810* (New York: Oxford University Press, 1998), 73–93.

66. *CAJ*, March 1 and 8, 1848.

67. Methodism evolved in social and economic milieux that were themselves evolving. The antebellum generation of Methodists made an accommodation with a corporate America still in its infancy. Postbellum capitalism, operating by a laxer moral code, posed a greater ethical test, which some Methodists, notably in the Book Concern, failed with distinction. John Lanahan, *The Era of Frauds in the Methodist Book Concern* (Baltimore: Methodist Book Depository, 1896).

68. E. M. Coulter, *William G. Brownlow: Fighting Parson of the Southern Highlands* (Chapel Hill: University of North Carolina Press, 1937); F. M. B. Hilliard, *Stepping Stones to Glory: From Circuit Rider to Editor and the Years in Between: Life of David R. McAnally, D.D., 1810–95* (Baltimore: Gateway Press, 1975).

69. *Richmond Christian Advocate*, March 1, 1855, quoted in Charles F. Deems, ed., *Annals of Southern Methodism for 1855* (New York: Gray Press, 1856), 359. The early years of the Nashville house were marked by insecurity and debt, exacerbated by the financial panic of 1857. H. C. Jennings, *The Methodist Book Concern: A Romance of History* (New York: Methodist Book Concern Press, 1924), 169–72.

10

"A Mere Calculation of Profits and Loss": The Southern Clergy and the Economic Culture of the Antebellum North

Kenneth Startup

Employing a vituperative cadence that suggested long practice, an anonymous writer for the *Southern Presbyterian Review* placed the Confederate success at the Battle of First Manassas—and the larger war—in a cosmic context: The still young war represented the unfolding of "long delayed [divine] wrath upon the unjust, the avaricious, the covenant-breaking, the usurping, tyrannical, and licentious, and God-defying, Sabbath-breaking, and Bible perverting nations." To be sure that no one misunderstood, the writer specifically indicted the North for "falsified faith, and covenant-breaking, and [a] sectional self-aggrandizement policy of seventy years" now culminating in a terrible war.[1]

For the purposes of this chapter, *avarice* is significantly prominent in the litany of malfeasance cited above. It was a venerable loadstone of antebellum Southern sermonology, often employed alongside the more generic, more commonplace term *mammonism*. These were words and ideas that had provided Southern clergymen with much of their pulpit and printed rhetoric during the decades immediately before the war. Few themes—excluding the general call to repentance and salvation—rivaled, in force or in quantity, the ministers' antimammon message. Usually, these antimammon jeremiads were aimed pointedly, emphatically at the ministers' immediate congregations and communities. Often the messages warned of God's impending judgment on the idolatrous mammonists and their society.[2]

The Southern ministers' indictment of mammonism derived from many sources but was grounded in one overarching reality, the transformation of the Western world, in the late eighteenth and early nineteenth centuries, to a thoroughly capitalist ethos and practice. It was a transformation, a revolution, touching every aspect of Southern life, especially after the turn of the nineteenth century.[3] Ministers marveled at the pace and force of the changes in their culture. Generally and unavoidably, the clergy recognized that the South was fully engaged in a global economic community and that the South reaped great benefits and faced potentially disastrous consequences as a result of its full absorption into Western capitalism. For Southern divines, the principal threat from capitalism appeared in the form of the popular, indiscriminate appetite for rapid material gain. Southerners knew—their ministers knew—that the booming international cotton trade, westward expansion, the easy availability of credit, and the revolutions in transportation and communication made economic success possible on a scale never before seen in the South. The "main chance" seemed well within the reach of most white southerners. But this potential and passion for gain alarmed many of the Southern preachers. They detected, despite popular, external religious affectations, that the economic enthusiasm of the day was leading to a deadly indifference toward higher, spiritual things. The ministers suspected that the men and women sitting before them in the pews each Sunday were more interested in commerce and cotton than salvation and saintliness.[4] "Even the most pious," according to one southern divine, "have persuaded themselves, that, to aim at worldly prosperity is to aspire to greater degrees of holiness."[5]

The Southern ministers stridently condemned this excessive appetite for gain as a destroyer of social and civil relations. Notably, the clergy believed that it was largely the mammonist mentality—the capitalist excesses—of masters that precluded the reformation of the slavery system. Greedy masters perverted their patriarchal calling by overworking slaves and denying them their proper portion of the material benefits their labor produced. This exploitation and oppression also appeared in the abuse of the white working poor in the South.[6]

Equally sinister to the ministers was their perception that the commercial, capitalist culture of the South promoted an obsession with accumulation and ostentation calculated to divide Southerners along economic class lines. Consequently, Southern ministers excoriated fashion and luxury as inimical to cherished republican ideals. Similarly, the clergy challenged southerners to reject indolence as yet another antirepublican, anti-Christian manifestation of the economic excitement of the age. Not that Southern ministers were levelers; they recognized and approved gradations in society. In the preachers' social ideal, the pious and discreet rich were as worthy of respect as the virtuous, diligent working poor.[7] The divines did not believe that men and women—white men and women—should simply accept a lowly or impoverished existence. Patient, honest, and humble effort might—and properly should—elevate a family economically. Again, capitalism had its benefits in the clergy's estimation, and the possibility of responsible economic improvement for the lower orders was numbered among the benefits most lauded by the ministers. Of course, the clergy believed that the greatest benefit of a prosperous, capitalist South resided in its potential to sustain denominational colleges, seminaries, orphanages, Sunday schools, and missionary endeavors. Still, these benefits were attended by extraordinary risks to Southern society. And so the ministers generally disparaged the economic *mentality* of their day and region and warned of God's judgment on such an offending people. The warning still resonated during the war years, with some ministers describing the conflict as a punishment for, and purgation of, the South's venal sins.[8]

Though they focused primarily on the South, the Southern clergy's war against mammonism—and by extension their critique of an excessive, aggressive capitalism—was never exclusively sectional. Occasionally, during the earlier decades of the era, the ministers leveled their jeremiadic fire at the larger national passion for gain, and more specifically, at Northern economic conventions.[9] As the era closed and secession and war became realities, the ministers' decades-long criticism of Southern economic attitudes and practices served as a filter, a medium through which the Southern clergy increasingly measured and criticized Northern economic life and thought. In the last two decades of the era, important ideas and events coalesced, adding intensity to the Southern clergy's hostile scrutiny of the Northern economic mind and Northern culture generally. The volume and pace of the Southern clergy's critique of the North reached its apogee after the mid-1840s. The fact that, for decades, they had censured their own communicants and neighbors for their illicit economic aspirations and attitudes only added weight to the clergy's negative assessment of Northern economic ideas and practice late in the era.

As in their depiction of Southern economic life, the slave state ministers, when addressing the Northern economic environment, rarely employed the language of formal, academic economic theory; for instance, they rarely used the word *capitalism*. Some Southern preachers were very well versed in Smith, Ricardo, Fourier, and Comte, but like their less erudite peers, even these professorial pastors usually framed their economic commentary in more accessible, more biblical language. Then, too, for the ministers, things economic were never primarily matters of abstract theory but were linked with immediate moral choices and circumstances. For the Southern ministers, historical and epochal economic forces were never more important than the individual's response to those forces—a response ideally based on straightforward biblical instruction derived from a largely literal reading of Scripture. Consequently, the Southern preachers presented their critique of the Northern economic mind in fairly simple, moralistic terms. In fact, and not surprisingly, the Southern clergy's discussion of Northern economic conventions reads far less like a measured, methodical assessment and more like a stilted, often strident, morality play that interlaced various strands of Northern economic, social, political, and religious culture.

Nothing in the antebellum era so sharpened the Southern clergy's cross-sectional gaze as the abolition movement. From the 1820s, Southern ministers had mounted a biblical defense of slavery in an attempt to deflect external and internal opposition to it. After the 1830s, that defense became an almost ubiquitous fixture of Southern evangelicalism.[10] The Southern clergy also defended slavery on constitutional, social, and—significantly—economic grounds. In this last regard, the ministers believed and asserted that slavery provided, *potentially*, an ideal means for securing individual and corporate economic improvement; they never doubted the capitalist character or economic value of the slavery system. But it was an ideal system to the divines *only* insofar as the institution operated, *potentially*, within the Biblical confines of a patriarchal system, including a proper regard for the needs, physical and otherwise, of the slaves. As Alabama Baptist Iveson Brookes asserted, slavery provided for a benign, peaceful social order while allowing Southerners, even slaves, to enjoy the benefits of moderate commercial capitalism.[11] The reality that many Southerners—as Brookes conceded—refused to act with appropriate biblical restraint and decency in their management of slaves, did not obviate, in the minds of Brookes and his peers, the social and economic benefits of the institution. In the late 1830s, one Southern divine publicly advocated the enslavement of the "hireling" classes of the whole world as a means of assuring better, more humane treatment for these brutalized, exploited workers.[12] Other Southern apologists, notably George Fitzhugh and James Hammond, offered a similar socioeconomic defense of slavery.[13]

It may be useful to add that the Southern clergy's defense of slavery, however energetic and strident, was never the center or ground of Southern evangelicalism; actually *evangelism*, the saving of souls, was the ground and center of Southern evangelicalism. And a more complex worldview grew out of this simple redemptive aspiration, a world view far larger than the plantation ethos.[14] That this larger Southern evangelical worldview served as a bulwark for slavery did not mean that the ministers had arbitrarily crafted their theological or social theory to sustain the institution. In much of his recent scholarship, Eugene Genovese has helped to reveal the complex, layered character of Southern evangelicalism. Genovese has clarified the fact that Southern ministers saw the world as more than a stage for a struggle between abolitionism and slavery; they regarded that contest as ancillary to the far more significant contest between *secular* and *sacred* worldviews and social orders. Their economic and social defense of slavery—resting on their belief in the biblical *truth* of slavery—was intended, in turn, to bolster their larger goal of biblical order and authority in human affairs and the progress of the gospel.[15]

Again, their vision of a biblical society subsumed the economic and social defense of slavery. This economically, socially rooted affirmation of slavery was subsequently complemented by the Southern ministers' attempt to link abolitionism with Northern greed and selfishness and with a peculiar, destructive distortion of Northern economic sensibility. The connection plotted by the ministers between abolitionism and Northern economic thought followed a tortuous path. Even so, over time, the Southern divines crafted and presented a critique that allowed them to conjoin several rhetorically powerful anti-Northern, antimammon themes.

In its most exotic form, the Southern preachers' challenge to abolitionism depicted that movement as a stalking horse for an international economic tyranny. This perspective appeared in 1842 in the *Southern Quarterly Review*.[16] In a detailed discussion of English abolitionism, the anonymous writer asserted that England's decision to free West Indian slaves a decade earlier was nothing more than a bid on the part of English capitalists to weaken West Indian economic competition in favor of burgeoning East Indian interests. The writer observed that West Indian abolition had occurred only after the economic value of the West Indies was determined to be less than that of England's Asian possessions. And beyond this, the writer asserted that the exploitation of Indian peasants was as brutal and thorough as had been the exploitation of West Indian slaves. The writer castigated the hypocrisy and gullibility of the English abolitionists for willingly commingling mammonism and Christian morality. For the writer, the whole West Indian abolition movement ultimately amounted to nothing more than a "mere calculation of profit and loss."[17] The writer supported his conclusions with a closely woven chronology of the connection between the economic development—exploitation—of India and the progress of West Indian abolition. The chronology and a substantial array of documentary evidence gave the article a tone of historical objectivity and authority. The writer went so far as to assert that the English abolition movement was part of a larger conspiratorial effort by English capitalists to engender hostility between the North and South as a means of fomenting an American civil war, thereby blunting American competition with England's expanding capitalist world empire.

Whether or not this writer was a clergyman is largely irrelevant because the conclusions drawn in the article were replicated by Southern ministers throughout the antebellum period. Just two years later, for instance, a writer for the *Christian Index* of Georgia asserted that American abolitionists had entered into an "alliance with a foreign people—a people that have never forgiven the assertion of our independence, and that would be gratified at nothing so much as our ruin."[18] In 1851, an article in the *Nashville and Louisville Christian Advocate* described English manumission schemes since the Revolution as nothing more than shrewd and cynical attempts to weaken economic competitors; the writer suggested that all who abetted such schemes were little better than traitors to the United States and adjuncts to a virulent and immoral form of imperial-industrial capitalism.[19]

In 1859, the South Carolina Presbyterian John Adger endorsed the prevalent Southern view that English capitalism, through the English government and press, was "stimulating the blind fury of Northern Abolitionists." England, Adger stated, "acts from the instinct of British abolition sentiment and from the instinct of British commercial jealousy."[20] In fact, the supposedly sinister connection between English abolitionists—and by extension English capitalism—and American abolitionism had been marked fully twenty years earlier in the *Christian Index*. An article explicitly condemned the English abolitionist George Thompson, at the time visiting in New York, for inciting the Northern public against slavery and the South.[21] Precisely how English bankers, industrialists, and merchants hoped to gain economically from the abolition of slavery was rarely explained in detail by the Southern ministers. At most, they suggested that a general weakening of the American economy—through the abolition of slavery—would leave England more fully in command of world trade and world industrial output; and then, too, there was the matter of eliminating Southern competition with the cotton producers of India.

The language of economic conspiracy gave place sometimes to even more dramatic and alarmist apocalyptic imagery. In 1857, a Baptist congregation passed an antiabolition resolution redolent with grim prophetic prose. American abolition, according to the Mattamuskeete, North Carolina, congregation, was only the "hand-maiden of mystery Babylon, gotten up in . . . London."[22] Potentially, it was a powerful nexus; mercenary international abolitionism might also be seen by Southerners—influenced and instructed by their ministers—as cosmically wicked and spiritually debauched.

Beyond their attempt to link the antislavery North with a sinister, possibly satanic and alien capitalist conspiracy, the Southern ministers offered a more common but slightly more obtuse critique of the Northern economic mind that merged its larger economic culture with the purportedly warped *philanthropic spirit* of the abolitionist movement. This line of criticism disdained what the Southerns ministers regarded as the peculiar penchant of abolitionists, and Northerners generally, for a cheap benevolence. It was a perspective that reflected the old and vivid image of miserly but economically astute Yankee culture, full of self-righteousness and selfishness. In the perspective of Southern preachers, the Northerners—led by abolitionist zealots—could morally condemn and possibly ruin the South economically, all the while continuing to aggrandize themselves financially, politically, and morally. Southern ministers saw a prospering Northern middle class and rich abolitionists like Lewis Tappan, in their fine houses and fashionable clothes, glibly giving away the South's economic future. As they warred against Southern prosperity, the Northerners prospered. For the southern ministers, the abolitionists' benevolence was not true benevolence for it cost the abolitionists—and their communities—nothing; all true benevolence required a legitimizing personal or corporate sacrifice. The Old South's most famous diarist, Mary Chestnut, concurred fully: She characterized abolitionism as the "cheapest philanthropy trade in the world—easy. Easy as setting John Brown to come down here and cut our throats in Christ's name. These people's obsession with other decent people's customs reduced to self-serving and sanctimonious nonsense."[23]

This image of a disingenuous and moralizing—though mercenary—North conformed to the Southern clergy's long-held suspicion of capitalism itself. The Southern ministers often inveighed against what they regarded as an intrinsic moral flaccidity in the capitalist ethos, a willingness to tolerate "sharp," selfish, and dishonest conduct as long as the trappings of conventional churchmanship, including charity, were maintained. They had seen this kind of hypocrisy in their own churches; they now saw it dominating Northern society. For many Southern ministers, the specious Northern benevolence they reviled was only a gloss, a mask for northern greed, a greed that among many other offenses, cruelly exploited Northern workers: "When our Northern brethren have freed their own land from these evils [the hypocritical abuse of the North's "laboring classes"] it will be fully time enough to think of interfering with the domestic institutions of the South." The writer pointedly condemned the Northern "proprietors in all their manufacturing and mercantile establishments" as agents of economic exploitation and oppression in their own region; indeed, Northern capitalists were so oppressive and calculating that some working women of the North were driven by neglect and ill treatment to "a house of infamy." This was a vivid triptych to be sure—Northern capitalism, spiritual hypocrisy, and sexual degradation.[24] Again, the Southern clergy may have been more inclined to condemn the abuse of Northern workers because they had long observed—and routinely condemned—the same greed-driven abuses in their own communities. Nevertheless, the shared sins of the two regions did not soften the Southern clergy's confrontational stance, especially in the face of the mounting Northern criticism of Southern culture, Southern virtue, and Southern churches.

Southern ministers were not hard-pressed to produce what, to their minds, were striking, discrete examples of Northern mendacity—this blending of moral and mercenary hauteur and hypocrisy. As chapter 9 by Richard Carwardine indicates, they found much that reinforced their perspective in the circumstances surrounding the sectional division of American Methodism in 1845. For the next fifteen years, the Methodist clergy of the South would disparage many of their former brethren in the North, not only as theologically unorthodox perverters of the *Discipline*, but also as self-righteous mammonists. The ground of this charge stemmed from the failure of the Northern-dominated General Conference to divide certain assets of the church with the departing Southern Methodists. The Southern Methodists blamed the abolitionists and their closest allies in the General Conference for the property crisis; and they wondered and worried over the apparent complacency of the larger Northern Methodist community, which tolerated what to the Southern clergy was outright fraud. The church property issue was bound to gain a wide audience in the South, given the simple reality that Methodism was the predominant evangelical denomination in the region.[25] The property struggle resulted in several litigations and ultimately reached the Supreme Court in 1854.[26] All of the major legal decisions confirmed for the Southern clergy and laity the perfidy of the Northern Methodist leadership and the larger Northern evangelical community.

The damaged relationship between Northern and Southern Methodists was certainly not mitigated by judicial pronouncements, even if most of those decisions sustained Southern claims and issued from Northern courts. Following a favorable legal decision, one Southern divine wrote sarcastically, "What a panoply of war, what awful nodding of Northern plumes, in defense of their possession of Southern money-bags, which they were determined never to give up, right or wrong. . . . Possession was 'nine points of the law,' and they resolved to maintain [their hold on Southern property] in utter defiance . . . of moral obligation."[27] Southern Methodist clerics angrily censured a Northern church that lamented injustices done to slaves while casually doing injustice to their former coreligionists. An article in the *Richmond Christian Advocate* stated succinctly that the Northern Methodists were dishonestly, selfishly, "unjustly holding [Southern Methodist] property," including resources intended for the benefit of Southern widows and orphans.[28] Such language implied that exploitative and excessive Northern capitalism was reaching southward to wound defenseless Southerners.

The image of Southerners struggling to preserve their rightful property against the selfish designs of grasping Northerners figured prominently in the subsequent justifications of secession and war by Southern ministers, politicians, and journalists. In the summer of 1861, the Rev. Dr. N. M. Crawford, Baptist and ardent Confederate, sometime president of Mercer University, and a man intimately connected with the cultural, political, and social elite of Georgia, stated, "We are fighting for our property which belongs to us, and the title to which none, and least of all the people of the North, have the right to question or gainsay."[29]

Once the war began, Crawford's perspective seemed confirmed by the fact that Northern evangelicals attempted to claim Southern church property by right of military conquest, as Daniel Stowell has recently documented.[30] In the wake of advancing Northern armies, influential Northern churchman enlisted powerful political and military leaders in their efforts to confiscate Southern church property. Secretary of War Edwin Stanton issued explicit orders in late 1863 and early 1864 for Union commanders to see that church property in occupied Confederate territory was transferred to the control of "loyal" ministers.[31] Such orders actually meant that Southern church property was to be ceded to Northern denominations. For example, acting on Stanton's 1864 order, "representatives of the ABHMS [American Baptist Home Missionary Society] seized about thirty buildings."[32] Southern clerics and many border state ministers condemned the confiscation policy as an unrivaled example of Northern greed and arrogance.

Perhaps the most aggressive opponent of the confiscation program was W. M. Leftwich, a lawyer who became a Methodist itinerant in Missouri.[33] Five years after the war's end, Leftwich still furiously denounced the confiscation effort. He asserted that the confiscators had no more right to the property of Southern churches than "South Sea Islanders."[34] And harking back to his legal training, Leftwich indicted Northern Christians as "particeps criminus" for their "appropriating" of Southern church property.[35] Leftwich was far from being dispassionate or objective in his observations and opinions. He was an adamant Southern partisan. Nevertheless, his views merit attention because they generally conform to the attitudes and assumptions of the antebellum Southern clergy, notably to the cynical disdain the preachers evidenced for the purported economic rapacity of Northerners, especially Northern Christians. Even Abraham Lincoln apparently recognized the potential of the church confiscation policy to deepen the white evangelical South's conviction that the Union's war effort masked pernicious plans for Northern aggrandizement. Lincoln, in fact, began "restoring the southern churches' property to them" months before the war ended.[36]

In the conflict over church property and related issues, Northern evangelicals had unwittingly added a powerful negative nuance to the South's already substantial store of sectional fear and suspicion. This nuance, that Northern churchmen were mendaciously selfish, was all the more potent since it complemented the Southern clergy's—and Southern culture's—larger, older perception of a calculating, mercenary North. And the antebellum and wartime recriminations over church houses and other denominational property in turn complemented bitter sectional jealousies over matters like the missionary prerogatives of free and slave state evangelicals. These jealousies, voiced loudly by Southern ministers, had occasionally been couched in economic language, playing again on Southern suspicions of Northern motivations, even of northern missionary endeavors.

In the last decade before the war, Southern Methodists pointedly accused Northern Methodists of attempting to insinuate their "avaricious" missionaries into the slave states.[37] In Arkansas, popular meetings were held to censure such intrusions.[38] In their bitter, tangled rejoinder to the intruders, the Southern ministers implied that the Northerners were practicing a kind of illicit spiritual capitalism—grasping for Southern souls and Southern money. In 1850, the venerable Bishop William Capers, aged missionary to the slaves, criticized Northern Methodist ministers who were coming south into the Louisville, Kentucky, Conference; he claimed that these selfish, "mischief working" preachers were taking black members away from the conference.[39] And one Southern minister wondered publicly, with a skeptical inflection, at the fact that in spite of Northern Methodism's abolitionist perspective, their missionaries and ministers in the slave states eagerly admitted slaveholders to their churches.[40] There was some suggestion that these Northern missionaries expected to benefit financially from their Southern sojourns. This last assumption rested on and reinforced another idea, common in the antebellum South, that every Northerner in the slave states was a man-on-the-make, sly and cynically mercenary. Of course, even Harriet Beecher Stowe's arch villain was the predatory Yankee mammonist and slave overseer Simon Legree.

What made the presence of these Northern immigrants more distressing to Southerners was the air of moral and cultural superiority some of the newcomers manifested; it was a superiority some Southern clergymen adamantly challenged. In the 1820s, the Rev. Thomas Griffin had issued such a challenge before a Baltimore audience: "Who ever heard of a Mississippi Yankee fancying a girl that owned no slaves? . . . [and] how came slaves to this country [in the first place]? Northern men, Northern money, and Northern ships brought them to America." Griffin noted that northerners, in their financial connection to slavery, invariably revealed their "peculiar" economic "shrewdness and foresight." Sarcastically, he continued his attack, stating that when slaves were no longer profitable in the South then, southerners, too, would become high-minded, "patriotic, moral, and religious," people. That Griffin had delivered these comments in the 1820s and then republished them nearly twenty years later in the *Natchez Free Trader* suggests the staying power, if not the deepening, of such attitudes among the southern ministers. Griffin had expressed the idea forcefully—Northern moralizing was only a veneer, a theatrical performance that was masking a distorted, aggressive self-interest.[41]

In attitude, tone, and content, Griffin's denunciation was comparable to the general perspective of the many Southern clerics who spoke to issues like the church property conflict and the intrusions of Northern missionaries. To be sure, Griffin reinforced Southern perceptions of a symbiotic relationship between Northern selfishness and northern moral arrogance. And his embittered dismissal of and disdain for the abolition crusade was echoed repeatedly by his peers. Griffin also suggested a marked connection between personal, private greed and abolitionist hypocrisy, though he put the matter in a historical context by describing the personal gain of the abolitionists' ancestors in the slave trade. Still, for Griffin, abolitionist hypocrisy was not merely a matter of cheap benevolence or regional economic aggrandizement but was also a matter of private gain at another's expense; the descendants of New England slave traders would continue to thrive off the inherited income of the slave trade while they reduced southerners to penury. Like the rest of his discourse, this aspect of Griffin's perspective was defensive and distorted, but it proved nonetheless to be an attractive line of attack for many Southern ministers. How much easier it was to repudiate abolitionism if the case could be made that abolitionists anticipated direct, personal economic gain from their antislavery attitude and from the eventual abolition of slavery.

Some Southern clerics, therefore, proposed an outright, individual entreprenuerialism on the part of abolitionists. In 1837, the *Christian Index* ridiculed an abolition meeting in New York, noting that the American Anti-Slavery Society intended to raise "$100,000" in the near future for the "support of [its] agents."[42] The writer clearly intended to imply that abolitionists were earning substantial—if not exorbitant—salaries while advocating the impoverishment of the South. It is interesting that this writer went on to make the old connection between American and English abolitionism: "Every friend of his country . . . ought to inquire seriously, how far the feelings of Old England are becoming the feelings of the New."[43] Only two months later, the same paper condemned the indifference of "English and Irish" ministers, who complacently enjoyed their "splendid" salaries in the midst of a "starving" population of Irish peasants.[44] Such a pointed indictment of individual, self-interested moral incompetence resonated repeatedly in the southern clergy's denunciation of Northern abolitionists. In 1844, the *Christian Index* carried the article "Abolition Hirelings," denouncing the "premiums" paid for "articles against slavery" in Northern religious papers. The article compared the "hireling scribblers" to the perverse, mercenary witnesses who had given testimony "against the martyr Stephen."[45] The "hireling" canard proved to be a durable and evidently powerful weapon. The same 1857 resolution of North Carolina Baptists that linked abolitionism to "mystery Babylon," also asserted that the abolition movement was sustained in the United States through the activities of "hired clergymen at the North."[46]

Perhaps what made the criticism of a personally, economically interested abolitionism so potent and compelling for the southern preachers was the ongoing struggle in the South over their own compensation. Although most evangelical denominations in the Old South endorsed the idea of a salaried clergy, the public and the ministers continually wrangled over the appropriate size of remuneration and some preachers and laymen even persisted in presenting poverty as the ideal economic and spiritual status for Southern divines. Pronouncements from the pulpits and in the religious press of the antebellum South reveal a deeply conflicted clerical community on the matter of compensation.[47] Given their own anxiety and guilt about any possible blending of ministry and mammon, the Southern preachers were especially sensitive to, and highly suspicious of, the private, commercial aspects of abolitionism. Their suspicions seemed confirmed with reports of Harriet Beecher Stowe's financial success. Indeed, Southern clerics presented Stowe as the very model of a mercenary moralist. In 1853, one of the most prominent and influential Southern religious periodicals claimed that Stowe was amassing "a large fortune . . . out of the woes of the South upon which she trades and denounces." The writer went on to characterize Stowe as a moral profiteer: "How much of her large income, we may ask, does she appropriate to the liberation of the slaves? Probably not one dime. No, no northern principle 'knows a tale worth two of that.' Money is a stronger God than morality, and more powerful even than slavery, one has aptly said in this respect, 'Dollars in the North fatten to full moons, but seen in the South—they count as picayunes.'"[48] Here was an important ancillary theme: Why did not more Northerners simply buy and liberate slaves? Surely, the Southern divines implied, it would be an act of high moral purpose—and instructive in its sacrifice—to see thousands of slaves bought and liberated by wealthy abolitionists in the North.

As an example of the personal moral duplicity and material selfishness they perceived in the Northern antislavery movement, the editors of the *Nashville and Louisville Christian Advocate* reported in 1853 that a member of a Methodist church in Kentucky—a church affiliated with Northern Methodism—sold an old, infirm slave. Such a mercenary act, the editors claimed, would have prompted the expulsion of so pronounced a mammonist from a Southern church.[49] Southern ministers clearly wished to portray Northern virtue and abolitionist idealism as conditional, subject to individual and corporate calculations of self-interest. In 1856, the Rev. W. A. Smith, president of Randolph-Macon College, emphasized this perspective when he asserted that Northern prosperity rested in large measure on the profits of colonial New England's participation in the slave trade. These profits, according to Smith, had provided much of the capital for the North's subsequent economic development. Moreover, the Northern economy continued to benefit from the productivity of slave-worked cotton fields. With a snide inflection, Smith suggested that a commitment to perfect moral consistency required abolitionists to repudiate, to rescind willingly some considerable portion of their region's and—by logical extension—their own wealth, even as they demanded the end of Southern prosperity.[50]

The Southern ministers' characterization of abolitionism, Northern culture, and Northerners was marked deeply by numerous moral and logical contradictions and rationalizations, not the least of which was the Southern clergy's own admission of the South's destructive, raging materialism. Again, ironically, it was probably the excesses and abuses of Southern capitalism and their fear and resentment of those excesses that gave Southern ministers a heightened sensitivity to the perceived illicit connection between Northern benevolence and Northern economic excesses and abuses. Convinced of their own society's mammonism—and concerned by their own appetite for middle-class remuneration—the Southern ministers were unwilling or unable to concede a selfless idealism to abolitionism or to the North. Just possibly, the Southern clergy's strident disdain for the abolitionists and the larger Northern culture reveals far more about these ministers and the South than it does about the Northern men and women they characterized so negatively. It may well be that a *capitalist* Southern clergy wished to deflect its own sense of shame, not over slavery, but over its own absorption into the capitalist ethos.[51] Moreover, is it not reasonable to consider that the Southern divines, unable to break mammon's hold on their beloved South, expended some of their embarrassed and frustrated fury by imputing and exaggerating mammonist selfishness and insincerity to their Northern protagonists?

But for all their misperceptions, exaggerations, and distortions—and despite what surely were mixed and suspect motivations—the Southern ministers possibly came close to some valid conclusions; scholarly inquiry has occasionally suggested as much. For instance, their attempt to link abolitionism to a highly aggressive and abusive form of capitalism bears some resemblance to Eric Williams's masterwork, *Capitalism and Slavery*, published originally in 1944.[52]

Controversial then and now, Williams's thesis has nevertheless continued to merit serious scholarly attention. Although critics have effectively repudiated some of his fundamental conclusions—notably his idea that West Indian slavery and the slave trade were economically untenable by the early nineteenth century—they have not completely addressed the underlying issue, that English abolitionists were successful in their campaign because the public and press regarded slavery as an outmoded, declining economic system. Like historians of economics—and everything else—some of Williams's critics miss the point that *perception* may be more important than reality.[53]

Williams's work focused on the English abolition movement in the West Indies, a subject that also captured the attention of Southern ministers. Like them, Williams claimed that the abolition movement was not ultimately sustained by a benign idealism. In two sentences calculated to provoke intense controversy, Williams stated, "The humanitarians [English abolitionists] were the spearhead of the onslaught that destroyed West Indian [slavery]. . . . But their importance has been misunderstood and grossly exaggerated by men who have sacrificed scholarship to sentimentality and . . . placed faith before reason and evidence."[54]

Williams then proceeded to describe the English abolition movement as an intellectually and morally superficial campaign that had ignored the massive social evils current in England at the time, evils intimately associated with industrial capitalism: "Wilberforce was familiar with all that went on in the hold of a slave ship but ignored what went on at the bottom of a mine shaft."[55] In Williams's estimation, the English abolitionists had also failed to recognize the powerful economic forces that ultimately secured the success of English abolitionism. He contended that English abolitionism actually provided a moral legitimization, or at least a useful moral veneer, for a rapacious form of imperial, industrial capitalism.[56] Still, his intention was not to defame the abolitionists but rather to see them as symbiotically—if *inconsistently*—linked to a dynamic, often destructive, economic system, a system that, nevertheless, provided them with the wherewithal to mount their moral campaign. Recent scholarship has qualified, but also reaffirmed, some of Williams's salient conclusions. For example, John Ashworth is among those recent historians who have marked a significant connection between American abolitionism and capitalism.[57]

William Freehling described a true entrepreneurial abolitionist when he traced the life and career of Pearl Andrews in *Road to Disunion*.[58] A native of Massachusetts, Andrews found abolitionism to be the perfect means for combining a powerful capitalist appetite with high idealism:[59] "Harnessing enterprise to emancipation came naturally to a seeker torn between Christ and capitalism."[60] Andrews tried to convert Texans to emancipation in the mid-1840s by linking abolitionism to economic prosperity. Slavery, according to Andrews, inhibited white immigration to Texas. Renounce slavery, Andrews advised, and white "immigrants would swarm to this lush prairie. Then rising land values would compensate slaveholders for lost slaves."[61] As a prominent speculator in Texas land, Andrews could win both morally and financially should his dream of a free Texas be realized. Ultimately, however, his "Christian entrepreneurship" failed completely.[62] But regardless of Andrews's success or failure, Freehling's portrait of the abolitionist land speculator tends to support the perspective of the Southern ministers.[63] Like these preachers, Williams and Ashworth suggested a bifurcation of the abolitionist-capitalist connection. They emphasized an institutional, national, *macro* relationship between capitalism and abolitionism; William Freehling clearly marked the private *micro* merging of morality and money.

This is not to say that Williams, Ashworth, or Freehling replicated the Southern clergy's perspective in any detailed or comprehensive way. In fact, these scholars moved the story in quite a different direction; all aptly described the profound self-interest of the proslavery advocates. Still, the scholars identified and described a connection, a synthesis between a confident, aggressive capitalism and an equally aggressive, economically interested abolition movement.

Other historians have suggested a less obvious if no less divisive manifestation of the purported selfishness of the Northern economic culture. In this

view, the melding of powerful economic, social, and moral forces in the antebellum North, as in England, led to a sort of cultural myopia, making the
subordination or destruction of other, inferior cultures—West Indian and
Southern slave society—more palatable and probable.[64] This was a recapitu-
lation of the Southern clergy's elemental accusation that Northerners, though
not necessarily animated by an appetite for direct economic gain at the South's
expense, were nevertheless casually and cruelly dismissive of the economic
sacrifice they required from Southerners. Economically confident and prosperous, Northerners refused to ameliorate, or even concede, the South's economic anxiety and the potential economic pain represented by abolitionism.

Gavin Wright offered a partial corroboration of this theme in *Old South,
New South*. For Wright, Northerners, entirely convinced of the economic
superiority of industrial capitalism which was based on wage labor, easily
adopted a general anti-South, antislavery attitude without considering the
economic consequences.[65] Such a mentality represented a kind of "cultural
imperialism," a phrase and concept Wright credited to Howard Temperly.[66]
Moreover, Wright asserted that Southerners, confronted with stark Northern
economic and cultural assumptions, "were able to see through the hypocrisies and injustices of northern society more clearly than most Northerners."[67]
To Southern ministers, Northern society was selfish, therefore, in a particularly degraded way. Northerners, led by their abolitionist high priests, had
made idols of their own economic and moral conceptions and were blind, in
their complacent materialism, to the sacrifice they demanded of Southerners.
Of course, the southern ministers had also proclaimed openly, incessantly,
that Southerners were also beguiled by a false priesthood—the rich, ostentatious plantation elite—and had *blindly* made their own idols of indolence,
cotton, and commerce.[68]

The Southern clergy's characterization of the abolitionist North was woe-
fully biased and simplistic, but by 1860 it carried real force, especially when
seen in the context of other long-perceived examples of selfish, abusive
Northern attitudes and practices—notably, high protective tariffs.[69] Early in
the war, a prominent Southern minister made just such an explicit connection between Northern war aims and Northern greed: "Not a gun would [Lin-
coln] fire if he could collect the Southern imposts [and still let the seceded
states go]. . . . [The South] must fight unless we are willing to pay any tax
which avarice and rapacity, armed with power, may assess."[70] Such an alle-
gation effectively compressed the sectional suspicions and hostility of the
evangelical Southern clergy into a simple, powerful image of an economically grasping and morally degraded North.

The Southern divines believed and proclaimed that Northerners, especially the abolitionists, had made a "mere calculation" that cost them nothing or, in some rare cases, made their fortunes. In their assertive, naive
economic and moral confidence, the Southern ministers were certain that
Northerners had made a disastrous miscalculation in failing to gauge
Southern determination to preserve the economic prerogatives and prosperity of the slave South.

Notes

I wish to thank Mark Noll for his assistance in the preparation of this chapter. His suggestions were uniformly helpful. I am particularly grateful for his advice concerning sources for the "guilt thesis."

1. "The Victory on Manassas Plains," *Southern Presbyterian Review*, January 1862, 606.

2. Typical of the genre are "The New Year," *Christian Index*, January 5, 1837; "Individual Responsibility," *Nashville Christian Advocate*, February 23, 1849; slightly more muted is W. A. Scott, *Daniel, as a Model for Young Men, a Series of Lectures* (New York: Robert Carter, 1854), 186–89. For a more complete discussion see Kenneth Startup, *The Root of All Evil, the Protestant Clergy and the Economic Mind of the Old South* (Athens: University of Georgia Press, 1997), esp. chaps. 1 and 2.

3. I will not attempt to discuss in this chapter the complex and often abstract arguments relating to the emergence of a capitalist South. I am convinced that such a South existed by the beginning of the antebellum era and matured rapidly during the period. My working definition of capitalism is taken directly from Neal Wood, though I do not agree with his assertion that capitalism requires "free wage earners" (19). I do agree with his definition that a capitalist is someone who "lives on the profits derived from the expropriation of surplus value produced by the workers. His economic behavior is motivated by a desire to maximize profit and minimize loss, and the necessity to accumulate capital." Neal Wood, *John Locke and Agrarian Capitalism* (Berkeley: University of California Press, 1984), 19–20. This definition is especially useful in describing the economic mentality that governed the slave South. Dated, but still compelling for its clarity and conclusions regarding the rise of *American* capitalism, is Louis M. Hacker, *The Triumph of American Capitalism: The Development of Forces in American History to the End of the Nineteenth Century* (New York: Simon & Schuster, 1940). Other works useful in describing or debating the issue of antebellum Southern capitalism—and helpful in placing Southern capitalism within the larger national and Western economic context are John Ashworth, *Slavery, Capitalism, and Politics in the Antebellum Republic Commerce and Compromise*, vol. 1 (New York: Cambridge University Press, 1995); Joseph Reidy, *From Slavery to Agrarian Capitalism in the Cotton Plantation South, Central Georgia, 1800–1880* (Chapel Hill: University of North Carolina Press, 1992); and Gavin Wright, *Old South, New South, Revolutions in the Southern Economy since the Civil War* (New York: Basic Books, 1986).

4. A more thorough discussion of the Southern clergy's concern with the economic preoccupation of Southern Christians is found in Startup, *Root of All Evil*, chap. 2.

5. Ibid., 27.

6. Ibid., chaps. 4 and 5. This note applies to the entire discussion of the clergy's fear that mammonism was corrupting the South's social and political structure.

7. Ibid., chap. 6. This note applies to the entire paragraph.

8. *Annual, Southern Baptist Convention*, 1863; "Our Servants," *Confederate Baptist*, June 1, 1863.

9. In the "panic year," the editors reviled the national appetite for quick riches in "Children and Money," *Christian Index*, March 2, 1837. Later that year, an article bitterly denounced economic injustice in the North: "Traveling Preacher, No. 3," *Christian Index*, August 31, 1837.

10. Anne C. Loveland, *Southern Evangelicals and the Social Order, 1800–1860* (Baton Rouge: Louisiana State University Press, 1980), 199–202; Donald G. Mathews, *Religion in the Old South* (Chicago: University of Chicago Press, 1977), chap. 4.

11. Iveson Brookes, *A Defence of the South against the Reproaches and the Incroachments of the North: In Which Slavery Is Shown to Be an Institution of God Intended to Form the Basis of the Best Social State and the Only Safeguard to the Permanence of a Republican Government* (Hamburg, S.C.: Republican Office, 1850).

12. "Traveling Preacher, No. 3."

13. For a thorough explication of Fitzhugh's views and the general socioeconomic defense of slavery, see Eugene Genovese, *The World the Slaveholders Made: Two Essays in Interpretation* (New York: Vintage Books, 1971); Hammond's socio-economic perspective is described in Drew Gilpin Faust, *James Henry Hammond and the Old South: A Design for Mastery* (Baton Rouge: Louisiana State University Press, 1982).

14. The complex culture of the antebellum South is too large a topic for consideration in this chapter. To be sure, Southern evangelicalism both shaped and was shaped by the currents of popular and practical culture. Complicating matters still more is the reality that a considerable, and not surprising, gulf existed between the ideal culture of evangelicalism as espoused by preachers and the real evangelicalism of most antebellum Southerners. A superb recent treatment of the antebellum Southern evangelical culture is Christine Leigh Heyrman, *Southern Cross, the Beginnings of the Bible Belt* (New York: Knopf, 1997). My reference to an evangelical "worldview" relates primarily to the clergy's aspiration and ideal.

15. See especially Eugene Genovese, *A Consuming Fire, the Fall of the Confederacy in the Mind of the White Christian South* (Athens: University of Georgia Press, 1998). Genovese makes the telling point that in the wake of defeat in 1865, the Southern ministers lamented that they must now "live under the very social system that they had condemned as un-Christian in tendency. Punished for their lapses they now found themselves enmeshed in a materialistic, marketplace society that promoted competitive individualism and worshiped Mammon" (102). Clearly, in Genovese's estimation, it was far more than just the loss of slavery that disconcerted and depressed the secessionist clergy after Appomattox.

16. *Southern Quarterly Review*, April 1842, 446–93.

17. Ibid., 465.

18. "Slavery No. 1," *Christian Index*, July 5, 1844.

19. "Slavery," *Nashville and Louisville Christian Advocate*, February 20, 1851.

20. Though Adger agreed with some of the conclusions contained in the "Majority Report," he opposed reopening the slave trade. His words are from John B. Adger, "Review of Reports to the Legislature of South Carolina on the Revival of the Slave Trade," *Southern Presbyterian Review* (October 1859), 108 and 112, respectively.

21. "Abolitionism," *Christian Index*, June 8, 1837.

22. "A Blast Against Abolitionism," *North Carolina Christian Advocate*, August 6, 1857.

23. Mary Boykin Chesnut, quoted in Elizabeth Fox-Genovese, *Within the Plantation Household: Black and White Women of the Old South* (Chapel Hill: University of North Carolina Press, 1988), 357.

24. "Northern Slavery," *Christian Index*, December 1, 1843, for quotations in this paragraph.

25. The size and influence of Southern Methodism and the process of sectional separation are discussed fully in Donald Mathews, *Slavery and Methodism: A Chapter*

in American Morality, 1780–1845 (Princeton, N.J.: Princeton University Press, 1965). For another excellent review of the disruption of American Methodism, Mitchell Snay, *Gospel of Disunion, Religion and Separatism in the Antebellum South* (Chapel Hill: University of North Carolina Press, 1997). Comparing the specific size of Southern Methodism, Snay asserts, "They were the largest denomination in the South in 1850, comprising 37.4 percent of the churches in the eleven states that would become the Confederacy, plus Kentucky and Maryland" (127).

26. The "property controversy" and the Supreme Court's decision are treated briefly in Snay, *Gospel of Disunion*, 134.

27. "Things as Strange as They Are True," *Nashville and Louisville Christian Advocate*, January 29, 1852.

28. "The Property Question," *Richmond Christian Advocate*, copied in *Nashville Christian Advocate*, April 20, 1849.

29. N. M. Crawford, "What We Are Fighting For," *Christian Index*, July 24, 1861.

30. Daniel W. Stowell, *Rebuilding Zion, the Religious Reconstruction of the South, 1863–1877* (New York: Oxford University Press, 1998), 29–31.

31. Ibid., 30.

32. Ibid.

33. W. M. Leftwich, *Martyrdom in Missouri, a History of Religious Proscription, the Seizure of Churches, and the Persecution of Ministers of the Gospel in the State of Missouri during the Late Civil War and under the "Test Oath" of the New Constitution*, vol. 2 (St. Louis: Southwestern Book and Publishing Company, 1870).

34. Ibid., 71.

35. Ibid., 78.

36. Stowell, *Rebuilding Zion*, 31.

37. "'Signs of the Times' in Arkansas," *Nashville Christian Advocate*, March 30, 1849.

38. "Southern Movements in Arkansas," *Nashville Christian Advocate*, August 24, 1849.

39. "Letter from Bishop Capers," *Nashville Christian Advocate*, February 8, 1850.

40. "Northern Interference," *Nashville Christian Advocate*, April 27, 1849.

41. All of Griffin's comments and perspectives cited are from a "copied" article. Indeed, the article containing Griffin's opinions is typical of the penchant of the antebellum religious press for copying articles from other papers. Griffin's opinions appeared first in the *Natchez Free Trader*, only to be copied by the *New Orleans Christian Advocate*, and copied again as "Tobias Gibson—Lorenzo Dow—the old meeting house in Washington—Mead and Poindexter—A Duelist—Dr. Winans—Miles Harper—Thomas Griffin—Original Letter—Slavery and Slaveholders—Northerners in the South," *Nashville and Louisville Christian Advocate*, October 12, 1854.

42. "Abolitionism," *Christian Index*, June 8, 1837.

43. Ibid.

44. "Traveling Preacher, No. 3."

45. "Abolition Hirelings," *Christian Index*, August 9, 1844.

46. "A Blast Against Abolitionism," *North Carolina Christian Advocate*, August 6, 1857.

47. In "Covetousness the Sin of the Baptist Denomination," an article copied from the *Cross and Baptist Journal*, readers were alerted to the fact that far too many Baptist preachers were "greedy for filthy lucre." *Christian Index*, July 20, 1837.

234 THE ECONOMICS OF SECTIONAL STRIFE AND REVIVAL

48. These specific quotations about Stowe are taken from "Key to Uncle Tom's Cabin," *Nashville and Louisville Christian Advocate*, May 28, 1853. In an attempt to bolster their clearly sectional perspective on Stowe, the article also included comments highly critical of Stowe that were taken from a review in the *Boston Courier*.

49. "An Incident Illustrative of the Abolitionist Conspiracy," *Nashville and Louisville Christian Advocate*, September 3, 1853.

50. William A. Smith, *Lectures on the Philosophy and Practice of Slavery, as Exhibited in the Institution of Domestic Slavery in the United States: With Duties of Masters and Slaves*, ed. Thomas O. Summers (Nashville, Tenn.: Stevenson & Evans, 1856), 160–61. Another example of this commonplace criticism of the North is "O South! South!!" *Christian Index*, October 16, 1859.

51. The "guilt thesis"—the idea that antebellum Southerners were defensive and confrontational because they were inwardly obsessed with personal and corporate guilt over slavery—has been challenged effectively by many scholars. For example, William Cooper and Anne Loveland respectively described the white South's reconciliation of democracy and evangelicalism with the slave system, a reconciliation sufficient to counter any inclination toward guilt. See William J. Cooper, Jr., *Liberty and Slavery, Southern Politics to 1860* (New York: Knopf, 1983); and Loveland, *Southern Evangelicals*. Cooper and Loveland are among many scholars cited by Gaines Foster in his excellent review of the topic, "Guilt Over Slavery: A Historiographical Analysis," *The Journal of Southern History* 56 (November 1990): 665–94. Foster especially recognized Eugene Genovese as having "provided the most explicit, sustained, and comprehensive attack on the guilt thesis" (683). Foster also cites Jack P. Maddex as a dissenter from that perspective (685). In 1979, Maddex presented one of the most formidable challenges to the guilt thesis. He concluded, in a closely reasoned analysis, that the defense of slavery—especially by Southern ministers—far from being a matter of guilt, derived from the conviction that slavery was entirely compatible with their "comprehensive world view." Jack P. Maddex, Jr., " 'The Southern Apostasy' Revisited: The Significance of Proslavery Christianity," *Marxist Perspectives* 7 (Fall 1979): 132–41, quotation 137.

52. Eric Williams, *Capitalism and Slavery* (Chapel Hill: University of North Carolina Press, 1944).

53. Seymour Drescher, "Review Essay," on *The Antislavery Debate: Capitalism and Abolitionism as a Problem in Historical Interpretation*, ed. Thomas Bender (Berkeley: University of California Press, 1992); *History and Theory* 32 (October 1993): 311–320. Drescher is among Williams's most effective and persistent critics. See especially Seymour Drescher, *Econocide: British Slavery in the Era of Abolition* (Pittsburgh: University of Pittsburgh Press, 1977); and Drescher, *Capitalism and Antislavery: British Mobilization in Comparative Perspective* (New York: Oxford University Press, 1987). Another powerful critique of Williams is Roger Anstey, *The Atlantic Slave Trade and British Abolition, 1760–1810* (Atlantic Highlands, N.J.: Humanities Press, 1975).

54. Williams, *Capitalism and Slavery*, 178.

55. Ibid., 182.

56. Ibid., see especially chaps. 10 and 11. This note applies to the entire paragraph.

57. John Ashworth, *Slavery, Capitalism, and Politics in the Antebellum Republic Commerce and Compromise, 1820–1850*, vol. 1 (Cambridge: Cambridge University Press, 1995).

58. William W. Freehling, *The Road to Disunion, Vol. 1, Secessionists at Bay, 1776–1854* (New York: Oxford University Press, 1990), esp. chap. 21 for a discus-

sion of Andrews's activity in Texas. This note refers to the entire discussion of Andrews's character and conduct.

59. Freehling, *Road to Disunion*, 373. Freehling describes Andrews as a man "dreaming of the double bliss of an emancipator who would turn materialistic gamble into a holy triumph."

60. Ibid.

61. Ibid.

62. Ibid., 376.

63. Ibid.

64. The idea of a sectional myopia is nothing new, of course, and certainly the astigmatism could be found to be obscuring both regional visions. Perhaps the finest discussion of this issue is Lewis Simpson, *The Dispossessed Garden: Pastoral and History in Southern Literature* (Athens: University of Georgia Press, 1975); see also Simpson, *Mind and the American Civil War: A Meditation on Lost Causes* (Baton Rouge: Louisiana State University Press, 1989).

65. Gavin Wright, *Old South, New South, Revolutions in the Southern Economy Since the Civil War* (New York: Basic Books, 1986), esp. chap. 2. This note applies to the entire paragraph and generally to the idea of an economic and cultural myopia in the North.

66. Ibid., 31.

67. Ibid., 32.

68. Startup, *Root of All Evil*, esp. chap. 4.

69. "The State of the Country," *Southern Presbyterian Review*, January 1861, 864. The clergy's perspective in this regard—that the North had penalized the South for its own economic benefit—was a prominent part of the secessionists' rationale for leaving the Union. For an especially cogent example of this view, see "Report of the Ordinance of Secession," (Georgia), January 29, 1861, in *Official Record of the War of the Rebellion*, ser. 4 (Washington, D.C., 1890–1901) 1:81–85. And in his first presidential address in Richmond, Jefferson Davis could not resist a reference to "70 years of taxation being in the hands of our enemies." *The Papers of Jefferson Davis*, ed. Lynda Crist and Mary S. Dix (Baton Rouge: Louisiana State University Press, 1992), 7:184.

70. Crawford, "What We Are Fighting For."

"Turning . . . Piety into Hard Cash": The Marketing of Nineteenth-century Revivalism

Kathryn T. Long

On Saturday, February 27, 1858, the most successful daily newspaper in the United States, James Gordon Bennett's *New York Herald*, turned its full attention to revival. Under stacked headlines, seven deep, the story received a full column of coverage on page one:

> Great Revival of Religion in New York.
> Progress of the Movement.
> Remarkable Conversions Among the Unrighteous.
> Sinners Brought to the Way of Grace.
> One of the "Forty Thieves" Repentant.
> Wonderful Manifestations of Penitence and Piety.
> Revivals Elsewhere.
> &c., &c., &c.,

For the first time in the nineteenth century in America, revivalism was front-page news in the secular press, appearing alongside stories of city politics, the activities of Congress, and sensational crimes. As the headlines indicated, the human interest aspect of the religious fervor was tailor-made for a mass-circulation daily like the *Herald*. There was emotion, large crowds, and the promise of "open confessions" by businessmen and politicians.[1]

Although Bennett took the lead in recognizing the news value of the "great revival," his competitors wasted little time in catching up. Through a series of adroit moves during the next few days, Horace Greeley, a long-time rival and editor of the powerful *New York Tribune*, outmaneuvered Bennett and positioned the *Tribune* as a kind of national clearinghouse for revival information. On the following Monday, Philadelphia papers also carried reports of the New York prayer meetings, and by the end of the week the *New York Times*, the *Boston Post*, and even the distant *New Orleans Picayune* had noted the religious excitement in the nation's largest city.

During the months of March and April 1858, the revival became a media event, covered by metropolitan dailies in all parts of the United States.[2] Newspapers, in fact, took the lead as catalysts for the spread of this revival's most familiar manifestation, the noon urban prayer meeting. The New York press, led by the *Tribune* and *Herald* but including other respected dailies such as the *Times* and the *Evening Post*, spurred the ongoing coverage. Newspapers in other cities reprinted their stories, emulated their writing style, and followed their practice of culling anecdotes from the many weekly religious papers.

Rank-and-file evangelical Protestants expected to read about revivals in denominational weeklies, but the appearance of such news in secular papers created a sensation. A letter from a correspondent in the upstate village of Troy to the *Tribune* captured the reaction: "The elaborate accounts of the great National religious awakening, which are just now being published in The Tribune, has [*sic*] created a great demand for the paper in this vicinity. Hoyt's news-room is thronged with applicants of all classes and conditions for your sheet."[3] Such commentary pointed to the commercial aspect of the secular press's discovery of the revival: It sold papers. With the advent of the penny press in the 1830s, metropolitan dailies had embraced the concept of news as a product for mass consumption. Although they also offered subscriptions, the *Herald* and the *Tribune* maintained their high circulations through "cash and carry" sales. The papers were hawked on the streets by newsboys for two cents a copy to a more middle-class audience than the people who bought the cheaper and smaller penny papers. Success depended on news that sold, and editors knew from day to day what stories appealed to a mass readership. At bottom, to editors such as Bennett and Greeley, the revival of 1857–58 was a commodity.[4]

The expanded sale of religious information as news was only one of the ways in which the "businessmen's revival" of 1857–58 marked a new phase in relationships among Protestants, money, and the market.[5] By setting those relationships of the late 1850s against the earlier history of revivalistic evangelicalism and its supporting agencies, it is possible to see where lines of influence between Protestants and the market grew, shrank, or reversed course. The same benefit arises from examining several other factors of the 1857–58 revival, including circumstances in which business methods or considerations of economic realities exerted a growing impact on the Protestant churches. For both the increased eagerness to market the revival as news and the deeper involvement of commercial values in the practices of the revival, the years just before the Civil War are unusually important. They summed up, even as they altered, much that had gone before in the Protestant approach to money.

Revivalism as Commodity

Historians have recently demonstrated that treating revivals as a commodity was not a new thing. Modern evangelicalism, Harry S. Stout argues, emerged in tandem with the commercialization of Anglo-American culture that began in the early eighteenth century. Along with other leisure-time activities, "religion increasingly represented a product that could be marketed," with revivalist George Whitefield as its most brilliant early entrepreneur.[6] Whitefield, for example, pioneered the practice of open-air preaching on weekdays in urban centers. There, in competition with commercial vendors and secular entertainments, he offered his "product," the gospel message, to a mass, popular audience. To encourage demand for his preaching, Whitefield used advertising and other public relations devices, sensing intuitively, as Stout has suggested, that "mass media and mass consumption were inseparable."[7] During his 1740–41 preaching tour in the colonies, the revivalist turned to colonial newspapers to publicize his activities.[8] A half century later, in the early Republic, a similar kind of market-centered mass revivalism was developed even more effectively and on a much broader scale by the Methodists, a group Nathan Hatch has described as "the prototype of a religious organization taking on market form."[9]

Whitefield and the many "peddlers" of popular Christianity who followed him in the free-market ethos of early nineteenth-century America were religious entrepreneurs. They were innovators who helped create a popular, religious "marketplace," using print media and early advertising techniques to promote their cause.[10] To be sure, what these evangelists sought was certainly not profits. Although employing the techniques and even the rhetoric of the marketplace to pursue evangelism, they maintained a distance between themselves and the greedy selfishness they saw as infecting the actual economic order.[11] Frank Lambert poses this relationship for Whitefield: "He was unsparing in his criticism of those who placed pursuit of earthly gain above concern for eternal salvation." At the same time, "Whitefield was a pioneer in the commercialization of religion, discovering within the market the very strategies and languages to attack its excesses." In short, Whitefield "proclaimed a message not of the market but in the market."[12]

Examples from the nineteenth-century reflect this same tendency among revivalistic Protestants to selectively appropriate market methods while rejecting certain market values. From the beginning, Methodists understood that their purpose—or in the words of the Methodist *Discipline*, their "business"— was "to save as many souls as possible."[13] God providentially had raised up Methodism, early proponents believed, "to reform the continent, and spread Scripture holiness over these lands."[14] With a corps of itinerants and a strategy of continual expansion, the movement cultivated the vision of a nationwide market for religion. Yet at the same time, Methodist expansion and skill at market penetration was fueled by an ethic of sacrifice and voluntary poverty rather than a desire for self-aggrandizement or a hope of financial gain. Francis Asbury scoffed at the pretensions of educated ministers who wished to settle in towns and sell their souls to wealthy congregations. In contrast, he challenged young men to become traveling preachers at a salary of $80 a year, a survival wage at best.[15]

As David Paul Nord (chapter 7 in this volume) shows, the same combination of market methods and market wariness attended the work of the great publishing societies. By 1829, the American Tract Society was printing more than 6 million tracts annually, and the American Bible Society was producing more than 300,000 copies of the Scriptures. Yet neither agency was clearing any profit. A comparison of the American Bible Society's receipts for the year 1829 with those Carey & Lea, the most prosperous commercial publishing house of the era, reveals that the commercial enterprise took in twice as much money in sales although the volume of production for the two organizations was roughly equivalent.[16] By the late 1840s, some commercial booksellers complained that the "charity presses" of the benevolent societies and major denominations were driving private publishers out of the book trade by undercutting their prices.[17] Once again revivalistic Protestants were proving expert at marshaling the techniques of industrialization while resisting the profit motive and other standard values of the marketplace.

Despite the correlation some scholars have suggested between Charles Finney's revivalism and middle-class entrepreneurial success, the same creative tension holds true in his case. On the one hand, Paul Johnson has noted Finney's appeal to the entrepreneurial classes of Rochester, New York.[18] Promotional technique, in fact, found a place in Finney's famous "new measures." In 1832, for example, when the revivalist held special meetings at the Chatham Street Chapel in New York City, he sent lay men and women door to door in the neighborhoods surrounding the church to distribute slips of paper advertising the services.[19] On the other hand, James Moorhead has also pointed out that Finney never was "completely comfortable with the ethos of commercial or industrial capitalism." In the wake of the panic of 1837, Finney attacked the credit system and rejected "the mass consumption necessary to sustain a modern economy."[20] The revivalist himself was so financially strapped in the fall of 1837 that he lacked the funds to support his family through the following winter.[21]

Beginning, then, with the colonial revivals that were centered around the preaching tours of Whitefield and throughout the religious fervor of the first half of the nineteenth century, pietistic Protestants "marketed" revivalism through their own, largely nonprofit print media, including religious periodicals, books, and tracts. Such established patterns held true during the winter months of 1858, as denominational newspapers began to publicize an unusual degree of religious interest in New England, in the Old Northwest, and in the New York City metropolitan area. By late February in New York City, evangelicals had accelerated local public relations efforts. Clerks and merchants belonging to the Young Men's Christian Association (YMCA) distributed thousands of printed cards that advertised the Fulton Street Prayer Meeting, a businessmen's gathering that was the epicenter of religious excitement. On February 20, the YMCA sent to secular papers throughout the country a circular letter—in effect, a press release—publicizing additional business-men's prayer meetings.[22] As on previous occasions, evangelical use of such techniques was geared toward heightening the religious effects of the revival rather than toward financial gain. Although it was able to pay for these pro-motional efforts, the New York City YMCA struggled with recurring debt during the years that followed.[23]

When the secular press, particularly the *Herald* and the *Tribune*, became the dominant players in coverage of the revival, the dynamic changed. Publi-cists for previous revivals had been the revival preachers themselves, their cadres of lay supporters, or the ministers who edited denominational news-papers. By contrast, during the spring of 1858, Bennett and Greeley, two editors of profit-making newspapers, became the most influential reporters on Protestantism in America. They introduced new styles of packaging revivalism for consumption in the mass medium of the popular press. More than a hundred years earlier, George Whitefield had used the colonial press to further his public relations purposes.[24] But in 1858, the balance of power shifted. The press would now use revivalism for its own ends. Packaging the story of a religious awakening for commercial consumption rather than for nonprofit religious edification had the long-term effect of pushing American revivalism toward a businesslike frame of mind. As evangelicals cooperated with and encouraged the mass circulation dailies, they began to narrow the distance between their values and the values of the economic order.

James Gordon Bennett is generally credited as the first to begin religious news coverage in a newspaper intended for a general audience.[25] The *Herald* was quick to publish news on religious subjects, especially when these stories involved sex or scandal. Whereas the Mormons, Millerites, and other religious groups outside the American mainstream furnished most of this mate-rial, Bennett could also raise the ire of the Protestant establishment, as he did, for example, with a series of articles in 1840 on the annual meetings of the Protestant voluntary societies. The articles were relatively inoffensive, but religious leaders were infuriated at finding themselves featured in a paper that flaunted its disdain for evangelical values by opposing temperance, printing advertisements from theaters and prostitutes, and publishing a Sunday edi-tion. Many Northern evangelical leaders maintained a strong dislike for the *Herald*, an attitude Bennett returned in kind, especially as he refused to embrace the evangelical idea of the secular press as an agent of Christian (Protestant) civilization "on the side of Virtue and Religion."[26] The *Herald*, claiming in 1858 a distribution of 62,000 copies, was "the most widely circu-lated daily in the United States." Because of Bennett's proslavery stance and regular coverage of the South, the paper enjoyed particular influence in that region.[27]

Greeley, on the other hand, embodied the ideals of those, including evan-gelicals, who deplored the *Herald*, and he positioned his *Tribune* as the secu-lar, daily counterpart of the religious family weeklies. To be sure, Greeley was more overtly political than most evangelical Protestants, and, as a Uni-tarian, religion was not the same driving force for him as it was for many evangelicals.[28] But even as Northern evangelicals recognized the *Tribune* as a secular paper, "without any fixed religious convictions," they nonetheless respected it as having "an honest purpose to promote the best interests of society."[29] Especially through its weekly edition, the *Tribune* was the dominant paper of the North.

By the early spring of 1858, Bennett's *Herald* and Greeley's *Tribune* were well positioned to take commercial advantage of the growing revival excitement. Both newspapers were respected—and sometimes hated—for their political reporting. They were also practiced in writing human interest and sensationalistic stories, news that moved far afield from traditionally serious subjects. Along with other businesses, both had suffered in the wake of the national financial panic of October 1857, with Greeley especially hard-hit. In February and March 1858, advertising inches (and revenue) were still down. With eight pages in their daily and weekly editions, twice those of most mass circulation journals, the papers had space to fill.[30] Revival stories provided much desired content. They had the added attraction of convenience since the most popular New York City prayer meetings were being held less than five minutes' walk from the newspaper offices.

Of the two editors, Horace Greeley had the best marketing strategy. From its initial, extensive coverage on March 1, 1858, to a special "Extra" edition five weeks later devoted entirely to the revival, the *Tribune* provided the most complete record of the nationwide awakening to appear in any single newspaper, secular or religious.[31] To package the news in a way that would sell, the *Tribune* employed the techniques of popular journalism and highlighted the human side of the religious awakening. Greeley devoted reams of newsprint to what he considered the chief "novelty"—the lead story—of the revival: interdenominational, noon prayer meetings "held in the center of the business circles of the city, and sustained largely by the most prominent business men."[32] In addition were stories of unusual conversions, the transformation of a Manhattan theater into a place of prayer, and eccentric but harmless characters who formed part of the revival scene such as a Methodist woman dubbed "Screeching Harry" and a ne'er-do-well, "boorish" Tommy Lloyd.[33] Greeley enjoyed a publishing coup with his sensational coverage of the dramatic conversion of Orville Gardner, a well-known bare-knuckle fighter and sometimes thug. Presented to readers in a series of *Tribune* articles, Gardner became one of the first "celebrity converts" in American history, certainly the first Christian sports celebrity.[34]

Such efforts paid off in increased sales for the *Tribune*. They also elicited gratitude from evangelicals and enabled Greeley to build readership among Northern Protestants, a key segment of his market.[35] After the revival, Charles Finney commented that the *Tribune* "was instrumental in doing very much to extend the work. All honour to Mr. Greeley for the honourable course he pursued."[36] Detractors scoffed at the *Tribune's* emphasis on religion as little more than Greeley's "knack of turning . . . piety into hard cash."[37]

Economic potential was not the sole reason for Greeley's interest in religion. The respectability of this revival also appealed to his Whig temperament. It was an orderly awakening with a high "moral tone . . . great sobriety, and a commendable freedom from undue excess."[38] Most of all, Greeley hoped, at least at first, that the revival's spiritual fervor would cause people to support his political agenda. Bennett suggested that the revival was taking place because the churches were leaving politics alone.[39] Greeley disagreed: "Simultaneously with the deep interest felt and expressed by the great body of the religious men of the North in the attempt to save Kansas from the evils and horrors of Slavery . . . we see a revival breaking out and extending over the country quite unprecedented in its character." Greeley expressed his hope that the revival would bring "a great increase" to the Republican party and "infuse into it . . . additional zeal, earnestness and sincerity."[40]

James Gordon Bennett, who geared his religion coverage to New York City readers unsympathetic with the revival, groused about Horace Greeley, "hard at work trying to turn a penny by issuing 'Revival Tribunes' full of pious information and godly news."[41] In his *Herald* articles, Bennett maintained high levels of skepticism and satire. Once, for instance, he printed tables that showed a total daily attendance at New York City prayer meetings of more than 6,000 but also other tabulations that reported daily attendance of more than 14,000 New Yorkers at the theater. He concluded: "It would seem that Satan still has the majority."[42]

Bennett also hijacked the rhetoric of religious fervor for secular purposes, celebrating in one headline the "Revival of Spring Business—The Weather, Opera and the Balls." A few other papers imitated Bennett's practice. A correspondent to the *Boston Evening Transcript* wrote from New York, "Nature has revived. . . . Fashion has also revived . . . music has revived . . . ruffianism has revived. . . . But what are all these revivals compared with that of Religion and of Business."[43] Although not widespread, such comments anticipated the commercialization of conversion rhetoric that became popular in American advertising and other arenas of public persuasion during the final decades of the nineteenth century.[44]

In the eyes of most evangelicals, however, the *Herald* and its followers were isolated hecklers in a national press that was displaying encouraging signs of sanctification. Revival proponents pointed to the generally sympathetic press coverage in the *Tribune* and other urban dailies as evidence of the supernatural origin of the awakening. Presbyterian Samuel Irenaeus Prime, a key revival supporter, described the newspaper coverage as "unprecedented," and he suggested that it might be "an evidence of the all pervading power of the Holy Spirit."[45] Religious weeklies gossiped in their columns that "in the editorial corps of our principal papers are a considerable number of pious and truly evangelical Christians."[46] Eagerness to view the involvement of the commercial press as the sign of a dawning millennium for American journalism had, however, a serious consequence. The efforts of the dailies were celebrated but not thoughtfully evaluated. Evangelicals had known for more than a century that revivals were news; now the popular papers were coming to the same conclusion. The day when the press would lend its "incalculable influence" to "the cause of Christ's kingdom" seemed close at hand; the revival was spreading; God was at work.[47]

Such a vision had no place for the possibility that by extolling press efforts to commercialize the revival, evangelicals were exposing their religion to corrosion from the market. They easily forgot that the commercial papers were businesses, primarily concerned with profits rather than with piety. One or two religious weeklies did acknowledge the influence of the marketplace on the mass-circulation dailies. The *Western Recorder* noted that "some of the daily journals of N. York devote several columns to revival intelligence, for which they find themselves amply remunerated by an increase of circulation."[48] Yet no religious paper considered how this profit motive might affect the way in which secular newspapers viewed the revival and presented it to their readers.

The only cautionary words came from the secular dailies themselves. The *New York Evening Post*, a Republican paper edited by William Cullen Bryant, for years had survived on a modest circulation even as it resisted the tactics of the popular dailies.[49] In an editorial on March 11, 1858, the paper defended its low-key approach to the revival: "The devotional exercises and the religious experiences and opinions of a people . . . are invested with a character of privacy that ought not to be violated for the purpose of affording public amusement or gratifying the curiosity of the skeptic."[50]

The *Evening Post*'s criticisms implied that the mass-circulation dailies were interested in the revival primarily as entertainment. Revival news no longer was primarily information or, as in the denominational weeklies, religious inspiration. Instead, it was a product to be consumed for pleasure. The popular press, led by Bennett and Greeley, realized that there was a limit to public interest in the standard summaries of revival meetings and conversion statistics as these had been traditionally reported in the religious press. Mass audiences wanted to read private dramas of salvation; they wanted to eavesdrop on the prayer meetings; they were curious about the involvement of the famous. One popular anecdote, "Jessie's Gold Ring," concerned Jessie Frémont, wife of the explorer and presidential candidate John C. Frémont. According to the story, Mrs. Frémont had attended a Sunday service at Plymouth Congregational Church during the revival. During the meeting, a collection was taken for the poor. Having no money with her, Jessie Frémont took off an engraved gold ring and placed it in the offering. Although it was a private gesture, Frémont's contribution soon became news.[51]

News about Protestants in the secular press before this time was usually limited to brief notices of church services or to the activities of clerical elites. Editors, except for Bennett's occasional jab, had treated Protestant Christianity with a measure of deference. The revival marked the first time that papers sensationalized coverage of the nation's churches. Despite the mild nature of revival news—compared to stories of "Dissipation, Degradation and Death," which were the staple of the genre—a new policy was beginning in the journalistic treatment of religion.[52] Some papers even printed revival jokes. With the informality of humor introduced into religious reporting, there was after 1858 no longer a reverent distance between secular reporters and popular religion. An occasional editor or correspondent would still profess horror at the idea of "Religion . . . being made the subject of 'reports,' anecdotes, jokes and 'profitable matter,'" but most were more than happy to profit from delivering such religious news.[53] "Serious" coverage of Protestant activities continued, but the attention to revivalism in 1858 marked the beginning of an ongoing commercialization of middle-class Protestantism in the secular press.

Not only did secular papers commercialize and sensationalize the revival, they also played a key role in shaping perceptions of it for both contemporaries and historians. The effects of that shaping were most obvious in the treatment of women. What was most newsworthy for the papers were the urban prayer meetings. Yet these meetings were held in the business districts, where women were usually invisible. Because men were in the majority at these meetings and because male interest in religion was relatively unusual in the "feminized" Christianity of the mid-nineteenth century, the papers fastened onto the noon prayer meetings to sell papers. At most of these meetings, even when women attended, they were not allowed to speak or to offer public prayers.

Such press coverage explains why this religious awakening is often called the "Businessmen's Revival," despite the fact that many women, such as Phoebe Palmer and Elizabeth Finney, were actively involved. Newspapers were fascinated with the conversion and piety of men. Paradoxically, however, objective statistics in the press documented the presence of thousands of women at prayer meetings and other activities. Women were, in fact, a central part of this business*men*'s revival. Nonetheless, men at prayer came to be the most enduring image. The merchandising of religious news as entertainment, in other words, meant that the revival was perceived as an event in popular culture rather than as a product of fairly specific theological and ecclesiastical traditions, which, because they were not newsworthy, have been lost to view.

The post–Civil War revival activities of evangelical Protestants expanded this commercialized approach. With Ira Sankey's solos, the sale of hymn books by vendors, and a corps of trained ushers, D. L. Moody's crusades more openly reflected the trappings of entertainment than had antebellum revivals. In the 1890s, revivalist Fay Mills enlisted advertising committees whose explicit purpose was to "see that [newspaper] reporters attend all services" and that "complete reports are printed in every issue [of the local papers]."[54] By the early twentieth century, the press's hunger for sensationalistic stories put enormous pressure on revivalists such as Billy Sunday no longer to preach but to "perform."[55] To maintain press coverage and to continue to draw crowds, urban revivalism had become both entertainment and a business. Revivals increasingly embodied the image and style the press had created for them. In addition, businessmen were often the key organizers of urban revivals.

The influence of the secular press in marketing revivalism was not, of course, the only factor contributing to the commercialization of mass evan-

gelism during the second half of the nineteenth century. Nor did the revivalists themselves completely capitulate to the attractions of the marketplace.
D. L. Moody did not profit directly from the commercialism that surrounded
his revivals, although after his rise to fame he enjoyed a very comfortable
income.[56] Throughout the final third of the nineteenth-century, revivalist Sam
Jones continued to preach that debt was as dishonest as theft and to warn
against playing the stock market. Nevertheless, at the close of his career he
boasted that he had made more than a quarter of a million dollars through
preaching, speaking, and writing.[57] The tension between proselytizing and
profit making had diminished appreciably since the early years of the century.

During the Revival of 1857–58, Protestants celebrated and encouraged
popular press coverage of religious activity as evidence of the Christianization
of a powerful social institution. Yet their dreams were illusory, given the
economics of mid-nineteenth-century mass communications. What looked like
Christianization to religious leaders was also a commercial exploitation of
popular religion. Editors such as Greeley and Bennett marketed revivalism
to a mass audience as entertainment, just as they sold other human interest
news. Evangelicals, for their part, used the commercialism of the press to their
own advantage but with little recognition that by doing so they were encouraging their audiences to become consumers, as well as converts.

The Business of Revival

Money and commercial culture affected the revival in more ways than just
newspaper coverage. Although most of the changes involved attitudes toward
businessmen and the world of commerce, clergymen were not immune. For
example, almost all observers regarded the revival's stress on interdenominational unity as a good thing. Most clerics also affirmed the lay leadership
of many revival activities, including the noon prayer meetings. Yet there were
also competing forces at work to pull the ministers back, if not to their denominations, at least to local congregations and to the ministerial role. One
of the most basic was money. Particularly in urban areas, the rapidly shifting
population created a revolving-door effect in many congregations, where
membership additions were offset by those leaving for churches elsewhere.[58]
This situation made it difficult for churches to sustain a stable financial base.
Yet at this very time of need, ministers found that church-centered revivals
not only improved piety but also loosened purse strings as well. A prime
example was New York City's Eighty-sixth Street Methodist Episcopal
Church, which had been almost abandoned in 1856. Under a new pastor in
the fall of 1857 a revival began, and by the end of 1858 church membership
was up to 185 and a new building was underway.[59] Also, from 1857 to 1858,
Henry Ward Beecher's Plymouth Church in Brooklyn witnessed an increase
of 350 members, accompanied by a 60% increase in pew rentals (from $16,300
to $26,052).[60] Participation in union meetings was fine, but for most minis-
ters membership growth in local congregations was the bottom line.

As the label of a businessmen's revival suggested, the awakening also celebrated a new sense of the compatibility between the worlds of religion and commerce. An indication of the closer Protestant-business ties that helped to set the stage for the changes associated with the revival was the rise of a different sort of Christian writing. As more and more Protestant men sought their fortunes in urban centers during the 1840s and 1850s, books began to appear, mostly written by ministers, that extolled the benefits of Christian character for achieving true business success.[61] The books also exhorted merchants to use their wealth wisely and suggested that Christianity and the marketplace could be linked in many positive ways. In 1855, Baptist Henry Clay Fish acknowledged "that [it] is now almost at the peril of his piety that a Christian young man embarks in business, especially if of a commercial or professional character." Yet Fish also assured his readers that there was "no real antagonism between business and piety." Otherwise, "why has God commanded us to be 'diligent in business,' and at the same time to 'grow in grace'?"[62] Harking back to the Calvinist idea of all work as a calling, Fish insisted that Christians could dedicate their businesses to God's service for "the sublime end of saving men from the power and dominion of sin and death." A person who pursued such a course could be "another bright example of a *business Christian*."[63]

In 1858, sympathizers interpreted the revival's inroads among businessmen in terms of a "masculine millennium." They believed that piety could invade and sacralize commercial space and its inhabitants: "We trust that since prayer has once entered the counting-room it will never leave it; and that the ledger . . . the blotting-book, the pen and ink, will all be consecrated by a heavenly presence."[64] Individual merchants were encouraged to exchange the "tricks of trade" for Christian principles, "to have the same religion for 'down town' which they had for 'up town' . . . the same for the counting-room as for the communion table."[65] Throughout the revival meetings, scattered stories reported the appearance of sanctified business practices, for example, men who repaid outstanding debts or rectified dishonest dealings. In addition to the prayer meetings at downtown churches, a few smaller groups were established in stores and even in a print shop.[66]

Yet in most instances, influence seemed to move in the opposite direction as business methods were put to use for stimulating religious fervor. The revival celebrated not so much the religious transformation of business or businessmen as the affinity between urban revivals and a nascent corporate culture. The noon meetings were viewed as an appointment with God. They were scheduled to fit into the rhythm of the business day and reflected the intense time consciousness of Victorian men.[67] Prayer was considered a productive activity. The uniform format for the proceedings resembled an agenda as much as a liturgy. The directions stipulated that the opening hymn, Bible reading, and leader's prayer were not to occupy more than twelve minutes; in addition, at precisely 12:30 P.M., the leader was to interrupt the proceedings and allow those who desired prayer to stand for thirty seconds; and finally, at 12:55 he should announce the closing hymn, "any one having the floor yielding

immediately." The newspapers commented on the "peculiar legislative ability" needed to direct the proceedings.[68] It was a businessmen's job.

In prominent public spaces, posters for the prayer meetings hung alongside advertisements for consumer goods or entertainments, positioning religion as an enterprise that belonged in the business districts. The initiative of
YMCA workers in New York City, Boston, and Philadelphia in distributing
fliers and hanging banners reinforced publicity in the press and made the revival, after national presidential campaigns, one of the nation's first mass-
advertising efforts.

In Boston, Theodore Parker, a Unitarian abolitionist who was critical of
the revival, pointed to the advertisements as evidence that it was purely "a
business operation."[69] Parker meant the comment as a criticism, but for many
evangelicals it felt like an accolade. Charles Finney, never shy about the use
of promotional methods, chided the New York clergy for not recognizing that
they had produced a businessmen's revival because they went about organizing it "like businessmen." With satisfaction, he noted, "They took pains
to give public notice of these meetings, as they would notify matters of business or politics. They used the appropriate means, and it was remarked almost immediately after, 'Now God is answering prayer.'"[70]

This businesslike approach to revival piety appealed to the evangelically
oriented merchants, clerks, and aspiring entrepreneurs who populated downtown business districts. Despite stereotypes of devout women and worldly
men perpetuated by rhetoric associated with an ideology of separate spheres,
the revival demonstrated the existence of significant numbers of urban men
who were sympathetic to evangelical piety. Although they did not make up
the majority of men in the cities, their status as members of the dominant
British-American commercial community constituted an influential subgroup,
one that until the revival had lacked a public religious identity.

The particular character of this business community, as well as the way in
which Northern Protestants in general were being shaped by market values,
is brought into sharper focus by seeing how the events of the 1857–58 revival
represented a clash between two visions of revival, each with implications
for the relationship between religion and commerce. One was a New England
tradition, rooted in seventeenth- and eighteenth-century American Puritanism, that stressed revivalist piety and the moral transformation of the community together.[71] The other was a more individualistic ideal that regarded
community renewal as only a by-product of individual transformation.

The nineteenth-century heirs of the New England tradition, such as Charles
Finney and Lyman Beecher, were convinced that revivals would accomplish
"the complete moral renovation of the world" by "elevating the intellectual,
spiritual, and social conditions of men."[72] When they spoke of "moral renovation" and social elevation, they envisioned the impact of revivals on entire
communities: "It is found . . . in instances almost innumerable, that a revival
had renovated . . . a community; has driven away vice; has encouraged industry; has given spring to intelligence . . . where chilling selfishness, or hateful
discord, or unblushing crime, seemed to have established a perpetual reign."[73]

The combination of zeal engendered by revivals and community perceived as a moral unit provided much of the impetus behind the formation and growth of the nineteenth-century's Protestant benevolent empire. The same combination also provided the basis for an alliance between northern evangelical Protestants and the Whig party as it emerged, during the second quarter of the century, in a common desire to improve people and society.[74] As a result, the New England pattern of combining revivals and reform became the most publicly influential model for how revivals should influence society. It was a pattern that assumed common social goals, and therefore social harmony, on the basis of a shared commitment to conversionist piety and unselfish service.[75] This ideal, although widely disseminated, found its most fertile soil in areas of Yankee migration, including upstate New York and the upper part of the Old Northwest, places where relatively homogeneous communities most replicated the Puritan village.[76]

The environment in urban centers such as New York City was much less conducive to moral regeneration through classic New England communalism. Explosive population growth, the separation of middle-class business and residential districts, heterogeneity reinforced by increasing immigration, and an almost constant population flux—all made the traditional stable community an illusive ideal.[77] Even the era's dominant Protestant churches suffered from the general instability. Fifty-one of the Presbyterian churches established in New York city during the nineteenth century were dissolved before the century was over; seventeen had a lifespan of less than five years.[78] Equally significant for the future of evangelical social reform were the tensions that arose, on the heels of revivalism and the prosperity of the benevolent empire, over the issue of slavery. The vision of revivalism and united community reform proved illusive as major Protestant denominations split, sectional identification increased, and the benevolent societies were beset with agitations. Socially active Protestants began to focus on political action rather than religious effort as the means to reform. The Liberty, Free Soil, Know Nothing—and especially the Republican—parties represented a fusion of the revivalistic, crusading spirit of the Northern evangelicals and specific social-political agendas.[79] In effect, as Yankee activists attempted to impose their vision of a godly republic on the nation, revivalistic politics were taking the place of religious revivals as agents of social transformation.

This transformation of the New England ideal of revival worked to the advantage of the socially conservative Reformed churches, which had historically propagated a more individualistic view of revival. From early in the nineteenth century, the New York City Presbyterian and Dutch Reformed establishment had resented efforts by New Englanders to impose a Yankee religious culture on the metropolis.[80] The change in Yankee interest from revival to political crusades turned out to strengthen this conservative position. The result was a broad shift in understanding religious awakenings that was quite apparent with the 1857–58 revival.

During the height of the revival, perhaps as many as twenty noon prayer meetings met throughout the city and a hundred more convened in churches and halls during the early morning or evening hours.[81] Yet, the face of the city remained unchanged. A leading Presbyterian, Samuel Irenaeus Prime, wrote in the *New York Observer*, the Old School Presbyterian paper he edited:

> Though in this city alone we may number the converts by thousands, and in the country by tens of thousands, we know that the numbers are so small compared with the vast multitudes remaining unchanged, that we have no right to expect any perceptible improvement in the masses of the community. . . . Crime is not likely to be checked because here and there a criminal has become an honest man. . . . Wall street will be as rife with fraud, and the Stock Board as full as gambling as it has been, and the whole world around us will move on as if the Providence and Spirit of God had not combined to arrest men in their mad career, and . . . had not turned their thoughts by force for a time heavenward.[82]

Prime represented a theologically moderate Presbyterian revival tradition with three emphases that distinguished it from the New England pattern: (1) an understanding of revivalism that centered on a revitalized church rather than a transformed secular community, (2) a concern for doctrinal purity over moral reform as essential to social stability, and (3) a certain biblical literalism that led Old School Presbyterians to reject such New England doctrinal innovations like "disinterested benevolence" as a basis for community renewal. Presbyterians were more inclined to identify revivals with a revitalized church than to automatically associate church and community renewal with each other.

The difference of opinion is well illustrated by the explosive issue of slavery. Unlike many New England Congregationalists, many conservative Presbyterians did not believe that slavery was a sin. The commitment of the influential Charles Hodge and his colleagues at Princeton Theological Seminary to a conservative approach to biblical exegesis further emphasized doctrinal correctness. Within the framework of such conservatism, Hodge argued that slavery was not a sin in itself nor slaveholding inherently evil since the Bible clearly allowed the institution.[83] On a popular level, this appeal to the commonsense interpretation of the written biblical text was translated into a more encompassing, socially conservative constitutionalism. Social harmony was based on a shared commitment to the rules, to the dictates of the written word. A stable society depended on respect for the law, whether the law of the land or the law of God.[84] Even revivals were bound by written rules, and these rules explained the celebrated unity of the 1857–58 revival, including its regulation about "No Controverted [controversial] Points Discussed."[85] Following that rule meant eliminating slavery from discussion.

This Old School Presbyterian revival tradition, although much less visible than the New England approach, nonetheless had affinities with the social and cultural setting of the merchants and bankers who constituted the middle class of urban centers like New York City. Church-centered revivalism was a good fit for the fluid urban community, and conservative biblicism had parallels with the growing contractual focus of commercial trade. A theological tradition that distanced itself from direct humanitarian reform also appealed to merchants with large financial interests in the South.

By the 1850s, only a few of the evangelical businessmen in the city, most notably New Englanders Lewis and Arthur Tappan, publicly opposed slavery. The more common view was expressed in a letter from a merchant to an abolitionist: "There are millions upon millions of dollars due from Southerners to merchants and mechanics alone, the payment of which would be jeopardized by any rupture between the North and South. We cannot afford, sir, to let you and your associates endeavor to overthrow slavery. It is not a matter of principles with us. It is a matter of business necessity."[86]

Instead, socially conservative New York Protestants expressed their benevolence through such organizations as the American Bible Society, American Tract Society, and American Sunday School Union. These organizations distributed literature with the goal of converting and teaching people, but they steered clear of slavery and other controversial issues that might threaten national unity or commercial interests. Each society maintained a concern for social stability and the preservation of the Union. If the churches could not preserve social stability by themselves, it could be promoted by united efforts in support of conversionist aims. But such stability came at the exclusion of direct efforts for social reform.

A similar approach was taken in the Revival of 1857–58. Jeremiah Lanphier, the lay worker credited with organizing the key noon prayer meeting at Fulton Street, originally had been hired by the North Dutch Church as a missionary to canvass the neighborhoods around the church, mostly populated by lower-class immigrants.[87] The new missionary, however, soon found his most receptive audience among the young clerks and business travelers who filled the boarding houses throughout the area. His success in filling the pews led Lanphier to begin the lunchtime prayer meeting. The focus of his work changed from being a missionary to the poor to acting as a chaplain for businessmen.

Part of the success and unity of the Fulton Street meeting as an outreach to businessmen derived from the exclusion of women until it had been well established. In addition, as a letter written to the editor of the New York Tribune in March 1858 indicates, the poor and African Americans were not welcome: The African-American writer explains how he and a female companion were ushered from the second floor prayer gathering to an empty room on the third floor. "The colored people have good meetings 'up' here," they were told.[88] Thus, people whose presence appeared to threaten social harmony found the atmosphere at Fulton Street less than congenial. Such incidents were not mentioned in denominational papers or in later books about the revivals. Instead, apologists emphasized the laudable characteristics of the Fulton Street gatherings: the harmonious, nonsectarian atmosphere; the emphasis on prayer; the high number of reported conversions. Instead of an ethical imperative or even the traditional concern for doctrinal purity, prayer had become a "new element of usefulness and power," the basis for Christian unity, and a source of direct, divine intervention to transform the nation.[89] According to conservatives, it was the "deep and all-pervading earnest piety" of the 1857–58 Revival, rather than political activism or social reform, that would "rouse up the national conscience, to arrest the national decay."[90]

The focus on prayer and personal piety struck a chord with some Baptists and Methodists. For such evangelicals, a stress on personal piety confirmed long-standing convictions that the only true path to social change came through individual conversions.[91] By the 1840s and 1850s, Methodists had become more politically active, and some were recognizing publicly that even evangelistic efforts exercised a "politically and socially constructive influence" on society. Still, many continued to believe that influence should be "indirect," that is, purely religious in intent and operation.[92] Prominent Methodist apologists of the 1857–58 revival—Abel Stevens, Nathan Bangs, and Phoebe Palmer—all represented middle-class or formalized New York City Methodism. Even for Methodists, the genteel atmosphere of the metropolis fostered "safe" opinions on subjects such as slavery, as well as a generally conservative social stance.[93]

The shift to the Presbyterian, pietistic type of revival did not pass unnoticed. The controversy that surrounded the annual meeting of the American Tract Society (ATS) in May 1858 illustrates the strain. At this gathering, a reform group and a socially conservative group debated the ATS policy on slavery and abolition. In 1857 a compromise had been attempted whereby the ATS would avoid the "political aspects" of slavery but would publish materials condemning the moral evils that could accompany the practice, such as cruelty by slaveowners. Shortly following the compromise, a tract was printed, entitled the *Duties of Masters*, that immediately angered Southerners. The society withdrew the tract in September 1857 after being inundated with objections. The morning of the annual meeting, May 12, 1858, more than 1,300 male members of the society, a majority of them New York City merchants, jammed the Dutch Reformed Collegiate Church in Lafayette Place.[94] The crowd clearly favored withdrawing the compromise and returning to a policy of strict silence on slavery. George Cheever—a New York City Congregational minister and an abolitionist who stood in the New England revival tradition—described how the attitude of the crowd changed from initial reserve to a tumult that was close to a "Christian mob."[95] Critics later charged that many merchants in the galleries supported the conservatives principally from economic motives. By May 1858, businessmen, both secular and Christian, were convinced that offending the South would spell disaster for commercial interests already weakened by the financial panic of 1857—an attitude they would maintain until the outbreak of the Civil War.

Observers at the time noted the contrast between harmony at prayer meetings and squabbling at the ATS. At a meeting of the American Congregational Union the following night, William T. Eustis, a New Haven minister, spoke in glowing terms of the revival in Connecticut. Eustis extolled "the catholicity of spirit which the Revival had everywhere manifested, and . . . its effects in breaking down walls of partition between Christian denominations." Then he turned to the subject of the ATS:

> When yesterday morning . . . visiting this city for the first time since the mani-
> festation of divine grace among you, I entered a convocation where I heard
> reverend men on the one side pressing the argument of worldly prudence, say-
> ing, It is expedient, it is profitable; and on the other side the . . . unanswerable
> declaration, It is true, it is right; and then the clamorous applause that went up
> for the trade argument . . . I was disposed to ask, "Where is the evidence of
> *your* revival of religion?"[96]

Abolitionist George Cheever spoke even more harshly. He poured disdain
on "an executive committee of the godliest, staid, conservative ministers and
elders in the community," who saw "their whole duty" as that of maintaining
"peace and union."[97] As the country had acquiesced to President Buchanan
in the betrayal of Kansas, so the church stood silent before the ATS: "God
speaks for the slave, they [the ATS] speak for the owner. What kind of piety
is this? What fruit of revival is this?"[98] Cheever and others lambasted the
society for establishing a Mason-Dixon line for sin: "South of such a line you
must not speak of such and such sins."[99]

By 1857–58, more evangelicals were moving up the social ladder to become middle class. As this occurred not only did the concept of social concern as an expression of revival begin to narrow but also its object became
more circumscribed. Instead of targeting the oppressed such as the slaves,
the destitute of the city, or social outcasts like prostitutes, evangelicals concentrated more and more on people like themselves—groups within the middle
class with special needs, such as young urban men and women or members
of the upwardly mobile, usually northern European "working poor." For an
example, John Wanamaker's "institutional church," Bethany Presbyterian, a
Philadelphia congregation that began as a Sunday school during the revival,
offered its range of social services primarily to those "with hopes of fulfilling
their aspirations to middle class status."[100] Henry Ward Beecher, ever the
weathervane of middle-class trends, devoted his time after the revival toward
helping his parishioners adapt, guilt-free, to the affluence that urban living
offered the successful.[101] Even some of the hereditary New Englanders were
finding their community in the Protestant middle class and retreating from
the community-wide ideal of the earlier revivalism. In so doing, their religious efforts became more compatible with the social conservatism of the city's
Protestant establishment.

Because of their involvement in the downtown prayer meetings and the
intense publicity given them by the press and clergy, businessmen were the
most visible participants in the urban revivals of 1858. This visibility, in turn,
signaled an important shift in perceptions of Protestant lay piety. During much
of the first half of the nineteenth century, churchwomen and ministers were
viewed as the stalwart Christians, whereas nonclerical men were identified
with worldly pursuits.[102] It was a division of responsibilities, some have suggested, that allowed Americans to embrace both laissez-faire capitalism and
the ideals of a virtuous republic: "Pious women would keep their husbands
and sons moral; productive men would work to become successful entrepreneurs in order to provide for their wives and children; and together they would

form Godly homes, the epitome of Christian progress."[103] By contrast, the revival popularized a different ideal: the Christian businessman as the modern Protestant layman, an image that would rival the powerful symbol of the praying mother as an icon of Protestant piety. Beginning in 1858 and continuing into the twentieth century, businessmen assumed an increasingly prominent profile as the mainstay of public Protestantism.[104]

During the revival, Prime and other conservative clergy emphasized the superior effectiveness of male piety. The traditional ideal of the praying mother portrayed a long-suffering woman whose prayers followed her husband and sons for years before they were answered.[105] Men, by contrast, brought the same attitude and expectations of success and efficiency to prayer as they did to commerce. As Prime made clear, "These are business men, and they address themselves to the great business before them."[106] Another cleric echoed Prime's description: "While there is no irreverence, there is a promptness, an earnestness, a directness, which allow no dragging, and show that men have come together for a purpose, and mean, with God's blessing, to accomplish that purpose."[107] The results demonstrated that when men prayed, God paid attention: "They were answered with a promptness and celerity never surpassed in the history of the church."[108] Men who previously had "given their country its preeminence in the daring, intense, and unexampled progress of worldly enterprise" now brought that same productivity to prayer. The use of the telegraph—an invention that represented male and, in evangelical eyes, millennial technology—to link the noon prayer meetings in various cities reinforced such views.[109]

A new sense of compatibility between religion and the commercial world resulted from the Revival. The Christian businessman stepped into place alongside the praying mother as an icon of Protestant piety. The methods of business were used to further the revival and Protestantism in general. But although the adoption of these methods most often seemed to bring efficiency and good returns on money "invested" in Protestant activities, little thought seems to have been given to harmful associations. Rather, the push was toward identifying the religious activism and individual responsibility required in an "earnest" Christian with the secular values of the Victorian business culture—hard work, initiative, and practicality.[110] Male business leaders became models of Protestant culture, at least in the newspapers, even if ordinary churchgoers were still most likely to be female.

The great revival of 1857–58 played an important role in the rapid growth of several Northern Protestant denominations.[111] It also played an important role in shifting traditional Protestant attitudes, at least in the North, toward money, the world of business, and the men who occupied that world. During the revival, religious news for the first time became a secular commodity that noticeably increased sales. During the revival some of the tension that most American Protestants historically had felt between the convictions of their faith and the practices of free-market capitalism became relaxed. In the context of the revival, the Christian businessman emerged as the model Protes-

254 THE ECONOMICS OF SECTIONAL STRIFE AND REVIVAL

tant layman. Other important changes would take place in Protestant attitudes toward money during the Civil War, then even more in the decades that followed. Even if the marketing of revivalism in 1857–58 recapitulated many strands of earlier Protestant history, the innovations that took place in those years left a mark on American Protestant religion that has not yet faded away.

Notes

This chapter uses material found in a different form in Kathryn Teresa Long, *The Revival of 1857–58: Interpreting an American Religious Awakening* (New York: Oxford University Press, 1998). I wish to thank Joel Moore for his assistance in adapting it for this book.

1. *New York Herald*, February 27, 1858. This account was not the first time the revival of 1857–58 had been mentioned in the secular dailies. For example, the *New York Tribune* had run a brief note, "The Hour of Prayer," on February 10, 1858; and under the headline, "Extensive 'Revival' in New York City," the *New York Evening Post*, February 11, 1858, had reprinted a two-paragraph report from the *New York Evangelist*. Indeed, Bennett had signaled his intentions with an editorial, "A Religious Revolution," and a few news notes in his Sunday paper, February 21. Nonetheless, the page-one position, flamboyant headlines, and extensive coverage on February 27 marked a new stage: the recognition of the revival as a major news story. Bennett's competitors, who rushed to follow his lead in covering the event, clearly got that message.

2. In addition to Eastern seaboard cities, coverage was most intense in the Old Northwest and the upper South—in such cities as Cincinnati, Louisville, Chicago, and St. Louis. But newspapers in the Deep South, including Charleston and Richmond, as well as New Orleans, also printed articles and editorials about the event.

3. "The Revival in Troy," *New York Tribune*, March 24, 1858.

4. For the *Herald* and the dynamics of the "cash and carry" press, see James L. Crouthamel, *Bennett's New York Herald* (Syracuse, N.Y.: Syracuse University Press, 1989), chap. 2; for the *New York Tribune*, see Frank Luther Mott, *American Journalism* (New York: Macmillan, 1950), 268–78. Michael Schudson, *Discovering the News: A Social History of American Newspapers* (New York: Basic Books, 1978), provides a broad overview of the consumer revolution in news brought about by the rise of the mass-circulation dailies.

5. For important earlier treatments of that revival, where, however, economic factors are not prominent, see Timothy L. Smith, *Revivalism and Social Reform: American Protestantism on the Eve of the Civil War* (Nashville, Tenn.: Abingdon, 1957), chap. 4, "Annus Mirabilis—1858"; and J. Edwin Orr, *The Event of the Century: The 1857–1858 Awakening*, ed. Richard Owen Roberts (Wheaton, Ill.: International Awakening Press, 1989).

6. Harry S. Stout, *The Divine Dramatist: George Whitefield and the Rise of Modern Evangelicalism* (Grand Rapids, Mich.: Eerdmans, 1991), 35. See also Frank Lambert, *Pedlar in Divinity: George Whitefield and the Transatlantic Revivals, 1737–1770* (Princeton, N.J.: Princeton University Press, 1994).

7. Stout, *Divine Dramatist*, 44, 99.

8. Lambert, *Pedlar in Divinity*, 52–67.

9. Nathan O. Hatch, "The Puzzle of American Methodism," *Church History* 63 (June 1994): 188. For a similar assessment of the Methodists, as well as an analysis of Christianity in American during the nineteenth and twentieth centuries that is based on a market model, see Roger Finke and Rodney Stark, *The Churching of America 1776–1990: Winners and Losers in our Religious Economy* (New Brunswick, N.J.: Rutgers University Press, 1992), 72–75 and passim.

10. For the populist religious entrepreneurs of the early nineteenth century, see Nathan O. Hatch, *The Democratization of American Christianity* (New Haven, Conn.: Yale University Press, 1989).

11. Michael Zuckerman, "Holy Wars, Civil Wars: Religion and Economics in Nineteenth-Century America," *Prospects: An Annual of American Cultural Studies* 16 (1991): 219–24, points to this distance as a characteristic of most American Protestants during the nineteenth century.

12. Lambert, *Pedlar in Divinity*, 8–9.

13. Quoted in Nathan Bangs, *A History of the Methodist Episcopal Church*, (New York: Carlton & Porter, 1840), 1:362–63.

14. *The Doctrines and Discipline of the Methodist Episcopal Church* (New York: Carlton & Porter, 1856), iv.

15. Hatch, *Democratization*, 83–88.

16. See also David Paul Nord, "Evangelical Origins of Mass Media in America, 1815–1835," *Journalism Monographs*, May 1984, 17.

17. *An Appeal to the Christian Public on the Evil and Impolicy of the Church Engaging in Merchandise; and Setting Forth the Wrong Done to Booksellers, and the Extravagance, Inutility and Evil-working of Charity Publication Societies* (Philadelphia: King & Baird, 1849), cited in Joan Jacobs Brumberg, *Mission for Life: The Story of the Family of Adoniram Judson* (New York: Free Press, 1980), 65.

18. Paul E. Johnson, *A Shopkeeper's Millennium: Society and Revivals in Rochester, New York, 1815–1837* (New York: Hill & Wang, 1978).

19. *The Memoirs of Charles G. Finney: The Complete Restored Text*, ed. Garth M. Rosell and Richard A. G. Dupuis (Grand Rapids, Mich.: Zondervan, 1989), 359; and William G. McLoughlin, *Modern Revivalism: Charles Grandison Finney to Billy Graham* (New York: Ronald Press, 1959), 87, 98, 138.

20. James Moorhead, "Social Reform and the Divided Conscience of Antebellum Protestantism," *Church History* 48 (December 1979): 428.

21. Finney, *Memoirs*, 388. A $200 gift from a friend supplied the need.

22. For a reproduction of the card advertising the Fulton Street meeting, see the *New York Herald*, February 21, 1858. The YMCA circular was reprinted in the *New York Evening Post*, March 11, 1858.

23. L. L. Doggett, *Life of Robert R. McBurney* (Cleveland, Ohio: F. M. Barton, 1902), 49–50.

24. For Whitefield's ability to "manipulate the press," see Stout, *Divine Dramatist*, 47.

25. Judith M. Buddenbaum, "'Judge . . . What Their Acts Will Justify': The Religion Journalism of James Gordon Bennett," *Journalism History* 14 (Summer–Autumn 1987): 55. On Bennett as such a pioneer, see his managing editor, Fredrick Hudson, *Journalism in the United States from 1690 to 1872* (New York: Harper, 1873), 453–54. See also Crouthamel, *Bennett's New York Herald*, 34–37, 42; and R. Laurence Moore, *Selling God: American Religion in the Marketplace of Culture* (New York: Oxford University Press, 1984), 33, 130–31.

26. Crouthamel, *Bennett's New York Herald*, 161.

27. For circulation figures, see the *New York Herald*, March 4, 1858. Crouthamel, *Bennett's New York Herald*, 69, has commented that the *Herald* was "the North's most prosouthern newspaper."

28. Daniel Walker Howe, *The Political Culture of the American Whigs* (Chicago: University of Chicago Press, 1979), 188, 355, n.26.

29. *Christian Advocate and Journal*, March 25, 1858.

30. During the first full weeks of February and March 1858, the *New York Tribune* averaged about two pages, six columns each, of advertising. By April, with spring sales underway, the average was close to three pages per issue. See the *Tribune*, February 1–6, March 1–6, and April 5–10, 1858. Bennett made a backhanded acknowledgment of similar problems in the *New York Herald*, March 18, 1858.

31. Apart from his publication strategy, one indication that Greeley had an intentional marketing plan was the questionnaire the *New York Tribune* apparently mailed to pastors in early March requesting revival information for the "Extra." See the *New York Tribune*, April 3, 1858.

32. *New York Tribune*, March 1, 1858.

33. For "Harry" Olney and Lloyd, see the *New York Tribune*, March 13, 1858.

34. For more on Gardner, see Long, *Revival of 1857–58*, 40–42.

35. The precise circulation impact is difficult to determine because of the common practice among editors of greatly inflating circulation figures. Scholars estimate that Greeley's *Weekly Tribune*, the most widely circulated of his various editions, went from sales of 15,000 in 1847 to 51,000 in 1853 to more than 100,000 during the presidential campaign of 1856, when it supported Republican John C. Frémont. Greeley's shrill opposition to slavery in 1857, combined with the financial panic, resulted in a heavy circulation loss, although probably not half of all subscribers, as James Gordon Bennett suggested. The revival and the Lincoln-Douglas debates contributed to the paper's recovery, and most analysts agree that by 1860 the *Weekly Tribune* had reached a high of 200,000. It is probable that the daily *New York Tribune* reflected similar fluctuations, though with a smaller total circulation. See Alfred McClung Lee, *The Daily Newspaper in America* (New York: Macmillan, 1937), 384; Glyndon G. Van Deusen, *Horace Greeley* (Philadelphia: University of Pennsylvania, 1953), 216, 229, 230; and Constance Rourke, *Trumpets of Jubilee: Henry Ward Beecher, Harriet Beecher Stowe, Lyman Beecher, Horace Greeley, P. T. Barnum* (New York: Harcourt, 1927), 314. James Gordon Bennett repeatedly complained about Greeley's ability to successfully capitalize on the revival. See "The Tribune Making Money Again," *New York Herald*, March 28, 1858, and "Weekly Newspapers," *New York Herald*, April 11, 1858.

36. See Charles Finney, *The Prevailing Prayer-Meeting* (London: Ward, 1859), 27.

37. *Boston Post*, April 9, 1858.

38. *New York Tribune*, March 1, 1858.

39. *New York Herald*, February 27, 1858.

40. *New York Tribune*, March 6, 1858.

41. *New York Herald*, March 28, 1858.

42. Ibid., March 26, 1858. See also tables published March 24 and 26.

43. *Evening Transcript*, April 6, 1858.

44. Quentin J. Schultze, "Keeping the Faith: American Evangelicals and the Mass Media," in *American Evangelicals and the Mass Media*, ed. Quentin J. Schultze (Grand Rapids, Mich.: Zondervan, 1990), 28.

45. *New York Observer*, March 25, 1858; see also April 1.

46. *Christian Advocate and Journal*, March 25, 1858; see a similar comment in the *Western Recorder* (Louisville), July 14, 1858, citing the *Boston Recorder*.

47. *Examiner*, April 8, 1858.

48. *Western Recorder*, April 21, 1858.

49. The paper's circulation was listed in the 1850s census as 1,500, although there were indications it had increased by late in the decade. The *New York Evening Post* had begun as a Democratic paper, switched to Free Soil in 1848, and then moved on to the Republican party. Although small, it was respected and influential. See Frank Luther Mott, *American Journalism* (New York: Macmillan, 1941), 257–58; and J. C. G. Kennedy, "Catalogue of Newspapers and Periodicals Published in the United States," in *Livingston's Law Register for 1852* (New York: John Livingston, 1852), 29.

50. *New York Evening Post*, March 11, 1858. Soon, however, the *Post* softened its stance toward revival coverage and embraced some of the same strategies it criticized here.

51. *New York Tribune*, March 8, 1858; see similar items in the *Chicago Daily Democrat*, March 17, 1858, reprinted from the *New York Evening Post*; and in the *Cincinnati Enquirer*, March 12, 1858.

52. Headline in the *Cincinnati Enquirer*, March 13, 1858. Even in the respectable *New York Tribune*, revival news never completely eclipsed such stories as "Alleged Wife Murder," March 5, 1858.

53. *Boston Evening Transcript*, April 6, 1858.

54. William G. McLoughlin, *Modern Revivalism: Charles Grandison Finney to Billy Graham* (New York: Ronald Press, 1959), 232, 333. In the early years of the twentieth century, R. A. Torrey expressed the concern that "a good deal of commercialism has been creeping into our work" (368).

55. Lyle W. Dorsett argues that from 1911 on, the secular press and Billy and Nell Sunday drove each other: the press seeking ever bigger stories, the Sundays ever more dazzling revivals. See Dorsett, *Billy Sunday, and the Redemption of Urban America* (Grand Rapids, Mich.: Eerdmans, 1991), 99–100.

56. James F. Findlay, Jr., *Dwight L. Moody: American Evangelist, 1837–1899* (Chicago: University of Chicago Press, 1969), 203, 365–67.

57. McLoughlin, *Modern Revivalism*, 309, 327.

58. For some examples of this trend in New York City churches, see a *History of the Stanton Street Baptist Church in the City of New York* (New York: Sheldon, 1860), 148. From September 1852 to December 1859 the church added 474 members, 204 by baptism, 152 by letter of transfer, and 118 by "experience and restoration." During the same period, 454 people left the church, 230 by letter, 29 through death, 46 by expulsion, and 149 by being dropped. In 1857, the Fourteenth Street Presbyterian Church received 65 new members but dismissed 39 with letters of transfer. "New York City. Fourteenth Street Presbyterian Church Session Minutes, 1851–1874," Philadelphia, Presbyterian Historical Society.

59. *Forty Years of Methodism in Eighty-Sixth-Street, City of New York* [pamphlet] (New York: Nelson & Phillips, 1877), 22. For membership statistics and the building dedication, see Samuel A. Seaman, *Annals of New York Methodism . . . 1766–1890* (New York: Hunt & Eaton, 1892), 311–12.

60. Clifford E. Clark, Jr., *Henry Ward Beecher: Spokesman for Middle-Class America* (Urbana: University of Illinois Press, 1978), 133. Although Beecher spoke at one widely publicized union prayer meeting in Burton's Theater, during most of the revival he directed his energies to his own church.

61. Richard Carwardine, *Transatlantic Revivalism: Popular Evangelicalism in Britain and America, 1790–1865* (Westport, Conn.: Greenwood, 1978), 255, n14, cites a number of such works, including James W. Alexander et al., *The Man of Business Considered in His Various Relations* (New York: Anson D. F. Randolph, 1857); Henry Clay Fish, *Primitive Piety Revived* (Boston: Congregational Board of Publication, 1855); and William Arthur, *The Successful Merchant* (London: Hamilton, Adams, 1852). There also were regular articles in the religious press. See, for example, "Men of Business," in the *Christian Advocate and Journal*, May 14 and July 2, 1857.

62. Fish, *Primitive Piety Revived*, 48–49.

63. Ibid., 56, 50; emphasis in original.

64. *Independent*, April 8, 1858, from a column written by Harriet Beecher Stowe.

65. Samuel Irenaeus Prime, *The Power of Prayer, Illustrated in the Wonderful Displays of Divine Grace at the Fulton Street and Other Meetings* (New York: Scribner, 1858), 179. Prime devoted chapter 14 of this book to urging men to "do their business" on Christian principles.

66. William C. Conant, *Narratives of Remarkable Conversions and Revival Incidents* (New York: Derby & Jackson, 1859), 366.

67. Edward Anthony Rotundo, "Manhood in America: The Northern Middle Class 1770–1920" (Ph.D. diss., Brandeis University, Boston, 1982), 270–2, discusses the urgency surrounding time and work among midcentury men; in contrast, the nature of middle-class women's activities in the home meant that their work continued to be more task-oriented.

68. *New York Tribune*, March 11, 1858. The *Tribune* reprinted the card that served as a guide for the John Street meeting. Most of the YMCA-sponsored meetings followed similar guidelines; others were somewhat more flexible. See also Conant, *Narratives*, 380–82.

69. Theodore Parker, *A False and True Revival of Religion* (Boston: William L. Kent, 1858), 10. For a description of the promotional activities in New York City, see the *New York Tribune*, March 2, 1858.

70. Finney, *Prevailing Prayer-Meeting*, 19, 25. Finney viewed the revival as a vindication of the principles of his *Lectures of Revivals* and signaled out S. I. Prime and others of his opponents for their unwillingness to acknowledge the importance of such "means."

71. Michael J. Crawford, *Seasons of Grace: Colonial New England's Revival Tradition in Its British Context* (New York: Oxford University Press, 1991), 50–51, 165.

72. William B. Sprague, *Lectures on Revivals of Religion* (Albany, N.Y.: J. P. Haven and J. Leavitt, 1832), 261. Although an opponent of new measures, Sprague was a New Englander and his social vision clearly reflected that tradition. See "Results of Revivals," chap. 9 of the *Lectures*.

73. Ibid., 266–67. Jonathan Edwards, *A Faithful Narrative of the Surprising Word of God*, in *The Great Awakening, Vol. 4, Works of Jonathan Edwards*, ed. C. C. Goen, (New Haven, Conn.: Yale University Press, 1972), had helped establish a narrative pattern that linked individual conversions to community reformation in his contrast between the "licentiousness" and "lewd practices" of the Northampton youth and the "glorious alteration in the town" during the revival of 1735 (146, 151). For more on New England narratives that follow Edwards's pattern, see Crawford, *Seasons of Grace*, 183–90.

74. Howe, *Political Culture*, 36.

75. See, for example, Charles Finney's comment that if young converts were "well grounded in gospel principles," they would have "but one heart and one soul in regard to every question of duty that occurs." Finney, *Lectures of Revivals of Religions* (1835), ed. William G. McLoughlin (Cambridge, Mass.: Harvard University Press, 1960), 426.

76. John Hammond studied these classic representatives of New England revivalism in *Politics of Benevolence: Revival Religion and American Voting Behavior* (Norwood, N.J.: Ablex, 1979).

77. From 1830 to 1860, the population in New York City increased from 200,000 to 800,000. Overall urban population in the United States during the same period grew from about 500,000 to 3.8 million, more than a sevenfold increase. Much of the increase came from German and Irish immigrants: During the peak famine years of 1847–54, 1.2 million Irish immigrants arrived. See Paul Boyer, *Urban Masses and Moral Order in America, 1820–1920* (Cambridge, Mass.: Harvard University Press, 1978), 67.

78. Theodore Fiske Savage, *The Presbyterian Church in New York City* (New York: Presbytery of New York, 1949), 31.

79. Richard J. Carwardine, *Evangelicals and Politics in Antebellum America* (New Haven, Conn.: Yale University Press, 1993), 320–22 and passim; Lori D. Ginzberg, *Women and the Work of Benevolence: Morality, Politics, and Class in the Nineteenth Century United States* (New Haven, Conn.: Yale University Press, 1990), chap. 4, chronicles the increasing shift from moral influence to political activism among Protestant reformers during the 1840s and 1850s.

80. For examples of early opposition, see Bertram Wyatt-Brown, *Lewis Tappan and the Evangelical War Against Slavery* (Cleveland, Ohio: Press of Case Western University, 1969), 60–68.

81. The lower figure cited by Smith, *Revivalism and Social Reform*, 64, is probably a good estimate for the number of noon meetings in the city at the height of the revival; the higher figure, from Prime, *Power of Prayer*, 46, could approximate the total number of prayer meetings held throughout the revival period in both Manhattan and Brooklyn.

82. *New York Observer*, April 22, 1858.

83. See Charles Hodge, "Abolitionism," *Biblical Repertory and Princeton Review* 16 (1844): 554. Hodge did acknowledge that the laws regulating slavery and their application could be evil (pp. 572–81). I am indebted to Robert Bruce Mullin for his analysis of the retreat from "moral considerations" by scholars at Princeton and Andover seminaries in their concern to protect the authority of the biblical text from the influence of Unitarian rationalism and the related social conservatism of their stances toward slavery. See Mullin, "Biblical Critics and the Battle Over Slavery," *Journal of Presbyterian History* 61 (Summer 1983): 210–26.

84. These "Unionist evangelicals" identified respect for law and the constitution with respect for the God-ordained American Union; see Carwardine, *Evangelicals and Politics*, 182–83.

85. Conant, *Narratives*, 380–82, reprints one such guide.

86. S. May, *Some Recollections of the Anti-slavery Conflict,* 127–28, quoted in Philip S. Foner, *Business and Slavery; the New York Merchants and the Irrepressible Conflict* (Chapel Hill: University of North Carolina Press, 1941), 14; see also Carwardine, *Evangelicals and Politics*, 180–81.

87. Talbot W. Chambers, *Noon Prayer Meeting of the North Dutch Church* (New York: Board of Publication, Reformed Protestant Dutch Church, 1858), 34.

88. *New York Tribune*, March 27, 1858. Of course, the *Tribune*'s editor, Horace Greeley, was sympathetic to blacks and antagonistic to the city's conservative establishment, including Prime, the *Observer*, and the businessmen of the Fulton Street meeting. It was not surprising he printed the letter. I have found no other evidence of such incidents at Fulton Street, although accounts of similarly rigid treatment of women lend it credibility.

89. Prime, *Power of Prayer*, 48, 52.

90. Ibid., 52, 55.

91. Carwardine, *Evangelicals and Politics*, 14, 124–26, has emphasized that although "conservative pietists" were common in these denominations, members of the Northern branches in particular represented a spectrum of social and political views. The successful urbanites most identified with the 1857–58 revival, such as Francis Wayland for the Baptists and Phoebe Palmer, Nathan Bangs, and Abel Stevens for the Methodists, did represent the conservative, pietist wing of these churches.

92. Russell Richey, "History as a Bearer of Denominational Identity: Methodism as a Case Study," in *Beyond Establishment: Protestant Identification in a Post-Protestant Age*, ed. Jackson W. Carroll and Wade Clark Roof (Louisville, Ky.: Westminster/John Knox Press, 1993), 276. The enduring affinity of many grassroots Methodists for the Democratic party, with its essentially laissez-faire philosophy, reflected the denomination's pietistic heritage and distrust of imperialistic Yankee Calvinists; see Carwardine, *Evangelicals and Politics*, 113, 273–76, and passim.

93. Smith, *Revivalism and Social Reform*, 192.

94. Women, as well as men, could either be lifetime "directors" of the organization for a contribution of $50 or more or lifetime members for $20. Women, however, were not allowed to vote. Because the stakes were so high, only voting members were given tickets to the 1858 meeting.

95. *National Anti-Slavery Standard*, May 22, 1858. The *Independent*, May 13, 1858, described the scene as "such uproar as is more appropriate to Tammany Hall than to a house of worship." Both sources, of course, were biased against the majority, but transcripts indicated that the meeting did get out of hand. Cf. *New York Tribune*, May 13, 1858.

96. *Independent*, May 20, 1858; emphasis in original.

97. George Cheever, *The Commission from God Against the Sin of Slavery* (Boston: J. P. Jewett, 1858), 22.

98. *Independent*, June 3, 1858. On May 4, 1858, Congress had passed a compromise bill that would admit Kansas to the Union if the state would ratify the Lecompton constitution. Kansas rejected the constitution in August, and the state was not admitted to the Union until 1861.

99. Cheever, *Commission from God*, 22. An anonymous author, thought by some to be Theodore Parker, made a similar allusion in "The American Tract Society," *Atlantic*, July 1858: 248.

100. William R. Glass, "Liberal Means to Conservative Ends: Bethany Presbyterian Church, John Wanamaker and the Institutional Church Movement," *American Presbyterian* 68 (Fall 1990): 189.

101. Beecher's novel *Norwood* (New York: Scribner, 1868), 209–14, contained a classic statement of his approach to this issue. See William G. McLoughlin, *The Meaning of Henry Ward Beecher . . . 1840–1870* (New York: Knopf, 1970), 112–17.

102. See Ann Douglas, *The Feminization of American Culture* (New York: Anchor, 1988).

103. Gail Bederman, "'The Women Have Had Charge of the Church Work Long Enough': The Men and Religion Forward Movement of 1911–1912 and the Masculinization of Middle-Class Protestantism," *American Quarterly* 41 (September 1989): 436.

104. Anne C. Rose, *Victorian America and the Civil War* (Cambridge: Cambridge University Press, 1992), 183.

105. For an illustration of such piety, see the request from "an anxious mother" in Prime, *Power of Prayer*, 157–58, who had prayed for her 35-year-old son "since his birth up to the present time."

106. Ibid., 69.

107. Chambers, *Noon Prayer Meeting*, 57.

108. Ibid., 307.

109. James H. Moorhead, "The Millennium and the Media," in *Communications and Change in American Religious History*, ed. Leonard I. Sweet (Grand Rapids, Mich.: Eerdmans, 1993); and Conant, *Narratives*, 394–96.

110. Findlay, *Dwight L. Moody*, 83.

111. See Long, *Revival of 1857–58*, app. B, "Membership Statistics and Church Growth Rates, 1853–1861," 144–50.

IV

GENERAL PERSPECTIVES

12

Protestant Reasoning about Money and the Economy, 1790–1860: A Preliminary Probe

Mark A. Noll

Large, unanswered questions remain for relationships between Protestants and the emerging market economy in the early United States, at least in part because there were so many levels to the question and because the changes of the era were so rapid. If the grand interpretations now available for these relationships are inadequate, as suggested especially in the historiographical chapters in this volume on Charles Sellers and E. P. Thompson, a better general interpretation will nonetheless make use of the productive research generated by these two outstanding scholars. Such an explanation will also, however, be especially alert to where historical actors did not sense the moral standoffs perceived by some modern historians between, on the one hand, communities, subsistence, production for local use, humane values, and religion as self-expression and, on the other hand, individuals, profit, production for market exchange, commercial values, and religion as self-control. Above all, a better accounting of Protestants and the market will not scruple to study religion as a motivating factor, as well as a reacting factor; it will be particularly alert for interpretive cues from the history and character of the religion at stake. Although an interpretation of Protestants and money arising from the history and character of Protestants will not necessarily be superior to an interpretation arising from the history and character of market transformations, such an attempt should contribute to a fuller, more historically faithful understanding of the complexities at issue.

Beginning a study of Protestants and money with religious considerations means paying attention especially to the two major transformations that shaped Protestantism in the United States' early history: Those two transformations were the rise of evangelicalism and the abandonment of church-state establishment.[1]

Framework: A Twofold Protestant Transformation

Evangelicalism arose in the North Atlantic region during the eighteenth century as a religion particularly well adapted to the turmoil of revolution, understood as rapid political transition, large-scale industrial and commercial change, and the movement of peoples on an unprecedented scale. The evangelical stress on the new birth, biblical authority, active personal holiness, and the cross of Christ represented a modification of previous Protestant emphases rather than an entirely new religion.[2] But this eighteenth-century modification was significant for breathing new life into stagnant forms of faith that had come into existence when the model of a unitary Christendom still prevailed, when personal freedoms were still quite restricted, and when lay people were still mostly silent. Evangelicalism amounted to a renovation of traditional Protestantism that shook up inherited religion in response to the eighteenth century's changing circumstances. It took off in the United States after evangelicals joined their fortunes to the cause of American independence; it flourished in Scotland after evangelicals recoiled from the threat of revolutionary France; it expanded rapidly in Ireland after evangelicals distanced themselves from the 1798 rebellion of the United Irishmen; in Canada evangelicalism grew rapidly after the failed American invasions of 1812–14 and again in the wake of the failed rebellions of 1837–38; in England the rise of evangelicalism was complexly related to the near revolutionary tumult of the early industrial age.[3]

Understanding that evangelicalism was preeminently a form of "experiential biblicism" suggests why it flourished in the wake of revolution. Successful revolutions, by their nature, destroy traditions. Unsuccessful revolutions, through the exertions of conservative countermobilization, subject traditions to nearly as much strain. In either case, revolutionary or counterrevolutionary, the past and its wisdom are set adrift. Both revolution and counterrevolution demand a search for anchors. In the North Atlantic revolutionary situations of the eighteenth and nineteenth centuries, an evangelical religion of biblical experientialism provided meaning to more persons and motivation for more groups than did political utopianism or the defense of a threatened establishment. Historians have expressed this connection in different ways. For Europe in general, W. R. Ward has written, "The great crisis of the French Revolution altered for ever the terms on which religious establishments must work, and in so doing it intensified everywhere a long-felt need for private action in the world of religion." For Northern Ireland, Finlay Holmes's judgment is similar: "In the aftermath of the tragedy of 1798 many found a new purpose and meaning in life through evangelical religion."[4] For the United States, Gordon Wood offers a similar explanation for the remarkable success of an evangelicalism that had mastered the tumults of its era: Evangelicals "developed and expanded revivalistic techniques because such dynamic folk-like processes were better able to meet the needs of rootless egalitarian-minded men and women than were the static churchly institutions based on eighteenth-century standards of deference and elite monopolies of orthodoxy."[5]

Evangelicalism was at its most effective in revolutionary situations because it communicated enduring personal stability in the face of disorder, long-lasting eagerness for discipline, and a nearly inexhaustible hope that the dignity of self affirmed by the gospel could be communicated to the larger community. The special American circumstances for this religion are well summarized by Nathan Hatch: The "nonrestrictive environment" of post-Revolutionary America "permitted an unexpected and often explosive conjunction of evangelical fervor and popular sovereignty. It was this engine that accelerated the process of Christianization within American popular culture, allowing indigenous expressions of faith to take hold among ordinary people, white and black. . . . The rise of evangelical Christianity in the early republic is, in some measure, a story of the success of common people in shaping the culture after their own priorities rather than the priorities outlined by gentlemen."[6] Such reasoning helps explain why the United States, the North Atlantic nation most affected by revolution, witnessed the most rapid growth of evangelical churches.

The social and economic implications of the new evangelical emphases were never entirely predictable. Evangelicalism per se did not necessarily mean an immediate radicalization of society or the economy. It is helpful to recall that Pietism on the Continent and evangelicalism in Britain began within the precincts of traditional Lutheran, Reformed, Anglican, and Presbyterian churches, and that in the colonies evangelicalism was first promoted by traditional Congregationalists and by middle-colony Presbyterians who would have created a traditional establishment if they had been able.

As the eighteenth century wore on, however, the logic of a religion defined by the self's own strategic choices pushed increasingly toward new religious forms, as well as new religious convictions. That logic worked out in effective secularization of the human institutions that virtually all Christians, Protestant and Catholic, had long held to be authoritative mediators of God's presence in the world. If new birth was a product of the Holy Spirit's immediate presence, if the Bible was the converted person's basic religious authority, and if these new realities trivialized the spiritual efficacy of institutions long considered necessary for mediating the new birth or interpreting Scripture, then the stage was set for important social change. The institutions undermined by this logic included centers of higher learning monopolized by church-state establishments and the monarchy understood as the primary fount of stable social order.

They also included established churches defined as authoritative communicators of divine grace through word and sacrament. In America the social and economic implications of evangelical religion were measurably extended when many Protestants made the strategic choice to give up deeply embedded conventions of Christian establishment and the heretofore widely accepted Constantinian ideal of a comprehensive church-state authority in order to promote new life in Christ.

Methodism, which became so important in the early United States, was an especially interesting variety of evangelicalism since its connectional system retained characteristics of an establishment (especially the human authority of Wesley or the bishops who succeeded Wesley). But when Methodism eventually took institutional shape, it did so as an alternative to church-state establishments. By the 1790s in Britain and even before then in the new United States, Methodists were clearly committed to the choice against mediated religion and the traditional aspirations of Constantinian Christians to exert formal control over society at large.

Several qualifications are in order at this point. The disestablishmentarian impulse was, first, always only a tendency among evangelicals at large, who continued to number many important leaders and followers who supported the social status quo. Several were major figures in the early evangelical movement, like Charles Wesley and the Countess of Huntingdon, as well as evangelical leaders in the Church of Scotland, like Thomas Chalmers in the early decades of the nineteenth century. Second, it must not be forgotten that there were nonevangelicals, like Jonathan Mayhew in America and Joseph Priestley in Britain, who also supported disestablishment.[7] Third, the willingness to give up the establishment principle did not mean giving up life in society. It meant, rather, renouncing the traditional mechanisms by which Christian churches (including almost all European Protestant churches) had protected their social prerogatives and inculcated their traditions. In the place of these mechanisms came voluntaristic ways of influencing society that did not require a formal establishment. Fourth, it is important to stress that there was a gradation of viewpoints: The disestablishmentarian impulse existed in pure form, in delicate balance with traditional establishmentarian instincts, and as a mere tincture, coloring a basic commitment to establishment. In addition, individuals and groups could display both establishmentarian and disestablishmentarian characteristics and change over the course of time with respect to such matters.

By the 1790s, the Protestantism that existed in America was a religion containing small but vigorous cadres of evangelical disestablishmentarians, along with probably larger, but also more lethargic, groups of evangelical establishmentarians and nonevangelical disestablishmentarians. Although few could recognize it at the time, the evangelical disestablishmentarians were the wave of the future.[8]

A number of reasons explain why evangelical Protestantism and the principle of disestablishmentarianism both expanded so rapidly in the early United States.[9] As a first instance, pragmatic considerations kept the framers of the Constitution from implementing the establishmentarian ideals that many of them still preferred. Among the founding fathers a majority probably favored some kind of church-state establishment but nothing close to a majority on its nature. The result was a pragmatic choice to keep the national government out of religion.[10] For their part, disestablishmentarian evangelicals also bene-fited from a tremendous libertarian surge generated by the Revolution, from a pragmatic agreement with the founding fathers on the morals necessary for republican government, and from an ability to exploit commonsense forms of reasoning in the vacuum created by discredited intellectual traditions.[11] They were the ones institutionally prepared for meeting the religious needs of a population that was fanning out into what was by European standards the incredibly vast spaces of the American West. They were also the ones who worked very hard and, like Francis Asbury, traveled very far to do what they accomplished.

The result was that for about four decades after 1790 the Protestants who were most thoroughly evangelical and disestablishmentarian (Methodists, Baptists, and Disciples) flourished; those that were ambiguously so (Presby-terians) did fairly well; those that tried to retain establishmentarian preroga-tives (Congregationalists and Episcopalians) advanced much more slowly. In the decades after 1790 the rising Protestants in the United States were those who had made a very specific, often self-conscious choice against the propri-etary model of institutional Christian authority that remained dominant in Europe.

Establishmentarian instincts made a comeback only in the 1830s, and then only partially. The domestication of Methodism, the increasing pressure on Protestants North and South to sanctify their own regions' social arrangements, the increasing systematization of voluntary social reform, and the very wealth created by the disestablishmentarians' success at self-discipline led to a strengthening of formalist positions. By that time, however, the pace of American developments—political and ideological, as well as constitutional—had rendered impossible a return to full-scale European establishmentarian-ism or even to the partial establishmentarianism practiced by the era's Cana-dian Protestants.[12]

Even before the eighteenth century had passed, the consequences of act-ing on disestablishmentarian principles could be observed. Evangelical dis-establishmentarianism shared a natural affinity, if often incomplete and unself-conscious, with other renunciations of traditional authority. Politically, for example, disestablishmentarian evangelicals were predisposed to principles of republican freedom, even though strong intra-Protestant debate continued on how best to define both republicanism and liberty.[13]

For economic questions, the spread of evangelicalism and disestablishment was almost certainly a factor in the new nation's embrace of the market. A move away from top-down monarchical, hierarchical, and colonial control in religion predisposed many evangelicals in the same direction economically, that is, toward localism and free trade. As antiestablishment evangelicals, these Protestants rejected close regulation of the public spaces in which they hoped to promote their religion, and they were predisposed in favor of situations in which individuals could make the choice for God freely. They were also confident in the Spirit-given ability to persuade free agents to move toward Christ. Such convictions about religion no doubt pushed many American evangelicals toward corresponding values in economic practice, including an acceptance of market reasoning.

At the same time, however, evangelicals retained a great deal of hereditary Protestant nervousness about the accumulation of wealth, suspicion about the seductive power of money, and caution about the corrupting influences of economic power.[14] For renovators of Protestantism, there was no contradiction in carrying some of these historic reservations about the accumulation of money into the much looser social environment of revolutionary and early national America.

There may also have been an indirect way in which the nature of evangelical Protestantism affected religious-economic connections in the early United States. European Protestants, who for the most part maintained the ideal of Christendom, had regularly thought in terms of all-encompassing models of life-as-a-whole, including economics, But now, since disestablishmentarian evangelicals had given up earlier ideals of Christendom, they often found themselves *reacting* to changes and circumstances in arenas, including economic arenas, over which they once had tried to exert self-conscious control. In their choice for voluntary spiritual suasion, they set aside self-conscious attention to the structures of society. American evangelicals largely turned aside from trying to construct complete worldviews; in practice their pietism drove them to a functional division of life into a sacred sphere, which received comprehensive and self-conscious attention, and a secular sphere, which did not.[15] The pietistic instincts of disestablishmentarian evangelicals put them in the position of seeking the transformation of persons and society through revival. What they sought, they often found. But by limiting the goals of their activity, they also increased the likelihood that dimensions of society they now neglected would influence them unselfconsciously.

Disestablishment evangelicals were historical actors who did not feel compelled to construct grand schemes of economic life under God, who maintained traditional reservations about the entrapping power of money, and who eagerly exploited new market conditions. Such a combination may look incoherent or mystifying to modern interpreters. It was, however, the end point to which the evangelical and disestablishmentarian transformations of the eighteenth century had led a majority of American Protestants by the beginning of the next century.

In an insightful book on religious-social connections in the era of the Great Awakening, Timothy Hall has noted what to modern eyes seems an irony in George Whitefield's aggressive use of practices associated with the new consumer consumption. "Itinerancy . . . paralleled commerce and embodied mobility, working hand in glove with both to challenge not only boundaries of space but those of society and self as well. The partnership with commerce was tinged with irony for preachers, including Whitefield himself, who often preached against the ostentatious use of consumer goods like 'jewels, patches, and gay apparel.'"[16] Hall describes this conjunction as an irony, probably because he knows that the Protestant acceptance of market practices eventually led to the privatization of religion and to social consequences that figures like Whitefield could not have approved.

But if Whitefield was numbered among those who, in effect, were giving up an ideal in which formal religious institutions aspired to control all aspects of life, his stance was not ironical. It was only an irony if market practices contradicted the most fundamental basis of his religion. If that most fundamental basis was no longer the goal of social uniformity under the guidance of an established church—if that goal had become, rather, a willingness to forsake all, including establishmentarian aspirations, in order to present the gospel—then there was no irony. Perhaps there was a miscalculation since it may be simply impossible to divide the world into public and private spheres, as disestablishmentarian evangelicals had attempted. But there was no irony. Instead, interpreted against the backdrop of Protestant history, especially the rise of evangelicalism, Whitefield's stance is about what we should expect. It certainly is what we find.

A Preliminary Probe

For a preliminary investigation into Protestant reasoning about money and the economy in the seven decades between 1790 and 1860, I have looked at about 130 separate publications that treated in some way markets, economic booms and busts, finance, insurance, benevolent contributions, attitudes toward money, and related subjects.[17] The most important conclusion that can be drawn from this survey is that Protestants regularly, consistently, and without sense of contradiction both enunciated traditional Christian exhortations about careful financial stewardship and simply took for granted the workings of an expanding commercial society. To use the words of Richard Pointer after his study of two kinds of Pennsylvania Presbyterians, "both groups sought to endorse, support, and protect the market economy without sacrificing the distinctions or integrity of their religious faith."[18]

This standard position appeared from many different kinds of Protestants in many places and in many circumstances. In a Boston sermon of 1813, for example, Joseph Emerson selected John 6:12 ("gather up the fragments that remain, that nothing be lost") to preach on "Christian Economy." According to Emerson, Scripture taught that humanity was to labor primarily for everlasting life, that wealth was fleeting, that money must be used to aid the poor, and that it was sinful to spend as much money as Americans did on liquor ($33 million annually, at a time when the expenditures of the national government totaled $32 million). Yet Emerson also proclaimed that nothing in the Bible sanctioned disregard for temporal existence and that it was proper to seek prosperity as long as temptation was avoided.[19]

Four years later, a Presbyterian sermon from the as yet unfamous Gettysburg, Pennsylvania, urged the faithful not to abuse prosperity by indulging in "ease . . . torpid insensibility and luxurious indolence . . . levity, and an excessive love of pleasure . . . vainglory, and presumptuous confidence." Yet after noting how many social, economic, religious, and political advantages God had given to Pennsylvania, the minister concluded by urging his hearers to learn "how to perpetuate prosperity, and enjoy it with a blessing." Using prosperity for benevolence was the way to keep it from being withdrawn.[20]

In one of the era's most widely read formal introductions to the subject, Francis Wayland of Brown University promoted the same conjunction. While expatiating at length on the necessity of freedom for economic growth, Wayland was careful to include a moral qualification: "It is almost superfluous, however, to add, that a free constitution is of no value, unless the moral and intellectual character of a people be sufficiently elevated to avail itself of the advantages which it offers."[21] A sermon published just before the outbreak of hard times in 1857 put succinctly what many commentators had said at much greater length during the previous half century: "Our object is to baptize the riches of men with the spirit of the gospel."[22]

An eccentric interpretation of the story of the rich man and Lazarus from Luke, chapter 9, was an exception proving the rule. In March 1828 a Unitarian sermon in Philadelphia denied that this story was intended to affirm anything at all about wealth or poverty. Rather, it was an allegory on the state of Jews and Gentiles. In fact, there was "nothing said in this parable to implicate the character of the rich man, or in favour of Lazarus."[23] Other religious voices also occasionally attempted to evade the Protestant majority's strict moral attitude toward money. In a sermon unusual because of its failure to urge self-sacrificing benevolence as the only way to sanction wealth, a minister in Salem, Massachusetts, in 1806 evoked humanity's natural principles of self-love and benevolence to argue that the primary duty of the rich was to set a good example in assisting others. Moreover, purchasing "elegancies" was not only good in itself but also useful to society since the only choice for using what one earned lay between spending on luxuries or using money to promote idleness, "which is dangerous to virtue."[24]

Much, much more common was the attitude of Adam Clarke in another sermon on the rich man and Lazarus, published in New York in 1812. As Clarke read the passage, the rich man was condemned as criminal. Even if he had gained his wealth legally, Clarke held that the rich man's soul was imperiled because he neglected benevolence by getting all he could and keeping all he got.[25]

Clark's interpretation was the norm. It sketched a fine line between accepting the legitimacy of wealth (also the means to gain wealth) and denouncing the abusive use of wealth for selfish ends. Only rarely did public speakers comment on the conventional means of gaining wealth, including market commerce. Rather, they were more concerned to surround economic life with injunctions about using wealth wisely, benevolently, and for the good of others.[26] Even the Restorationist leader Alexander Campbell, who in his extreme antiformalism was so contrary in so many ways, agreed: "There is nothing incompatible with diligence in business and fervor of spirit in serving the Lord."[27]

Although the growing commercialization of American society exerted a gradually more obvious influence on the nation's religious life, the basic Protestant stance throughout the antebellum decades remained an uncomplicated acceptance of commercial society alongside an extraordinary elaboration of scruples concerning how the wealth engendered by modern commerce should be used. The religious contribution to this combination could be regarded, in Marx's terms, as one of those "bourgeois prejudices, behind which lurk in ambush just as many bourgeois interests."[28] It is more in keeping with main developments of the evangelical Protestant history that so influenced American religion in this period to regard it as a continuation of historic Christian, Protestant, and Reformed attitudes as adapted to a social setting in which Protestants had mostly given up aspirations for establishmentarian control.

Thus, in line with their inherited theological traditions, Protestant attitudes toward money repeated many long-standing Christian convictions. In line with their status as self-denying disestablishmentarians, the same Protestants by and large did not attempt to define comprehensive religious norms for economic life. There is little evidence that Protestant thinking was directly coopted by market reasoning, and much evidence that market reasoning remained subordinate to intrinsically religious convictions. This situation may have changed somewhat in the years immediately before the Civil War, as the discussion below on "systematic benevolence" indicates. Yet even systematic benevolence remained mostly within the lines of traditional Protestant concerns. Finally, it is important to note that although most American Protestants shared many common attitudes toward money and commercial society, significant differences also made for lively, sometimes embittered strife.

Economic Christian Traditionalism

Protestant references to economic matters during the period 1790–1860 were to a large extent indistinguishable from traditional Christian teaching on the need for humans to act as stewards of wealth "owned" by God, the dangers of being ruled by mammon or the love of money, and the Christian necessity to use money for benevolent purposes.[29]

During these decades, Protestants were called on to give to many causes, with most of the appeals phrased in completely altruistic terms. To be sure, appeals for benevolence sometimes referred to the good that would accrue to the giver.[30] Toward the end of this period, ministers also seemed to take greater care in preaching about the need for Christians to support their ministers. By at least the 1850s, such appeals were also stressing the long-neglected duty "of giving, as a minimum, one-tenth of our income to the treasury of God."[31]

Very occasionally an assertion could be heard about the need to care for oneself and one's own before extending benevolence afar, such as the suggestion in Boston's *Christian Review* from 1849 that "it is just as much charity to feed, clothe, and educate our children, pay our taxes to the civil magistrate and our pew-rent where we worship, as it is to give for the support of the gospel in foreign lands, or to feed the hungry."[32] But this kind of reasoning was extremely rare. Much more commonly, Protestants were simply urged, without consideration of reward, to give for the general spread of Christianity;[33] for the support of orphans;[34] for the spread of Christian literature, especially Bibles;[35] for education;[36] for relief to victims of disasters;[37] for support of the poor;[38] for missions to Indians,[39] to the West,[40] to free blacks,[41] and to slaves;[42] and (as was then widely regarded as a civil, as well as religious, duty) for keeping Roman Catholics at bay.[43]

Just as traditional in discussions of economic questions were the lists of virtues recommended and vices denounced. An endless parade of exhortations promoted industry and frugality while denouncing idleness and luxury.[44] Comments, for example, early in the century on "the superior sensibility of females" provided the occasion for promoting "the active exercise of mercy."[45] Protestants also knew how to ring the changes in support of general benevolence[46] and against greed,[47] covetousness,[48] gambling,[49] and speculation.[50] The logic behind this torrent of moral exhortation was never in doubt. It was stated with economy by Theodore Dwight Woolsey, the Congregationalist president of Yale College, in 1855: "Philanthropy must degenerate without the support of piety. . . . [But] piety will give tone and power to philanthropy."[51]

Antebellum Protestants were also traditional in ascribing to God most of the responsibility for the presence or absence of wealth. Active injunctions to use money benevolently—such as that from the Methodist minister who urged his fellow itinerants to get by with less so that more could be given to the poor[52]—did not undercut the primary conviction that God was ultimately responsible for the economic condition of human beings.[53] To be sure, the lessons drawn from the belief in God's economic sovereignty did vary over the course of these decades. A regular theme was that since many were poor through no fault of their own, believers should be active in charity toward them.[54] Others pointed to divine control over economic matters as a way of urging the poor to seek consolation in God and the rich to flee from the wrath to come[55] or they used the presence of the poor to show how God intended benevolence to "cement mankind together"[56] or they pointed to wealth as a reminder of its corruptions.[57] By the end of this period, ministers were drawing lessons that would have sounded strange at the start of the century, as did the writer in *The Presbyterian Magazine* for 1857 who after consideration of actuarial science and a reminder of the clear scriptural duty to provide for one's own family, approved the use of life insurance.[58] But the general conclusion was clear: Even in an age of expanding markets, God still ruled economic life, as well as all other spheres of human existence.

Theoretical Minimalism

Because of its great familiarity, most such traditional commentary wavered between pious sincerity and pious cliché. Protestant thinking more directly attuned to America's changing economic circumstances did exist, but, except perhaps in the South, it was never particularly innovative.[59] Even the major textbooks of the period by the Unitarian Francis Bowen of Harvard and the Baptist Francis Wayland of Brown, which were consequential works, mostly popularized distillations of political economy as articulated by British Christian moralists. Both, for instance, stressed the analogy between moral philosophy and political economy. For Wayland, it was a "law of our nature" that attached labor to means of sustenance, and it was further in the very nature of things that every form of vice—he mentions intemperance specifically—was connected with "economic evils."[60] Bowen held that economic structures resembled the operations of the mind since "through their general effects upon the well-being of society, [are] manifest the contrivance, wisdom, and beneficence of the Deity." On the specific issue of when it was appropriate for middlemen ("the corn dealer") to raise and lower prices, Bowen concluded that "the corn dealer" was "a mere instrument in the hands of Him who brings good out of evil, and who turns the little passions of man to the purposes of His own benevolence and wisdom."[61] The point to be made about such commentary is not that it was superficial, however much it would fall out of favor in the age of Darwin and the social Darwinists. The point is rather that these respected economic discussions were carried out within a framework provided by Christian and Enlightenment convictions about the reasonableness of all nature, not a framework provided by deliberations on markets directly.

Most Protestant commentors on theoretical economics were less sophisticated than Bowen and Wayland, but they nonetheless regularly positioned commercial questions in the same traditional framework of Christian moral reasoning. In 1818, a Vermont minister, Joshua Bates, preached a discourse on honesty in dealing, in which he opined that a completely free market was as immoral as following the principle that "every man has a right to make as good a bargain as he can" or "to buy as cheap as he can and sell as dear as he can." Bates's basis for this conclusion was rooted securely in an overarching religious vision: "If [a monopoly that drives up prices] does not violate the laws of the land, it certainly does the laws of christian love."[62] Similar reasoning informed a Providence, Rhode Island, report on female wages in 1837; Richard Furman's definition of money and its workings, offered the following year; and an 1851 justification for churches going into debt: Recommendations for specific economic actions grew, not from foundational beliefs about how economics worked, but from Protestant convictions about how God held economic behavior to account.[63] In general, these examples of off-the-cuff economic theorizing represented spontaneous moral responses to the workings of the marketplace. They are what one would expect if Protestants had turned against the older practices of establishment and had, in effect, forsworn the kind of broad religious reasoning promoted by the Puritans and other establishmentarian Protestants of earlier generations.

The one major exception to conventional reasoning was provided by the minority of artisan thinkers who felt that a major economic reconstruction was required.[64] Henry Van Amringe's 1845 tract on association and Christianity was a good example of such work. In phrases from its subtitle, it

launched a root-and-branch attack on church complicity in "the Social Relations in Present Christendom," urged "Industrial Association" along with
"Christian Brotherhood and Unity," and went on to depict the present order
as ruled by selfishness, pride, mammonism, and cruelty. As a remedy, Van
Amringe urged preachers to proclaim not charity to the poor but a just distribution of wealth to the laborers who produced it. Ministers who did not do so
were only "hireling[s] in the pulpits of the rich" who should "take up their
cross and preach without money and without price." Van Amringe acknowledged his debt to Charles Fourier for the plan of "Associations and Phalanxes,"
and denied that they were influenced by Emmanuel Swedenborg. In his view,
slavery and the desperate situation of the working classes were national sins
that manifested rank ingratitude for God's bounty.[65] Two things were unusual
about such work: its opposition to the main course of American economic
development and its self-conscious effort to reason about economics from the
ground up. In the latter effort, Van Amringe represented an establishmentarian
instinct that most of his Protestant contemporaries had given up.

Uncoopted: Economics Subservient to Politics

Evidence that market reasoning had not penetrated deeply into the thinking
of antebellum Protestants is suggested by many circumstances. One is the fact
that economic realities were often perceived as functioning within a moral
framework that might be called "Christian republicanism." That is, the meaning of money was related more to a traditional republican assessment of power
than to self-conscious reasoning about the market.

Thus, an early appeal to provide money for young men going into the
ministry made wealth subservient to the standard calculus of Christian republicans: Ministers support religion; religion is essential for a moral social order;
therefore, giving to support ministers supports a moral social order.[66] The same
argument in slightly different form appeared in an 1832 defense of Sunday
Schools. Since these schools were one of "the great levelling institutions of
our age," support for Sunday School teachers amounted to a "more effectual
way to banish aristocracy from among us."[67] Although ministers seem to have
emphasized the republican significance of money less frequently as the years
passed, the fact that economic reasoning was regularly embedded in republican reasoning indicates that market assumptions were not affecting Christian
reasoning as directly as did political assumptions.[68]

Uncoopted: Metaphors Only

The same conclusion may be drawn from the relatively few occasions when
commercial tropes entered directly into theological discourse. Since such
occasions always seemed self-conscious and metaphorical rather than instinctive and literal, they reflected more the rhetorical skill of speakers than a fundamental effect on the speakers' religion.

An early use of economic metaphors reached the United States from Britain in 1819, a year of financial distress. William Huntington's American audience, therefore, may have been skeptical to hear his testimony that "I found God's promises to be the Christian's bank note; and a living faith will always draw on the divine banker."[69] At an address to the Brothers Charitable Society in Providence, Rhode Island, in November 1829, the speaker urged his listeners to imitate Jesus in their care of the poor and then employed what must have been a telling metaphor in one of America's leading centers of overseas trade: "All our good, happiness, pleasure, and possessions . . . are blessings which we have only upon credit, for which we are to account on an awful pay day!"[70] Less than two years later, *The Christian Index* of Philadelphia, edited by the irenic Baptist W. T. Brantley, printed a short note on the "Best Investment." Brantley suggested that "capitalists," who were complaining about low rates of interest, should "invest it [their money] in a stock that will not fail to produce the highest rate of interest. Lend it to the Lord."[71] In 1838, when Alexander Campbell appealed for funds to help a prospective minister attend college, he used a similar figure of speech: "The Bank of Heaven has many and various offices open on earth, in which men can secure some property in heaven." Campbell announced the opening of another branch by appealing for others to join him in funding the lad's education.[72] In 1857 a pamphlet reprinted from London repeated W. T. Brantley's metaphor almost word for word: A "loan" to Christ paid the highest rate of interests, that is, "saved and glorified souls," as well as treasure in heaven for the lender.[73]

These usages illustrate, once again, the thorough intermingling of religious and economic spheres. At the same time, such usages were comparatively infrequent, and when they appeared they were obviously metaphors used for effect. By comparison to the intermingling that was occurring among other spheres in these same decades—for instance, between Christian and republican language in both the political sphere and the writing of theology or between theology and commonsense reasoning in moral philosophy—the effect of commerce on religious thinking seems slight.

Uncoopted: In the Face of Economic Crisis

Reactions to financial crises further confirm that for most antebellum Protestants economic reasoning remained superstructure and religion remained deep structure. In response to the panic of 1837, for example, a number of leaders spoke out forcefully, but mostly by accentuating elements of the standard Protestant picture: God controlled economic life; the workings of the economy did not need to become a subject for specific religious consideration; and financial crisis provided a useful occasion to underscore the need for Christian economic virtues, to stress the ongoing requirements of benevolence, and to drive home a range of moral lessons.

A sermon at New Haven's Center Church, preached twice in May 1837 by the orthodox Congregationalist Leonard Bacon, illustrated the typical performance.[74] Bacon thought he could see God's hand at work in the bank failures and commercial panic.[75] Equally clear was the origin of the crisis in the hasty pursuit of money.[76] He hoped that the crisis could be improved by having his listeners flee "luxurious and profligate habits of expenditures"[77] and by overcoming the moral decay promoted by partisan politics, the spoil systems, and slavery. Bacon also reminded his listeners how important it was to keep up their benevolences since to stop giving for only a single year would set back the missionary cause by a century. Bacon was silent about economic causes and effects as such, though some of his contemporaries stated explicitly that the complexities of commerce along with a minister's duty to provide moral guidance rendered economic analysis pointless.[78]

Charles Finney's response to the crisis was unusual, not because of his basic standpoint, but because of his characteristic moral urgency. After defining "debt" as the failure to repay contracted obligations when due, Finney not only called such debt a sin but also proceeded to the extreme conclusion that "the whole credit system, if not absolutely sinful, is nevertheless so highly dangerous that no Christian should embark on it." Never at a loss for over-statement, Finney then went on to say that those who repudiated their debts had no more place at the communion table than "whoremongers, or murderers, or drunkards, or Sabbath-breakers, or slave-holders." In such injunctions, Finney hardly sounded like one of Charles Sellers's "Moderate Lights," eager to sanctify the marketplace and pacify workers. Rather, Finney's concerns linked him with most other Protestants, who also urged readers to keep the needs of churches, missions, and ministers ever uppermost in their minds, despite the hard times.[79]

As chapter 11 by Kathryn Long has indicated, Protestant accommodation to market reasoning did become more noticeable as the years went by. During the financial crisis of 1857–58, for example, Protestants seemed somewhat more accepting of economic distress, somewhat more concerned for the financial viability of their own enterprises, and somewhat duller in their moral indignation. Nonetheless, many of the same elements of the basic Protestant stance reappeared. New York's *American Missionary*, for example, ascribed the crisis to an inordinate love of gain (especially the failure of Christians to tithe), recommended frugality, urged generous benevolences to all religious enterprises, and reminded readers that the magazine also needed support.[80] The Episcopal *Church Review* drew on the publications of Francis Bowen and Henry Carey to offer a somewhat more complex explication of the economic forces at work, but it soon returned to a very traditional moral analysis: "The pursuit of wealth, if not essentially and universally a selfish pursuit, is, nevertheless, so nearly related to selfishness, and places one in such a dangerous proximity to the polluting gangrene of the moral nature of man, that but few if any escape the effects of its poisoning virus."[81] For his part, Alexander Campbell was content to list nine reasons for why giving should be even more liberal in hard times than in good.[82]

Throughout 1857 the *Oberlin Evangelist* showed how time was dulling the Finneyite edge, as well as the fact that Finney's earlier positions on economic matters still carried weight. It devoted more space to reminding readers about the financial needs of Oberlin's professors and of the *Evangelist* itself than had been done twenty years earlier (March 5, March 18, November 11, and December 9), but it repeated traditional injunctions to care for ministers (September 16) and others who were promoting the gospel (May 27 and July 22). It portrayed the good businessman as one who cut his own pay before harming workers (November 25), and it urged the standard appeals against covetousness (May 27), lotteries (September 30), and luxury (November 25). The *Evangelist* used the contemporary clamor for "systematic benevolence" (see below) to chide Americans and Britons for spending much more on the theater and on liquor than on benevolences and to claim that "the contributions to a single heathen temple in Calcutta are nearly as much as the united contributions of Britain and America" (October 28). In the most dramatic change from Finney's strictures in 1839, the *Evangelist* printed the testimony of a debtor who feared that his failure to meet his obligations would reflect ill on the Christian faith but who yet expressed great comfort in God, "who is infinitely rich" and who would deliver him eventually from his troubles (July 22, 120).

In a word, Protestants reacted to the financial crises of 1837–38 and 1857–58 with the same set of responses that they displayed in ordinary times: They ascribed to God the ultimate control of the economy, pressed home the need for industry and frugality, used economic conditions to defend traditional morality, and continued the appeal to use wealth benevolently. Concern for the Protestants' own enterprises seems, from a limited sample, to have been greater in 1857, though differences between the two crises appear to be mostly in emphasis rather than essence.

Systematic Benevolence

A more complicated situation is presented by the Protestant concern for "systematic" benevolence. The interpretive question is whether an increasingly well-organized philanthropy did not indicate that commercial reasoning—with its stress on regularity, efficiency, and maximum profit—was altering the religious meaning of benevolence itself. The question is important for focusing on one of the clearest instances when the era's commercial reasoning entered into the substance rather than simply the form of religious reasoning.

Specific appeals for charitable giving to be handled systematically arose at the same time as the voluntary societies themselves. In 1810, for example, the Bible Society of Philadelphia proposed an elaborate plan for publicity, fund-raising, and the use of gifts that anticipated some of the programs developed only a few years later by the American Bible Society.[83] Soon it was common for benevolent appeals to come backed with extensive statistics and further proposals for systematizing the collection of funds.[84] By the 1830s and 1840s, proposals were multiplying to describe "The Benefits of System in Our Religious Charities," to give not by passion but by system, to explain the ideals of "Systematic Benevolence," and to show why "Benevolence Should be

Conscientious and Systematic."[85] The character of such writing was fully on display in the *Quarterly Christian Spectator* of 1837, where an appeal was published to recruit agents for the benevolent societies. It argued that the work of these societies needed to "be performed by men especially set apart to this service," just as agriculture, manufacturing, and business needed specially assigned workers. Ministers could not manage benevolences effectively while discharging their proper duties. Among the thirteen qualifications listed for the agents were the religious traits of piety, circumspection, liberality, and devotion to benevolence generally and to the special aim of the specific society. But agents were to have marketplace virtues as well: a certain measure of talent; a decent appearance; a knack for making and carrying out plans; and good business habits, especially honesty and the ability to function as a "good financier . . . strictly accurate in all pecuniary concerns, and scrupulously and punctiliously so, in collecting funds and accounting for them."[86]

In the decade before the Civil War, the flurry of publication in support of systematic benevolence reached its peak. In 1850, the American Tract Society published in one volume a trio of substantial works, at least some of which were reprinted again in 1859, 1860, and 1864. All three made much of I Corinthians 16:1–2 ("Now concerning the collection for the saints, as I have given order to the churches of Galatia, even so do ye. Upon the first day of the week let every one of you lay by him in store, as God hath prospered him, that there be no gatherings when I come"). Edward Lawrence's *The Mission of the Church; or, Systematic Beneficence* criticized those who retired early from business for their selfishness and recommended assigning one person in a church to canvass on behalf of each voluntary society.[87] Samuel Harris's *Zaccheus; or, The Scriptural Plan of Benevolence* offered considerable reassurance to potential donors. In his view, systematic benevolence tended "to secure God's blessing on business, and to enlarge the means of giving . . . it is not to be supposed that systematic benevolence will insure wealth. . . . But there are various ways in which systematic beneficence tends to promote prosperity" by making one a better manager of money, identifying the giver with Christ's interests, and reaping the reward of scriptural promises.[88] Parsons Cooke's *The Divine Law of Beneficence* rehearsed traditional Protestant demands for the moral use of wealth: "A Christian infidel is no more a contradiction in terms, than a Christian without charity."[89]

Three long essays were published together in 1856 in New York as the results of a prize contest. Their agreement on the need for efficiency in organizing benevolence did not prevent the essayists from making contrasting comments about current conditions. Abel Stevens thought that "radical defects" existed in the "financial policy of the Church," whereas Lorenzo White felt that the church actually enjoyed more spirit of consecration than current efforts at benevolence had been able to tap.[90] Similar appeals continued throughout the decade.[91]

Criticism of systematic benevolence was predictable, especially from antiformalists who feared a Money Power riding roughshod over local religious concerns. Nonetheless, in keeping with a half century of consistent religious-fiscal accommodation, antiformalist complaints were considerably calmer than had earlier been the case. From the confessional side, the Old School Presbyterian James W. Alexander expressed the opinion in 1858 that associations, subscriptions, and systematic charity were at best necessary evils, "machinery which intervenes between us and the Saviour, to whom we would minister in his poor members."[92] Alexander Campbell's *Millennial Harbinger* featured a complaint two years later against the payment of salaries to ministers.[93] But these minority reports were gently stated. They represented a far cry from Campbell's rousing words in the first issue of an earlier publication, *The Christian Baptist*, from 1823. Those words took the form of a commentary on II Peter 2:2 ("and through covetousness shall they with feigned words make merchandize of you") in which Campbell denounced the sale of pews; reprehended all principles of establishment; railed against the overpaid clergy of England, Wales, and Ireland; criticized the money-raising activities of the voluntary societies; and in general warned against the great evil when money was "drained from the people" by "the itinerant beggars of this age."[94]

Modern observers, with sensibilities shaped more by discussions of the merits of capitalism than the merits of Protestant voluntary societies, might bring a different range of potential criticisms. Systematic benevolence could, for example, be seen as a clear case of adapting historic Christian attitudes to America's commercial society. The fact that the advocates of systematic benevolence did not feel a need to justify the application of commercial efficiency to religious benevolence suggests that market instincts were beginning to edge into the basic instincts of the Christian moralists. Even if this is true, however, the critical interpretive issue remains ambiguous. To use terms from Joel Carpenter, it is possible to see in systematic benevolence a case of justifiable "cultural responsiveness" since it may represent another instance where the God of the Incarnation again honors "earthen vessels by making them carriers of his greatest gift." On the other hand, it may represent a case of "too readily making peace with their new [commercial] surroundings."[95]

Carpenter's kind of commentary, because it brings to bear standards from within the Christian tradition that antebellum Protestants represented, is a proper place to begin an assessment of religion and money in early American history. It is not as if modern debate over whether Protestant moralism was used to humanize or to excuse the spread of market capitalism is misspent energy. Nor is it the case that nothing is gained from deliberating on how highly competitive products flooding an unregulated market increased the gross domestic product of religious goods and services. It is rather that rooted historical inquiry takes seriously not only the words of historical actors but also where those words come from. In this case, religious words were being used to regularize, excuse, dignify, mystify, ennoble, and occasionally resist the workings of an expanding market capitalism. But the words performing those functions had a long and complex history. They were words loaded with republican, revivalistic, Enlightenment, Puritan, Calvinistic, Reformation, and

medieval freight. And they were words that, at least in many instances, had
not completely escaped an even more ancient anchorage: "You cannot serve
God and mammon"; "The love of money is the root of all evil"; "There was
a certain rich man . . . and there was a certain beggar named Lazarus"; "What
is a man profited if he shall gain the whole world, and lose his own soul?"
For a better account of Protestants, money, and the market in the early United
States, historians must also read these words for all they are worth.

Notes

1. My interpretation of eighteenth-century Anglo-American Protestant history
relies especially on W. R. Ward, *The Protestant Evangelical Awakening* (New York:
Cambridge University Press, 1992); Harry S. Stout, *The Divine Dramatist: George
Whitefield and the Rise of Modern Evangelicalism* (Grand Rapids, Mich.: Eerdmans,
1991); David Hempton, *Religion and Political Culture in Britain and Ireland from
the Glorious Revolution to the Decline of Empire* (Cambridge: Cambridge Univer-
sity Press, 1996); and the early chapters of two colloquia: Mark A. Noll, David W.
Bebbington, and George A. Rawlyk, eds., *Evangelicalism: Comparative Studies of
Popular Protestantism in North America, the British Isles, and Beyond, 1700–1990*
(New York: Oxford University Press, 1994); and George A. Rawlyk and Mark A.
Noll, eds., *Amazing Grace: Evangelicalism in Australia, Britain, Canada, and the
United States* (Montreal and Kingston: McGill-Queen's University Press, and Grand
Rapids, Mich.: Baker, 1994).
2. See especially the useful definition provided by David W. Bebbington, *Evan-
gelicalism in Modern Britain: A History from the 1730s to the 1980s* (London: Unwin
Hyman, 1989), 2–17.
3. For fuller accounts of evangelical expansion in these areas and of "experiential biblicism," see Mark A. Noll, "Revolution and the Rise of Evangelical Social
Influence in North Atlantic Societies," in *Evangelicalism*, 113–36.
4. W. R. Ward, "The Religion of the People and the Problem of Control, 1790–
1830," in *Popular Belief and Practice, Vol. 8, Studies in Church History*, ed. G. J.
Cuming and Derek Baker (Cambridge: Cambridge University Press, 1972), 237. Finlay
Holmes, *Our Irish Presbyterian Heritage* (Belfast: Presbyterian Church in Ireland,
1985), 100.
5. Gordon S. Wood, "Evangelical America and Early Mormonism," *New York
History* 61 (October 1980): 371.
6. Nathan O. Hatch, *The Democratization of American Christianity* (New Haven,
Conn.: Yale University Press, 1989), 9.
7. For sorting out connections between religious belief and political convictions
in England, see J. C. D. Clark, *English Society, 1688–1832* (New York: Cambridge
University Press, 1985); and for the United States, Alan Heimert, *Religion and the
American Mind from the Great Awakening to the Revolution* (Cambridge, Mass.:
Harvard University Press, 1966). On the radicalism of some nonevangelical Protestants, see Isaac Kromnick, "Religion and Radicalism: The Political Theory of Dissent,"
in *Republicanism and Bourgeois Radicalism: Political Ideology in Late Eighteenth-
Century England and America* (Ithaca, N.Y.: Cornell University Press, 1990), 43–70.
8. Thomas Jefferson and Achille Murat who, in 1822 and 1832, respectively, pre-
dicted the soon dominance of Unitarianism in the United States were poor judges of

current events. Had they made these predictions in the 1790s, however, they would not have looked so foolish. Thomas Jefferson to James Smith, December 8, 1822, in *Jefferson's Extracts from the Gospels*, ed. Dickinson W. Adams, *The Papers of Thomas Jefferson*, 2nd ser., ed. Charles T. Cullen (Princeton, N.J.: Princeton University Press, 1983), 409; and Murat, *Esquisse morale et politique des Etats-Unis de l'Amérique du Nord* (1832), as found in Milton B. Powell, ed., *The Voluntary Church: American Religious Life, 1740–1860, Seen Through the Eyes of European Visitors* (New York: Macmillan, 1967), 54–55.

9. Besides the factors listed here, see Jon Butler's account of rising American Protestantism in terms of an expansion of authority among the denominations, *Awash in a Sea of Faith: Christianizing the American People* (Cambridge, Mass.: Harvard University Press, 1990), 268–82.

10. Thomas J. Curry, *The First Freedoms: Church and State in America to the Passage of the First Amendment* (New York: Oxford University Press, 1986), 193–222; and John F. Wilson, "Religion, Government, and Power in the New American Nation," in *Religion and American Politics*, ed. Mark A. Noll (New York: Oxford University Press, 1990), 77–91.

11. Respectively, Gordon S. Wood, *The Radicalism of the American Revolution* (New York: Knopf, 1992); John R. West, Jr., *The Politics of Revelation and Reason: Religion and Civic Life in the New Nation* (Manhattan: University Press of Kansas, 1996); and Mark A. Noll, "The American Revolution and Protestant Evangelicalism," *Journal of Interdisciplinary History* 23 (Winter 1993): 615–38.

12. See, for example, John Webster Grant, *A Profusion of Spires: Religion in Nineteenth-Century Ontario* (Toronto: University of Toronto Press, 1988), chap. 9, "Affairs of State."

13. On these debates, see West, *Politics of Revelation and Reason*; Hatch, *Democratization of American Christianity*; and Daniel Walker Howe, *The Political Culture of the American Whigs* (Chicago: University of Chicago Press, 1979). For the situation in Britain, where republican freedom tended to be embraced by Dissenters but resisted by Anglicans, see James E. Bradley, *Popular Politics and the American Revolution in England: Petitions, the Crown, and Public Opinion* (Macon, Ga: Mercer University Press, 1986); Bradley, *Religion, Revolution and English Radicalism: Nonconformity in Eighteenth-Century Politics and Society* (New York: Cambridge University Press, 1990); Jacob Ellens, *Religious Roots to Gladstonian Liberalism: The Church Rate Conflict in England and Wales, 1832–1868* (University Park: Pennsylvania State University Press, 1994); and Timothy Larsen, *Friends of Religious Equity: Nonconformist Politics in Mid-Victorian England* (Woodbridge, Eng.: Boydell, 1999).

14. On that inheritance, see, for example, W. Fred Graham, *The Constructive Revolutionary: John Calvin and His Socio-Economic Impact* (Atlanta, Ga.: John Knox, 1971), 65–94; and Gordon Marshall, *Presbyteries and Profits: Calvinism and the Development of Capitalism in Scotland, 1560–1717* (Oxford: Clarendon, 1980), 115–26.

15. The contrast with Britain is instructive. There is in the United States very little of the self-conscious effort to link theology and political economy described for the late eighteenth and early nineteenth centuries by Boyd Hilton, *The Age of Atonement: The Influence of Evangelicalism on Social and Economic Thought, 1785–1865* (Oxford: Clarendon, 1988), and A. M. C. Waterman, *Revolution, Economics and Religion: Christian Political Economy, 1798–1833* (Cambridge: Cambridge University Press, 1991). By contrast, self-conscious and comprehensive religious reflection on political economy in the United States did not begin until the 1830s and 1840s. For

such comparisons, I am indebted to Stewart Davenport and insights from his Yale University dissertation-in-progress, "Churchmen and Economists . . . 1820–1860."

16. Timothy D. Hall, *Contested Boundaries: Itinerancy and the Reshaping of the Colonial American Religious World* (Durham, N.C.: Duke University Press, 1994), 131–32, which draws on the insights of T. H. Breen, for example, "'Baubles of Britain': The American and Consumer Revolution of the Eighteenth Century," *Past and Present* 119 (1988): 73–104; and "The Meaning of Things: Interpreting the Consumer Economy in the Eighteenth Century," in *Consumption and the World of Goods*, ed. John Brewer and Roy Porter (London: Routledge, 1993), 249–60.

17. The following pages depend heavily on a bibliography of works by Protestants on these subjects prepared by Robert Wood Lynn and augmented by Bryan Bademan, as well as on research assistance from Mark Sharrod and especially Rachel Maxson. Full citations to the works annotated in brief are arranged chronologically in a separate list of sources at the end of this chapter. The sources consulted for this chapter are overweighed toward the appeals of voluntary societies, toward New England and New York, and toward representatives of white formalist organizations. Broader research might find significant variations to conclusions drawn from this sample.

18. See chapter 8, p. 184, in this volume.

19. Emerson (1813).

20. McConaughy (1817), 13–14, 15–16.

21. Wayland (1841), 113.

22. McFarlane (1857), 673.

23. Fisk (1828), 12.

24. Prince (1806), 20.

25. Clarke (1812).

26. For a few out of many other possible examples, see Anon. (1792), Bird (1795), Ely (1810), A. B. (1818), "On Making Friends" (1824), Woods (1827), "Christian Stewardship" and "Duties of Private Christians" (1833), Harris (1837), Keen (1851), Frothingham (1852) and Boardman (1859).

27. Campbell (1860), 400.

28. Karl Marx, *Communist Manifesto*, trans. Samuel Moore (Chicago: Regnery Gateway, 1954), 35.

29. For one account of those more general traditions, see "Social Thought, Catholic," *New Catholic Encyclopedia* (New York: McGraw-Hill, 1967), 13:341–50. Without having engaged in systematic comparisons, it is nonetheless my impression that the biblical texts that appear most frequently in antebellum Protestant discussions of money are the same texts that had appeared most frequently in earlier Christian discussions.

30. Thatcher (1795), New Jersey Bible Society (1810), Stark (1825), Stevens (1849), Shepard (1858).

31. "Salaries of the Clergy" (1857), 230. Other appeals for ministerial support include Peck (1855), which also stresses the tithe; Benedict (1855); and "Claims of the Gospel Ministry," (1839).

32. "Law of Giving," (1849), 426.

33. Anon. (1794), Bradstreet (1794), M'Corkle (1795), Hollinshead (1798), Gardiner (1803), Hampshire Charitable Society (1815), Bacon (1832), Nevins (1848), Constable (1855).

34. Buist (1795).

35. Anon. (1802), Albany Bible Society (1810), Bodwell (1813), Coffin (1817), Fish (1858).

36. Wright (1814).

37. Sabine (1818), Seward (1827).

38. "Duty of Churches" (1849), Parker (1849), Kirwan (1856), Mason (1857).

39. Thayer (1801).

40. Van Vecten (1829).

41. Patten (1808).

42. Riley (1821), "Contributions of the Colored People," (1847).

43. Akers (1851).

44. Niles (1801), Burkitt (1804), "Review of the Third Report," (1819), Finney, "Lecture XX" (1839), Chalmers (1852).

45. Dana (1804), 16. See also Puffer (1816).

46. Stansburg (1816), "Unjust Steward" (1833).

47. *Reward of Avarice* (1810).

48. Dick (1856).

49. *Wonderful Advantages* (1813), Weems (1818).

50. *Speculation* (1840).

51. Woolsey (1855), 16, 18.

52. MacLeod (1805).

53. Foster (1799), *Christian Monitor* (1806), Prince (1806), James (1825), "Law of Giving" (1849), Dick (1856).

54. Baker (1806), Brazer (1836).

55. Clarke (1812).

56. Parker (1803).

57. Bennett (1823).

58. "Life Insurance" (1857).

59. For exploration of regional differences, see Eugene D. Genovese, *The Slave-holders' Dilemma: Freedom and Progress in Southern Conservative Thought, 1820–1860* (Columbia: University of South Carolina Press, 1992); and Kenneth Startup, chapter 10, in this volume.

60. Wayland (1841), 122, 308.

61. Bowen (1856), 27, 483.

62. Bates (1818), 8, 15.

63. *Report on Female Wages* (1837), Furman (1838), Ramsay (1851).

64. On these artisans, see the works mentioned in note 78 in the Introduction to this volume. For a brief comment on H. H. Van Amringe, who is discussed here, see Jama Lazerow, *Religion and the Working Class* (Washington, D.C.: Smithsonian Institution Press, 1995), 129, 132.

65. Van Amringe (1845), quotation 17.

66. Congregational Charity Society (1792).

67. Seward (1832), 210.

68. For other examples, see Hurd (1799), Bodwell (1813), Green (1828).

69. Huntington (1819), 72.

70. Trumbull (1829), 10.

71. Brantley (1831), 62.

72. Campbell (1838).

73. McFarlane (1857), 685.

74. Bacon (1837), quotation below 9.

75. As could also Whitman (1837), 11

76. For a similar assessment, see Frothingham (1837), 11; Whitman (1837), 6.

77. As did also Peabody (1837); Whitman (1837), 14.

78. Frothingham (1837), 11; Whitman (1837), 4.

79. Finney (1839). The difficulty in drawing religious lines in antebellum America on the basis of economic position is suggested by the fact that Alexander Campbell's *Millennial Harbinger* reprinted Finney's lecture "Being in Debt" five months after it appeared in the *Oberlin Evangelist*. Finney's preemptory condemnation of debt repudiation contrasts with other responses to the panic, for example, Andrew Peabody (1837) of New England, who held that patience between debtor and creditor would not only restore economic health but also promote public spirited patriotism.

80. "Financial Distress" (1857).

81. "Political Economy" (1857), 49.

82. Campbell (1858).

83. Bible Society of Philadelphia (1810).

84. For statistics, Beecher (1815), and for an appeal for systematic self-denial, "Miscellaneous Notice" (1819).

85. Respectively, "Benefits of System" (1830), Griffin (1834), "Systematic Benevolence" (1834), and Church (1846).

86. "Necessity and Qualifications" (1837), 259.

87. Lawrence (1850).

88. Harris (1850), quotation 37.

89. Cooke (1850), 57.

90. Stevens (1856), White (1856).

91. Hawes (1858).

92. Alexander (1858), 353.

93. Epps (1860).

94. Campbell (1823).

95. Joel A. Carpenter, "Revivalism Without Social Reform," *Books and Culture: A Christian Review*, November–December 1998, 27.

References

The sources consulted for this chapter are listed here chronologically. British reprints appear under the date of American publication, American editions usually under the date of first publication.

1792

[Anonymous.] *Advice in Order to Prevent Poverty; to Which Is Added Directions to a Young Tradesman, Written by an Old One.* Windsor, Vt.

Congregational Charitable Society. *Reverend Sir . . . July 25, 1792* (broadside). Boston.

1794

[Anonymous.] *The Love of Christ the Source of Genuine Philanthropy. A Discourse on ii Cor. Chap. V. Ver. 14, 15. Occasioned by the Death of John Thornton, Esq. Late of Clapham, Surry: Containing Observations on His Character and Principles.* Providence, R.I.

Bradstreet, Nathan. *Two Sermons on the Nature, Extent, and Motives of Charity.* Newburyport, Mass.

1795

Bird, Jonathan. *A Sermon on Covetousness; Delivered at Southold (L.I.), February 1, 1795.* Sag-Harbor, N.Y.

Buist, George. *An Oration Delivered at the Orphan-House of Charleston, South-Carolina, October 18th, 1795, Being the Sixth Anniversary of the Institution*. Charleston.

M'Corkle, Samuel Eusebius. *A Charity Sermon*. Halifax, N.C.

Thatcher, Peter. *A Sermon, Preached in Boston, February 12, 1795, in the Audience of the Massachusetts Congregational Charitable Society; When a Collection Was Made for the Relief of the Indigent Widows and Orphans of Deceased Ministers*. Boston.

1798

Hollinshead, William. *The Gospel Preached to Every Creature: A Sermon, Delivered on Wednesday, the 21st March, 1798, before the Incorporated Baptist Church, in Charleston, South-Carolina, Being Their Charity Anniversary for the Assistance of Pious Youth, in an Education for the Gospel Ministry, and Published at Their Request*. Charleston.

Morse, Jedidiah. *A Sermon Delivered before the Grand Lodge of Free and Accepted Masons of the Commonwealth of Massachusetts, at a Public Installation of the Officers of the Corinthian Lodge, at Concord, in the County of Middlesex, June 25, 1798*. Leominster, Mass.

1799

Foster, John. *A Sermon, Preached before the Roxbury Charitable Society, at Their Anniversary Meeting, September 16, 1799 by John Foster, Pastor of the Third Church in Cambridge. Published at the Desire of the Society*. Boston.

Hurd, Isaac. *A Discourse Delivered in the Church in Brattle Street, in Boston, Tuesday, June 11th, 1799, before the Humane Society of the Commonwealth of Massachusetts by Isaac Hurd, A.M., fellow of the Mass. Medical Society*. Boston.

1801

Niles, Samuel. *A Sermon, Delivered Before the Massachusetts Missionary Society, at Their Annual Meeting in Boston, May 26, 1801*. Cambridge.

Thayer, Elihu. *A Sermon, Preached at Hopkinton, at the Formation of the New-Hampshire Missionary Society, September 2d, 1801*. Concord, N.H.

1802

[Anonymous.] *An Address to Christians Recommending the Distribution of Cheap Religious Tracts, with an Extract from a Sermon by Bishop Porteus, Before the Yearly Meeting of the Charity Schools, London [1782]*. Charlestown, Mass.

1803

Gardiner, John S. J. *A Sermon before the Humane Society, of the Commonwealth of Massachusetts, at their Semiannual Meeting, June 14, 1803*. Boston.

Parker, Samuel. *Charity to Children Enforced, in a Discourse, Delivered in Trinity Church, Boston, before the Subscribers to the Boston Female Asylum . . . at Their Third Anniversary*. Boston.

1804

Burkitt, William [1650–1703]. *The Poor Man's Help and Young Man's Guide*, 32nd ed. Albany, N.Y.

Dana, Daniel. *Discourse Delivered May 22, 1804, before the Members of the Female Charitable Society of Newburyport, Organized June 8, 1803*. Newburyport, Mass.

Nott, Eliphalet. *A Discourse, Delivered in the Presbyterian Church in the City of Albany; Before the Ladies' Society for the Relief of Distressed Women and Children, March 18th, 1804.* Albany, N.Y.

1805

MacLeod, Roderick. *A Sermon. Dedicated, Prinicpally, to the Clergy: The Profits of Which Will Be Applied to the Necessities of the Poor of this City.* New York.

1806

Baker, Luther. *An Address delivered to the Philanthropic Society.* Warren, R.I.

The Christian Monitor: A Religious Periodical Work (by "a society for promoting Christian knowledge, piety, and charity") 1, no. 1. Boston.

Prince, John. *Charity Recommended from the Social State of Man. A Discourse, Delivered before the Salem Female Charitable Society, September 17, 1806.* Salem, Mass.

1808

Patten, William. *A Sermon Delivered at the Request of the African Benevolent Society.* Newport, R.I.

1809

Emmons, Nathanael. *The Giver More Blessed Than the Receiver: A Discourse, Addressed to the Congregation in Franklin.* Boston.

1810

Albany Bible Society. *Address of the Managers of the Albany Bible Society, to the Public.* Albany, N.Y.

Bible Society of Philadelphia (William White, President). *Address of the Bible Society of Philadelphia to the Friends of Revealed Truth in the State of Pennsylvania.* Philadelphia.

Ely, Ezra Stiles. *A Sermon for the Rich to Buy, That They May Benefit Themselves and the Poor.* New York.

New Jersey Bible Society. *Address of the New Jersey Bible Society to the Publick.* New Brunswick.

The Reward of Avarice, or Abdalla and the Iron Candlestick. A Turkish Tale. To Which is Added The Story of Whang the Miller. Windsor, Vt.

1812

Clarke, Adam. *A Discourse on the History of the Rich Man and the Beggar: From Luke XVI.19–31.* New York.

1813

Bodwell, Abraham. *A Sermon, Delivered at the Request of the Female Cent Society, in Sandborn, New-Hampshire, December 23, 1812.* Concord, N.H.

Emerson, Joseph. *Christian Economy. A Sermon, Delivered before the Massachusetts Missionary Society, at Their Fourteenth Annual Meeting, in Boston, May 25, 1813.* Boston.

The Wonderful Advantages of Adventuring in the Lottery. Philadelphia.

1814

Wright, Chester. *A Sermon, Preached before the Middlebury College Charitable Society, at Middlebury, Vt., August 16, 1814.* Middlebury.

1815

Beecher, Lyman. *On the Importance of Assisting Young Men of Piety and Talents in Obtaining an Education for the Gospel Ministry.* New York.

Hampshire Charitable Society. *The Address, and Regulations, of the Charitable Society, Instituted by the Central Association of Ministers in the County of Hampshire. And the Address of the Society's Committee.* Northampton, Mass.

1816

Puffer, Reuben. *The Widow's Mite. A Sermon, delivered at Boylston, before the Boylston Female Society for the Aid of Foreign Missions, Jan. 8, 1816.* Worcester [MA].

Stansbury, Arthur Joseph. *A Charity Sermon, Delivered before the Dorcas Society, at the Presbyterian Meeting-House in the Village of Newburgh, January 28, 1816.* Newburgh [NY].

1817

Coffin, Charles. *A Discourse Preached before the East Tennessee Bible Society at Their Annual Meeting in Knoxville, April 30th, 1817.* Knoxville.

McConaughy, David. *The Duties and Dangers of Prosperity: A Sermon, Delivered in the Presbyterian Church in Gettysburg, on Thursday the 20th of November, 1817, Being a Day of Thanksgiving, on Account of the General Plenty and Prosperity.* Gettysburg, Pa.

1818

A. B. "On the Deceitfulness of Riches." *The Panoplist and Missionary Herald* 14, no. 12 (December): 546–50.

Bates, Joshua. *A Discourse on Honesty in Dealing; Delivered at Middlebury, on the Annual Fast: April 15, 1818.* Middlebury, Vt.

Sabine, James. *A Sermon, in Commemoration of the Benevolence of the Citizens of Boston, Who, on the Occasion of the Dreadful FIRES of the 7th and 21st of November, 1817, in St. John's, Newfoundland, Sent Down Gratuitous Supplies for the Relief of the Sufferers During the Inclement Season of Winter.* Saint John's.

Weems, M[ason] L[ocke]. *God's Revenge Against Gambling. Exemplified in the Miserable Lives and Untimely Deaths of a Number of Persons of Both Sexes, Who Had Sacrificed Their Health, Wealth, and Honor, at Gaming Tables*, 4th ed. Philadelphia.

1819

Huntington, William. *God the Guardian of the Poor, and the Bank of Faith: Or, a Display of the Providences of God, Which Have at Sundry Times Attended the Author.* Baltimore (from the 7th London ed.).

"Miscellaneous Notice, Relative to the Cause of Religion and of Missions." *The Panoplist* 15, no. 3 (March): 130–32.

"Review of the Third Report of the Directors of the American Society for Educating Pious Youth for the Gospel Ministry." *The Panoplist and Missionary Herald* 15, no. 14 (April): 145–61.

1821

Riley, William. *An Address Delivered Before the Young Men's Missionary Society of South Carolina, at Their Second Anniversary, Held on Monday Evening, May 14, 1821.* Charleston.

1823

Bennett, James. *Duty of Supporting the Gospel Ministry. An Abridgement of a Sermon Preached at Sheffield (Eng.), April 25, 1821.* Hillsborough, [?].

Campbell, Alexander. "A Text Illustrated by Facts." *The Christian Baptist* I, no. I (July 4): 33–36.

1824

"On Making Friends with the Mammon of Unrighteousness." *The Evangelist* (Hartford) I, no. 5 (May): 169–73; cont'd. I, no. 6 (June): 208–13.

1825

James, Robert W. *A Sermon Preached Before the Bible Society of Darlington District, South Carolina, May 3, 1824.* Charleston.

Stark, Andrew. *Charitable Exertions and Evidence of a Gracious State: A Sermon, Preached in the Associate Presbyterian Church in the City of New-York, on Sabbath Evening, January 2, 1825.* Albany.

1827

Seward, William Henry. "For Greece" (preached February 26, 1827). In *The Works of William H. Seward,* vol. 3. New York.

Woods, Leonard. "Duties of the Rich. A Sermon Delivered in Newburyport. Feb. 18, 1827, on the Occasion of the Death of Moses Brown, Esq." In *The Works of Leonard Woods,* Vol. 5. Andover, Mass.

1828

Fisk, T. *The Rich Man in Hell: A Discourse at the Lombard Street Church, in the City of Philadelphia, on Sunday Evening, March 16, 1828.* Philadelphia.

Green, Beriah. "Anniversary Address." *The Home Missionary* I, no. 2 (June I): 25–28.

1829

Trumbull, Henry. *Address on the Importance of Charity. Delivered Before the Brothers Charitable Society, at the Second Congregational Church, Providence—On the Anniversary of Said Society, Wednesday, November 18, 1829.* Providence, R.I.

Van Vecten, J. "Anniversary Address." *The Home Missionary* 2, no. 2 (June I): 20–22.

1830

"The Benefits of System in Our Religious Charities." *The Spirit of the Pilgrims* 3, no. 2 (November): 567–72.

1831

Brantly, W. T. [ed]. *The Christian Index* (Philadelphia) 4 (January 22).

1832

Bacon, Leonard. *The Christian Doctrine of Stewardship in Respect to Property. A Sermon, Preached at the Request of the Young Men's Benevolent Society, of New Haven, Conn.* New Haven.

Seward, William Henry. "Sunday School Celebration" (preached on Staten Island, July 4, 1832). In *The Works of William H. Seward,* vol. 3. New York.

1833
"Christian Stewardship." *The Spirit of the Pilgrims* 6, no. 5 (May): 261–68.
"Duties of Private Christians." *The Spirit of the Pilgrims* 6, no. 10 (October): 578–91.
"The Unjust Steward." *The Spirit of the Pilgrims* 6, no. 11 (November): 633–41.

1834
Griffin, E. D. "Address to persons assembled at the Monthly Concert. Considerations
    on Giving to Charitable Objects—Extracted Chiefly from a Sermon by Dr. Griffin,
    on Heb. xiii.16." *The Home Missionary* 6, no. 9 (January 1): 153–56.
"Systematic Benevolence." *The Home Missionary* 6, no. 11 (March 1): 200–2.

1836
Brazer, John. *The Duty and Privilege of an Active Benevolence. Address, Delivered
    before the Seamen's Widow and Orphan Association, on Christmas Evening,
    1835, in the Tabernacle Church, Salem, Mass.* Salem.

1837
Bacon, Leonard. *The Duties Connected with the Present Commercial Distress. A
    Sermon, Preached in the Center Church, New Haven, May 21, 1837, and Re-
    peated May 23.* New Haven, Conn.
Frothingham, Nathaniel Langdon. *The Duties of Hard Times. A Sermon, Preached
    to the First Church, on Sunday Morning, April 23, 1837.* Boston.
Harris, John. *Mammon: or Covetousness the Sin of the Christian Church.* New York
    (reprinted from the London ed.).
"The Necessity and Qualifications of Public Agents for Benevolent Societies." *The
    Quarterly Christian Spectator* 9, no. 2 (June): 255–63.
Peabody, Andrew P. *Views of Duty Adapted to the Times. A Sermon Preached at
    Portsmouth, N.H., May 14, 1837.* Portsmouth.
*Report and Proposal to the Public, on the Subject of Female Wages. By a Committee
    of the Female Benevolent Society* (First Congregational Church). Providence, R.I.
Whitman, Jason. *The Hard Times. A Discourse, Delivered in the Second Unitarian
    Church, and also in the First Parish Church, Portland, Sunday, January 1st,
    1837.* Portland, Me.

1838
Campbell, Alexander. "A Good Speculation." *The Millennial Harbinger*, new ser. 11,
    no. 9 (September): 431.
Furman, J. C. *Christian Missions Entitled to Support; a Discourse Preached before
    the Welsh-Neck Baptist Association at Their 7th anniversary, Held with the
    Mispeh Church, Darlington District, South Carolina.* Raleigh.

1839
"Claims of the Gospel Ministry to an Adequate Support: An Address of the Presbytery
    of Elizabethtown to the Churches Under Its Care." *Biblical Repertory and Princeton
    Review* 11, no. 2 (1839): 180–201.
Finney, Charles G. "Lecture XIII: Being in Debt." *The Oberlin Evangelist* 1, no. 17
    (July 31): 129–31.
Finney, [Charles G.]. "Lecture XX: How to Prevent Our Employments from Injur-
    ing Our Souls." *The Oberlin Evangelist* 1, no. 24 (November 6): 185–187.
"Ye Cannot Serve God and Mammon." *The Oberlin Evangelist*, 1, no. 12 (May 22): 93.

1840

Speculation, or Making Haste to Be Rich. The Story of William Wilson, the Whistling Shoemaker. Boston.

1841

Wayland, Francis. *The Elements of Political Economy*, 4th ed. Boston.

1845

Van Amringe, H[enry] H[amlin]. *Association and Christianity, Exhibiting the Anti-Moral and Anti-Christian Character of the Churches and the Social Relations in Present Christendom, and Urging the Necessity of Industrial Association, Founded on Christian Brotherhood and Unity.* Pittsburgh.

1846

Church, Pharcellus. "Benevolence Should be Conscientious and Systematic." *The Southern Baptist Missionary Journal* 43, no. 3 (August): 52–55.

1847

"Contributions of the Colored People." *The Southern Baptist Missionary Journal* 2, no. 4 (September): 94–95.

1848

Nevins. "I Will Give Liberally." *The Southern Baptist Missionary Journal* 2, no. 11 (April): 269–70.

1849

"The Duty of Churches Toward their Indigent Members." *The Missionary* (American Lutheran Church, Pittsburgh) 2, no. 8 (August): 1.
"The Law of Giving." *The Christian Review*, ed. E. G. Sears. (Boston) 14, no. 57 (September): 460–71.
Parker, Theodore. "A Sermon of Poverty" [preached at the Melodeon, Sunday, January 14, 1849], in *Speeches, Addresses, and Occasional Sermons*, vol. 1. Boston (published 1871).
Stevens, L. C. *Church Finances: Or, God's Law Providing for the Public Expenses of Religion.* Gardiner, Me.

1850

Cooke, Parsons. *The Divine Law of Beneficence; Zaccheus, or The Scriptural Plan of Benevolence; and the Mission of the Church, or Systematic Beneficence.* New York (published with Samuel Harris and Edward Lawrence by the American Tract Society).
Harris, Samuel. *Zaccheus, or, The Scriptural Plan of Benevolence.* New York (published with Parsons Cooke and Edward Lawrence by the American Tract Society).
Lawrence, Edward. *The Mission of the Church; or, Systematic Beneficence.* New York (published with Parsons Cooke and Edward Harris by the American Tract Society).

1851

Keen, T. G. (Pastor, St. Francis Street Baptist Church, Mobile, Ala.). "Elements of a Church's Prosperity." In *A Collection of Original Sermons, Contributed by Ministers of Different Denominations, to Raise Means for the Erection of a*

Protestant Female College, in Greensburg, Kentucky, Thomas P. Akers, ed. Louisville.

Ramsay, William. *Church Debts; Their Origin, Evils, and Cure*. Philadelphia.

1852

Chalmers, Thomas. "Christian and Civic Economy of Great Towns." *The Presbyterian Quarterly Review* 1, no. 3 (December): 433–51.

Frothingham, N. L. "Hiding the Lord's Money." In *Sermons, in the Order of a Twelvemonth*. Boston.

Frothingham, N. L. "The Annual Thanksgiving. Prosperity Under the Reign of Wisdom." In *Sermons, in the Order of a Twelvemonth*. Boston.

Report of the American Board of Commissioners for Foreign Missions, Presented at the Forty-third Annual Meeting, Held in Troy, New York, Sept. 7–10, 1852. Boston.

1855

Benedict, Amzi. "Ministerial Support." *The American National Preacher* 29, no. 6 (June): 129–40.

Constable, Henry. "An Essay on the Measure of Christian Liberality." In *Gold and the Gospel: Prize Essays on the Scriptural Duty of Giving in Proportion to Means and Income*, ed. Jesse T. Peck. New York.

Peck, Jesse T., ed. *Gold and the Gospel: Prize Essays on the Scriptural Duty of Giving in Proportion to Means and Income*. New York.

Sherwood, J. M. "The Leaven of the Kingdom." in *The American National Preacher* 29, no. 6 (June): 117–28.

Woolsey, Theodore. *The Danger of Separating Piety from Philanthropy*. New Haven, Conn.

1856

Bowen, Francis. *The Principles of Political Economy Applied to the Conditions, the Resources, and the Institutions of the American People*, 4th ed. Boston.

Dick, Thomas. *An Essay on the Sin and the Evils of Covetousness; and the Happy Effects Which Would Flow from a Spirit of Christian Beneficence. Illustrated by a variety of Facts, Selected from Sacred and Civil History, and Other Documents*. Philadelphia.

Fry, Benjamin St. James. "Property Consecrated; or, Honoring God with our Substance." In *Systematic Beneficence: Three Prize Essays*. New York.

Kirwan, Walter Blake. "Seeking Another's Wealth." In *History and Repository of Pulpit Eloquence*, vol. 1, ed. Henry C. Fish. New York.

Stevens, Abel. "The Great Reform: a Prize Essay on the Duty and the Best Method of Systematic Beneficence in the Church." In *Systematic Beneficence: Three Prize Essays*. New York.

White, Lorenzo. "The Great Question; or, How Shall I Meet the Claims of God upon My Property? A Prize Essay." In *Systematic Beneficence: Three Prize Essays*. New York.

1857

"Financial Distress—Hopefulness and Duty." *The American Missionary* 1, no. 11 (November): 250–51.

"Life Insurance." *The Presbyterian Magazine* 7, no. 5 (May): 214–21.

Mason, John M. "The Gospel for the Poor." In *History and Repository of Pulpit Eloquence (Deceased Divines)*, vol. 2, ed Henry C. Fish. New York.

McFarlane, John. "Altar-Gold; or, Christ Worthy to Receive Riches." In *Pulpit Eloquence of the Nineteenth Century*, ed. Henry C. Fish. New York. (Preached in London, 1855.)

Oberlin Evangelist, new ser., vol. 14: "Oberlin College. No. 2" (March 4): 35; "Oberlin College. No. 3" (March 18): 41–42; "The Grammar of the Lord" (April 1): 54; "Covetousness" (May 27): 84–85; "A Hint to the Rich" (May 27): 88; "The Communion of Saints" (July 22): 120; "Experiences of One in Debt" (July 22): 120; "Paying Ministers' Salaries Promptly" (September 16): 150; "Lotteries" (September 30): 158; "Small Offerings to God" (October 28): 174; "The Oberlin Evangelist for 1858" (November 11): 182; "Where Shall We Begin to Retrench?" (November 25): 189; "Nothing can Afford at a Glance a Clearer Insight into the Universal Prevalence of Luxury in the United States" (November 25): 189; "Friends of the Evangelist" (December 9): 199.

"Political Economy and the Future." *The Church Review* 10, no. 1 (April): 48–67.

"Salaries of the Clergy." *The Church Review* 10, no. 2 (July): 217–36.

"What the Free Church System Requires." *The Church Review* 10, no. 1 (April): 88–105.

1858

Alexander, James W. "Daily Service of Christ. A Charity Sermon." In *Discourses on Common Topics of Christian Faith and Practice*. New York. (Preached in New York June 13, 1858).

Cambell, Alexander. "The Hard Times." *The Millennial Harbinger*, ser. 5, 1, no. 1 (January): 31–32.

Fish, Henry C. *Christ the Great Want of the Soul. A Discourse Preached before the American and Foreign Bible Society, at Its Annual Meeting Held in the Tabernacle Baptist Church, Philadelphia, May 13th, 1858*. New York.

Hawes, Joel. "The Widow and Her Two Mites; Or, Christian Benevolence Exemplified." *The National Preacher and Village Pulpit*, new ser., 1, no. 4 (April): 101–11.

"Organized Christian Emigration." *The Oberlin Evangelist* 20, no 23 (November 10): 181.

Shepard, George. *The Moral Discipline of Giving: A Sermon Before the American Board of Commissioners for Foreign Missions*. Boston.

1859

Boardman, H. A. *The Bible in the Counting-House: A Course of Lectures to Merchants*. Philadelphia.

1860

Campbell, Alexander. "Short Sermons for Business Men—No. 5." *The Millennial Harbinger*, ser. 5, 3, no. 7 (July 1860): 399–402.

Epps, J. "Is It the Duty of Christian Churches to Support Their Pastors?" *The Millennial Harbigner*, ser. 5, 3, no. 7 (July 1860): 374–89. (Reprinted from the British *Millennial Harbinger*.)

13

Afterword

Daniel Walker Howe

This volume has undertaken a historical inquiry into the attitudes of Christians toward money, looking specifically at Protestants in the antebellum United States. Despite its importance, there has been little written on the topic, even by professional historians of religion. The two most potentially embarrassing subjects to discuss are sex and money, and it has taken longer for historians of American religion to bring themselves to address the latter. The clergy know well that their congregations often show even more reluctance to hear about money from the pulpit than to hear about sex. Scholars seem to have shared some of this sensitivity. For them to raise the subject of money in the context of religious history, they evidently feared, would betray a cynical, debunking attitude, perhaps even a vulgar, Marxist ideology. The work Mark Noll has collected here should demonstrate, however, that the subject can be addressed in a constructive and objective manner.

This book has discussed what Christians living in a particular time and country thought about money and its religious uses. What Christians have thought about money or how they have used it is not the same thing as Christian doctrine about money—that is a subject for theologians, concerned to derive implications for attitudes and behavior from authoritative traditions and biblical texts. But these authors, as historians, have instead been concerned to find evidence for what some Christians said and did, regardless of whether all Christians everywhere should share those ideas—and, indeed, fully recognizing that antebellum American Christians were influenced by many factors besides Christian faith. Nevertheless, Mark Noll is quite right to insist, as he does at the end of chapter 12, that Christian doctrine should be the initial frame of reference within which to analyze this subject.

The period treated by this volume was one in which American evangelicalism became increasingly commercial in its methods. The disestablishment of religion, which occurred in most states around the time of the Revolution but not until the first third of the nineteenth century in parts of New England, opened up a nationwide free marketplace in religion. Just as the deregulation of several industries in the late twentieth century led to business expansion and greater competition, so the "deregulation" of religion in the early national period was followed by a multiplication of sects and increases in church membership. Revivalists undertook to market religion to its potential customers with all the efficient organization and media technology at their disposal. Each denomination, like each product for sale, was one among many; each one had to be marketed effectively to customers if it were to survive. If the result was sometimes crass and corrupt, marketplace logic was by no means entirely a negative influence on American religion. It kept the churches adaptable, dynamic, and growing.

In many respects the great evangelical movement of the nineteenth century was in the forefront of the commercial and industrial expansion that is sometimes called the market revolution. In spreading the message of the gospel, Christian missions helped conquer space and time. Itinerant preachers on the frontier blazed the trails that commerce would follow, just as overseas missionaries showed the way to the merchant traders who followed them. When movable type had been invented in the fifteenth century, the Bible had been the first book printed. In our own day, evangelicals have been quick to recognize the potential of television and the Internet. So in the nineteenth century, Christian evangelists were also among the first to exploit new communication technology: improvements in printing, cheaper paper, post roads, steamships, and government subsidies for the post—all facilitated the early development of evangelical magazines and periodicals. Well might the devout inventor of the telegraph, Samuel F. B. Morse, send as his first signal in 1844 the pious message "What hath God wrought?"

By the same token, evangelicals pioneered the techniques of business organization. Their religious movement was well organized at a time when little else in America was. They learned how to enlist the energies of Christians from various denominations in common enterprises. They operated on a national level in a period when there was no nationwide business corporation except the Bank of the United States and no government bureaucracy but the U.S. Post Office. Indeed, they even engaged in international undertakings with fellow Protestants in other countries. One of these was the effort to stamp out the slave trade in Africa and substitute other forms of commerce as an economic basis for the region.

Despite all the affinity and cooperation between Protestant activities and the growing commercialization of society in the nineteenth century, we would not be justified in concluding that the evangelical preachers were tools of capitalist employers. Neither the chapters in this volume nor other recent historical research bear out such an interpretation. As David Hempton and John Walsh show, E. P. Thompson's portrayal of the English Methodists as agents of class exploitation is not persuasive for England, much less the United States. As suggested at many points in this book, Anthony Wallace's hostile depiction of the evangelicals of industrializing America as authoritarian, presumptuous, and militant is considerably nuanced if not entirely refuted.[1] The chapters in this book have presented more evidence of ways in which the market revolution was turned to the advantage of evangelical Protestantism than of ways in which Christianity served the interests of capitalism.

In the years treated by this volume, American Christians were divided between two rival political parties, Democrats and Whigs, which had sharply contrasting political programs. The Democrats, whose leading politicians included Andrew Jackson, Martin Van Buren, and John C. Calhoun, favored laissez-fair economic policies, states' rights, and white supremacy. The Whigs, led by John Quincy Adams, Henry Clay, and Daniel Webster, favored a protective tariff and other aid to businesses along with a strong national government. The Whigs also endorsed a number of moral and humanitarian social reforms like temperance, improving the conditions of insane asylums or prisons, and (sometimes) antislavery. Although they did not actually disavow white supremacy, the Whigs were less eager to take land away from the Indians and less enthusiastic about expanding and perpetuating slavery. When states debated their voting requirements, Whigs were more willing to permit voting by nonwhite men (women being disfranchised in all states). On the other hand, the Democrats showed more sympathy for the difficulties of European immigrants and the white poor than did the Whigs.

Members of certain religious denominations and ethnic groups voted predominantly for one party or the other. Roman Catholics from Ireland voted almost unanimously Democratic; Congregationalists, Quakers, and Unitarians generally voted Whig. In those states that allowed black men to cast ballots, African Americans voted Whig or, if the opportunity was present, for a minor, antislavery party. White Protestant denominations were sometimes divided between evangelical Whigs and confessional Democrats. (Evangelicals showed a greater interest in reforming the larger society; confessional Christians made doctrinal purity and rigor their top priority.) Among Presbyterians, the evangelical adherents of the New School voted predominantly Whig, whereas the confessional Old School Presbyterians inclined to the Democrats. Similarly, evangelical Low Church Episcopalians were mostly Whig but confessional High Church Episcopalians included many Democrats. Within some evangelical groups like the diverse Baptists, a distinction can be drawn between "formalists" and "antiformalists," with the latter being less interested in an educated ministry, less supportive of interdenominational missionary and reform movements, and more likely to vote Democratic. The Democratic party often insisted that religion had no proper place in politics. Probably for this reason, men who were indifferent or hostile to religion in general usually found the Democratic party more congenial. But not all religious labels implied a likely political preference; that a man was a Methodist, for example, gives us no indication of how he voted unless we also know that he lived in a community where Methodists voted as a block. On the whole, there was more correlation between religious groups and political views in the North than in the South.[2]

What is remarkable in all this, however, is that the religious bodies took stands on some kinds of public issues but not on others. They took public stands on issues involving moral values, not usually on matters of economic policy. Of course, the slave trade and the liquor trade were big businesses and seriously affected by moral reform crusades. But the economic side of these issues was seldom overtly addressed by the reformers. The spokesmen for churches had comparatively little to say about the swings of the business cycle except to warn that periodic downturns were the providentially ordained retribution for speculative avarice. They had even less to say about the wisdom of free trade as opposed to tariff protection. Although the political parties embraced both economic and moral issues, it was the moral issues and not the economic ones that energized their religious supporters. The political divisions among the sects did not seem to create corresponding divisions over economic policy.

The capitalist system within which Americans lived was taken for granted by antebellum Christians, except for a few communitarian groups outside the mainstream, such as Shakers and the early Mormons. Preachers did not so much celebrate or justify capitalism as teach their congregations how to cope with it and how to reconcile it with Christian charity. When antebellum preachers addressed the subject of economics, it was usually microeconomics, not macroeconomics. They offered advice on the place of money in private life rather than public life. And the advice offered was much the same whether the congregation was middle class or working class, evangelical or confessional, Northern or Southern, black or white.

The economic teachings the clergy imparted to their congregations typically went something like this: Serve God in your calling. Work hard, be thrifty, save your money, and don't go into debt. Be honest in business dealings; don't screw down the wages of those who work for you to the lowest possible level. (In that society, we should remember, even people of very modest means might have a "hired man" who helped on the farm or a "hired girl" who helped in the kitchen.) If you manage a surplus, the message concluded, be faithful stewards of your bounty; that is, be generous to the church and other good causes.

All in all, it was not bad personal advice, then or now. The message presupposed private property and a commercial order but not ruthless competition. It synthesized Christian personal morality with conventional American attitudes of the time. It attempted to infuse the marketplace with moral meaning. In the South, the ministers admonished masters not to overwork or abuse their slaves and to respect their family ties. In other respects their teaching was much the same as in the free states.[3]

The chapters in this volume have indicated something of the complicated interrelationship between American Christianity, especially the evangelical movement, and the expanding market economy of the period between the Revolution and the Civil War. Mark Noll sets out a fascinating range of questions at the outset, in his introduction. This is followed by a survey of the economic history of the period and examples of the historiographical debates that have energized the field, showing how historians are struggling to liberate the subject from various kinds of antievangelical bias. Noll returns at the end of the book with a "preliminary probe," which emphasizes the way in which an autonomous religious culture sought to tame the excesses of market capitalism while at the same time exploiting its resources for the benefit of organized Christianity.

Between these general statements, we are presented with six specific cases that explore the ironies and complexities of the dialogue between economic means and ideal ends. David Paul Nord illustrates the ambiguities of the era in his discussion of the publication of Bibles and tracts. Although the evangelical publishers made use of modern business methods and technologies, they regarded commercial culture itself, with its cynical willingness to exploit human weakness and hedonism for financial gain, as the "enemy."

Two chapters illuminate the sectional bitterness of the Civil War era. Richard Carwardine shows how the schism between Northern and Southern Methodist churches was immeasurably embittered by disputes over the sharing of financial assets. He thereby enables us to understand why the denomination remained divided for so long and could not promptly reunite once the abolition of slavery had rendered moot the original reason for the split. Kenneth Startup shows how the resentment of Southern Protestants against what they took as the self-righteous hypocrisy of Northern antislavery reformers manifested the antimercenary values of Christian tradition. These values also prompted Southern clergymen to criticize much of what they saw around them in their own section.

Richard Pointer's chapter on the similar economic views of Old School and New School Philadelphia Presbyterians offers a salutary warning to historians. We may well have overdone the polarization of evangelical and confessional wings of Protestantism in our desire to explain party politics and stands on the slavery question. Pointer shows that both kinds of Presbyterians were involved in the temperance movement, and both regarded the industrialization of their city with mixed feelings. In general, Presbyterian clergymen "were both powerfully drawn and occasionally repulsed by emergent capitalism." Neither theology nor party politics was unchangeable, as Ezra Stiles Ely, minister of Seventh Presbyterian Church in Philadelphia, illustrates. Ely was at first associated with the Old School confessionalists. In 1827, he famously called on all good Christians to support Andrew Jackson for president against the Unitarian John Quincy Adams.[4] By the early 1830s, however, Ely was showing New School theological sympathies, and he eventually became a Whig in politics.

The closeness and yet the ambiguity of the relationship between Protestantism and capitalism is a recurrent theme in the chapters here. Kathryn Long, with a fine sense of irony, shows how the New York City press exploited the "businessmen's revival" of 1857 for sensational news whether the publishers sympathized with the aims of the revival or not. David Hempton, writing on

Anglo-American Methodist finances, displays a profundity seldom achieved in a brief study. All too much Methodist history has been written in the jeremiad form, deploring a fall from the grace of a heroic past. Hempton shows clearly why the saintly, solitary circuit-rider had to give way to the settled minister, with a family and formal training, as Methodism—like the individual Methodists themselves—became more prosperous, influential, and stable. By the end of the nineteenth century, American and British Methodists were erecting giant gothic-revival houses of worship "not only to accommodate burgeoning numbers" but also to demonstrate "order, stability, and respectability." Hempton's conclusion is applicable to far more than just the history of Methodism: "Money is not simply a necessary and neutral commodity for getting things done but rather carries with it a symbolic revelation of the values for which it was collected and appropriated."

An "afterword" should look forward, as well as back; that is, it should consider not only what has been said but also where the conversation might go from here. Church historians sorely need a history of the pew proprietor and pew rental system that formed the basis for congregational financing in most American denominations before the adoption of the present system of annual pledges. Why and when that former system came into being, how it functional, its advantages and disadvantages, and what explains its eventual disappearance is a story waiting to be told. Until it is, we will not have a clear picture of nineteenth-century finances. For the colonial period, it would be good to have a clearer picture of how the Anglican and Congregational religious establishments were funded. In fact, there is no period of history, not even the present, for which reliable data on church finances are easily accessible. Information on individual congregations is usually poorly preserved. Information on central denominational treasuries is sometimes closely guarded, as in the case of Mormons and Roman Catholics. And as is well known, the fiscal malpractices of some evangelists, on television and otherwise, have been exposed as scandals.

The finances of religious education also deserve exploring. For the period covered by this volume, religious institutions bore much of the responsibility of educating young Americans. Most church-connected schools levied tuition charges, which were often means-tested so that those who could afford to do so subsidized the poor. (In the same way, those who paid the American Bible Society for their Bibles subsidized their gift to the poor.) It seems unfortunate that we should know so little about how education in the United States used to be funded before the rise of the present system of public education. At a time when the proposed introduction of vouchers has reopened the whole issue of educational funding, our discussion should be historically informed.

Finally, we should want to examine the relationship between Protestantism and American economic development in such a way as to illuminate what is going on in the larger world today. In the late twentieth and early twenty-first centuries, much of the world is undergoing economic and social transformations analogous to those the United States underwent in the late eighteenth and early nineteenth centuries. Among these transformations are industrialization, urbanization, population explosion, and the novel experience of political independence. In many developing countries of what we call the "third world," Protestant missions and religious revivals are experiencing amazing success. Worldwide Protestantism and Christianity in general are ceasing to be predominantly European and becoming much more cosmopolitan. The experience of American Christians with the market revolution of the nineteenth century ought to speak to the needs of Christians in the market revolutions of the twenty-first century.

Notes

1. Edward Palmer Thompson, *The Making of the English Working Class* (London: Gollancz, 1963); Anthony F. C. Wallace, *Rockdale* (New York: Knopf, 1978).

2. There are many studies of this subject; see especially Mark Noll, ed., *Religion and American* Politics (New York: Oxford University Press, 1990); and Richard Carwardine, *Evangelicals and Politics in Antebellum America* (New Haven, Conn.: Yale University Press, 1993).

3. Besides the chapters in this volume, see Nathan Hatch, "The Second Great Awakening and the Market Revolution," in *Devising Liberty*, ed. David Thomas Konig (Stanford, Calif.: Stanford University Press, 1995), 253–64; William R. Sutton, *Journeymen for Jesus: Evangelical Artisans Confront Capitalism in Jacksonian Baltimore* (University Park: Pennsylvania State University Press, 1998); and Kenneth Moore Startup, *The Root of All Evil: Protestant Clergy and the Economic Mind of the Old South* (Athens: University of Georgia Press, 1997).

4. Ezra Stiles Ely, *The Duty of Christian Freemen to Elect Christian Rulers* (Philadelphia: William F. Geddes, 1828). The speech that was printed had been delivered in 1827.

Index

abolition, 12, 13, 60, 195, 250–51. *See also* antislavery; slavery
  and Charles Sellers, 68
  in England, 221–22, 226, 228–29
  link to Northern economy, 220, 226–27, 229, 230
  and Methodism, 206–07, 211, 225
  Southern clergy critique of, 220–23, 228
Adams, John Quincy, 89–90, 297, 299
Adger, John, 221
African Americans, 13, 55, 68
African Methodist Episcopal Church, 13
Albany, New York, First Church
  and benevolences, 47–48
  and community of Albany, 48–49
  Deacon's fund, 44, 46, 47
  land rentals/sales of, 46, 47, 48
  Middle Church, 45, 46, 48
  North Church, 44, 45–46, 48
  as representative of Protestants, 43, 49
Alexander, Archibald, 179
Alexander, James W., 174, 177, 180, 281
Allen, Richard, 13
American Antislavery Society, 11, 226
American Baptist Home Missionary Society (ABHMS), 224

American Baptist Publication Society, 163, 168n.32
American Bible Society, 9, 11, 163, 250, 279
  auxiliary system and, 160–62
  income of, 52n.27, 239
  and manufacturing, 155
  as publishing force, 41–43, 147, 156
  retail versus charity sales, 147–49, 157–60, 300
American Board for Foreign Missions, 11
American Congregational Union, 251
American Missionary, 278
American Revolution, 3, 6, 19, 38, 43–45, 269, 296
American Society for the Promotion of Temperance, 11
American Sunday School Union, 155, 156, 250
American Tract Magazine, 153
American Tract Society (ATS), 250, 280
  auxiliary system and, 161–62
  colporteurs, 163
  and the market revolution, 164–65
  and printing, 155–56, 239
  retail versus charity sales, 159–60
  and slavery, 251–52
American Tract Society of Boston, 156

303